Optical Imaging and Sensing

Optical Imaging and Sensing

Materials, Devices, and Applications

Edited by Jiang Wu and Hao Xu

WILEY-VCH

Editors

Prof. Dr. Jiang Wu
Institute of Fundamental and Frontier Sciences, University of Electronic Science and Technology of China, No.4, Section 2, North Jianshe Road, 610054, Chengdu, P.R. China

Prof. Dr. Hao Xu
School of Physics, University of Electronic Science and Technology of China, No.2006, Xiyuan Ave, West Hi-Tech Zone, 611731, Chengdu, P.R. China

Cover Image: © Yuichiro Chino/Getty Images

Library of Congress Card No.: applied for

British Library Cataloguing-in-Publication Data
A catalogue record for this book is available from the British Library.

Bibliographic information published by the Deutsche Nationalbibliothek
The Deutsche Nationalbibliothek lists this publication in the Deutsche Nationalbibliografie; detailed bibliographic data are available on the Internet at <http://dnb.d-nb.de>.

Print ISBN: 978-3-527-34976-0
ePDF ISBN: 978-3-527-83518-8
ePub ISBN: 978-3-527-83519-5
oBook ISBN: 978-3-527-83520-1

Typesetting Straive, Chennai, India

Contents

Preface

To enable conversion between optical and electrical signals, optoelectronic devices that rely on light–matter interactions are gaining much attention and are ubiquitous in modern society. Typically, they can be fabricated using either conventional Si and compound semiconductors (like GaAs, InP, and GaN) or emerging materials, such as graphene, transition-metal dichalcogenides (TMDs), and perovskites, or even using a combination of these materials. Optoelectronic devices for imaging and sensing are integral components of many photon-involved applications, such as optical communications, public security warning, smart city monitoring, clinical imaging, and personal healthcare checking. Considering the rapid development of the Internet of Things, 5G-based techniques, and beyond, optoelectronic imaging and sensing devices are ever increasingly important for these applications. Thereupon, they are facing tough challenges and high demands in terms of device properties and parameters.

High-performance optoelectronic imaging and sensing devices, such as optical sensors, photodetectors, light-emitting diodes (LEDs), lasers, and flexible devices, are indeed desired for practical use. For instance, photodetectors and optical sensors in the near-infrared (IR) and mid-infrared (MIR) regions have a huge number of applications, ranging from telecommunications to molecular spectroscopy. Nowadays, the quest for high photoresponsivity, high detectivity, high photogain, low dark noise, fast photoresponse, and complementary metal-oxide semiconductor (CMOS)-compatible room temperature photodetectors in these spectral regions is ongoing. Moreover, recently there have been significant developments in on-chip integration using emerging two-dimensional (2D) materials and/or plasmonic structures to extend the wavelength range of silicon-based photodetectors beyond 1100 nm. There has also been a lot of interest in building MIR detectors based on 2D material design. In order to achieve further developments based on traditional materials and device structures, new concepts regarding emerging materials and device configuration design can also be adopted. Accordingly, this book covers topics including nanomaterial-based photodetector arrays for imaging, plasmonic photodetectors, optical resistance switches for optical sensing, optical interferometric sensing, novel materials for super-resolution imaging techniques, and nanomaterial advances and on-chip integration.

Chapter 1 describes the material systems and heterostructures for optoelectronics, device components, challenges, and prospects of nano-optoelectronic devices. 2D materials are an expanding family of atomically thin crystals, represented by semimetallic graphene, semiconducting black phosphorus, metallic and semiconducting transition-metal (di)chalcogenides, insulating boron nitride, and so on. They exhibit fascinating characteristics, including tunable bandgap, high carrier mobility, efficient light absorption, good structural stability, and mechanical flexibility. In particular, their excellent photodetecting properties render them attractive for optoelectronic devices. Different strategies have been investigated to further enhance the photodetection capabilities of 2D materials, such as chemical doping, surface modification, defect and strain engineering, and plasmonic nanostructure assisting. Moreover, 2D van der Waals heterostructures have been developed to enrich the optical functionalities, such as fast response over a broad spectral range. Advanced synthesis techniques are desired to realize industrial-scale integration. Chapter 2 focuses on the design and performance of 2D material-based photodetectors for imaging applications.

IR photodetectors and imaging focal plane arrays (FPAs) are critical devices for sensing and imaging applications. Various techniques have been developed to enhance the performance of IR photodetectors and FPAs, including resonant cavities, surface gratings, surface plasmonic resonances (SPRs), optical antennas, and plasmonic perfect absorbers. Chapter 3 reviews the concepts of SPR, its excitation (i.e. resonant conditions), dispersion relations, and near-field profiles and enhancement. The applications of SPR resonance in IR photodetection will also be discussed. Specifically, various SPR structures are discussed, including the metallic 2D sub-wavelength hole array (2DSHA) and localized SPRs in metallic circular disks and nanowires. Plasmonic perfector absorbers (PPA) and their enhancement effect on IR photodetectors will be discussed as well. In Chapter 4, representative publications regarding optical switches for optical sensing applications have been studied. The main properties of the structures are given. Such features are the setup and topology of the optical devices, their mechanism operation, dimensions (2D/3D), and isolated waveguides. Also, whether the time-domain simulations are performed has been investigated. In summary, based on the mentioned properties, these structures have the potential to be used as optical sensors.

In Chapter 5, the nonlinear interferometer is discussed, including experimental implementation of phase locking, enhancement of phase sensitivity, experimental realization of entanglement enhancement, and quantum noise cancellation (QNC). Meanwhile, other types of nonlinear interferometers are described, including a nonlinear Sagnac interferometer, a hybrid interferometer consisting of a nonlinear four-wave mixing (FWM) process and a linear beam splitter, a phase-sensitive FWM process acting as a nonlinear beam splitter, and interference-induced quantum-squeezing enhancement. Afterwards, a nonlinear interferometric SPR sensor has been theoretically proposed and its sensing advantages were demonstrated by using sensing parameters such as degree of intensity-difference squeezing, estimation precision, and signal noise ratio. In Chapter 6, recent achievements of this emerging and fast-growing field have been reviewed. The diffraction limit substantially impedes the resolution of the conventional optical microscope. Under traditional illumination, the high-spatial-frequency light corresponding to the subwavelength information of objects is located in the near-field in the form of evanescent waves and thus not detectable by conventional far-field objectives. Recent advances in micro/nanomaterials and metamaterials provide new approaches to break this limitation by utilizing large-wavevector evanescent waves with the spatial frequency shift (SFS) method. The current super-resolution imaging techniques based on evanescent-waves-assisted SFS method, using nanomaterials, photonic waveguides, wafers, and metamaterials, are illustrated. They are promising in investigating unobserved details and processes in fields such as medicine, biology, and material research.

Recent advances in monolithically integrated multisection semiconductor lasers (MI-MSSLs) have propelled microwave photonic (MWP) technologies to new potentials with a compact, reliable, and green implementation. Much research has examined that MI-MSSLs can realize the same or even better MWP functions compared to discrete lasers by taking advantages of enhanced light–matter interactions. Here, we review these recent advances in this emerging field and discuss the corresponding photonic microwave applications. Three main kinds of MI-MSSL structures are demonstrated in general, including passive feedback laser, active/amplified feedback laser, and monolithically integrated mutually injected semiconductor laser. The focus of this paper is on MWP techniques based on the nonlinear dynamics of MI-MSSLs. The primary MWP applications considered in this paper cover from electro-optic conversion characteristics enhancement, photonic microwave generation, MWP filter, to multiwavelength laser array for wavelength division multiplexing radio-over-fiber (WDM-RoF) networks. Especially, the four special dynamic states of free-running oscillation, mode-beating self-pulsations (MB-SPs), period-one (P1) oscillation, and sideband injection locking are considered and demonstrated in detail for photonic microwave generation. In Chapter 7, the authors take a look at the future prospects of the research directions and challenges in this area.

We acknowledge all the contributors and sincerely hope this book can help readers better understand materials, devices, and applications for optical imaging and sensing.

Chengdu, China
28 December 2022

Hao Xu
Jiang Wu
University of Electronic Science and Technology of China

1

Introduction of Optical Imaging and Sensing: Materials, Devices, and Applications

Qimiao Chen[1], Hao Xu[2], and Chuan S. Tan[1]

[1] *Nanyang Technological University, School of Electrical and Electronic Engineering, Singapore 639798, Singapore*
[2] *University of Electronic Science and Technology of China, School of Physics, Chengdu 610054, China*

1.1 Optoelectronic Material Systems

Typically, optoelectronic devices for sensing and imaging can be fabricated by either group IV (Si, Ge, Sn, and their alloys) or compound semiconductors (like GaAs, InP, and GaN). Recently, there have been significant developments in on-chip integration using emerging two-dimensional (2D) materials and/or plasmonic structures to extend the wavelength range of silicon-based photodetectors beyond 1100 nm.

1.1.1 Si Platform

Si is not only the dominant material for the modern Information Age (also known as Silicon Age), which is driven by Si electronics, but also one of the most promising platforms for photonics to construct sensing or imaging systems. Si photonics, which is compatible with the current Si electronic industry, will share the benefits of Si technologies in terms of monolithic electronic–photonic integration, scaling, and low cost, which may result in a Si-based optoelectronic revolution in the future. Si photonics have been developed for various applications in the near-infrared (NIR) and shortwave infrared (SWIR) ranges.

Si photonics is a rapidly developing field. Recently, it has achieved a breakthrough with Ge-related materials (SiGeSn). As shown in Figure 1.1, the SiGeSn material system has some interesting optical properties that are ideal for constructing optoelectronic devices for sensing and imaging: (i) the lattice constant and bandgap can be tuned independently to form high-quality heterostructures to control the carriers; (ii) bandgap can be converted from indirect to direct; (iii) coverage of NIR, SWIR, and mid-infrared (MIR) wavelengths; (iv) the compatibility of growth temperature (<400 °C) with complementary metal oxide semiconductor (CMOS) process; and (v) high mobility, low Auger, and nonpolar.

Optical Imaging and Sensing: Materials, Devices, and Applications, First Edition.
Edited by Jiang Wu and Hao Xu.

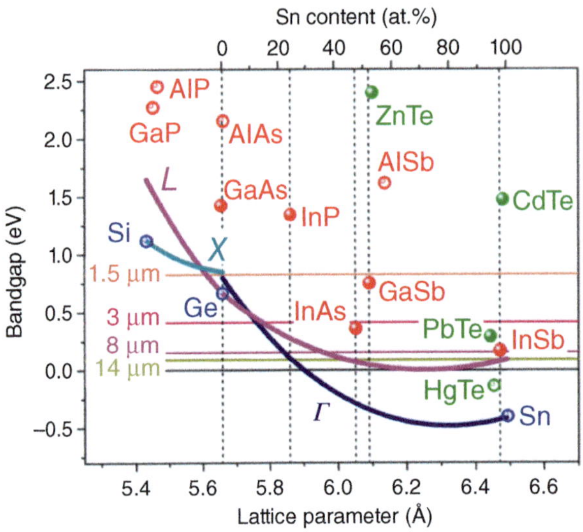

Figure 1.1 Bandgap versus of lattice constant of group IV, III–V and II–VI semiconductors. Source: Reproduced with permission from Moutanabbir et al. [1].

Early proposals to utilize SiGeSn materials for photonics date back to the 1990s. However, the actual development of practical devices has been hampered by the technical challenges of growing high-quality SiGeSn materials. While the thermal expansion coefficient mismatch between the epitaxial layer and Si substrates showed a positive effect on the epitaxial Ge layer, the large lattice mismatch between the elements (4.2% for Ge and Si, 14.7% for Sn and Ge, and 17% for Sn and Si) makes the grown layer tend to be defective and rough. For example, when the Ge or GeSn layer is epitaxially grown on Si substrates, misfit will inevitably be generated at the interface and propagate to the surface of the epitaxy layer as threading dislocations, which degrades the material quality and hinders the practical application of Ge or GeSn optoelectronic devices. Besides, it is challenging to incorporate Sn into Ge due to the extremely limited solid solubility of Sn (<1% in Ge), and Sn is prone to segregation at higher concentrations and growth temperatures. Therefore, GeSn alloys are unstable and require nonequilibrium growth. Fortunately, the device-level-quality SiGeSn materials have been grown recently by molecular beam epitaxy (MBE) or chemical vapor deposition (CVD) with various Ge and Sn precursors. We will introduce the growth of GeSn alloys and Ge by commercial CVD systems, which are of higher material quality and compatible with the industry for mass production.

To solve the first issue of lattice constant mismatch between epitaxial layers and Si substrates, relentless efforts have been made. Using a fully relaxed, thick-graded buffer layer can reduce the lattice constant mismatch between the Ge epitaxial layer and Si substrate. However, this method suffers from the issue of wafer bow as the graded buffer layer is usually too thick (~10 μm), so it could not be grown by industrial CVD tools. The second approach is the selective area growth (SAG) technique. The Si substrate is patterned, followed by epitaxial lateral overgrowth. Threading dislocations do not slide parallel to the growth direction, and they can slide to the window sidewalls and annihilate. The third more practical approach is the two-step growth method. A thin Ge seed layer was grown at low temperatures to restrict the mobility of Ge adatoms, thereby preventing 3D nucleation of Ge. Afterward, a thick Ge film is grown under high-temperature conditions for a higher growth rate and better Ge crystallinity.

For the growth of (Si)GeSn alloys, currently, the main Sn precursors are tetramethyl tin [$Sn(CH_3)_4$] and stannic chloride ($SnCl_4$). The main Ge precursors are GeH_4, Ge_2H_6, and Ge_3H_8. There are two approaches to obtaining GeSn layers with high Sn concentrations: (i) the spontaneous relaxation-enhanced (SRE) Sn incorporation process and (ii) the GeSn virtual substrate approach. When GeSn is grown on a Ge buffer using a fixed Sn concentration recipe, the Sn concentration will increase from the GeSn/Ge interface to the GeSn surface due to the strain relaxation. Thus, the grown layer will normally show two distinct layers, with the first strained layer being defective and the second relaxed layer having low defect density. Using the GeSn layer obtained by the SRE method as the virtual substrate to grow GeSn layers, an Sn concentration of up to 22.3% has been achieved. In addition to GeSn or Ge on Si substrates, some advanced engineered substrates, such as Ge(Sn) on insulator (GOI) and Ge on SiN (GON), have been proposed for the benefits of the smaller device area, increased packaging density, and high mobility.

Since the breakthrough in material growth, many devices have been developed for silicon photonics, including light emitters, amplifiers, photodetectors, waveguides, modulators, couplers, and switches. Progress has also been made in the most challenging components of Si light sources, and a complete set of components is being delivered for silicon photonics.

1.1.2 Two-dimensional Materials and Their van der Waals Heterostructures

Graphene was discovered in 2004 and the discoverers won the Nobel Prize due to its attractive 2D properties. Since then, many atomically thin 2D layered materials have been discovered, ranging from metallic (graphene and $TaSe_2$), semiconducting (WSe_2 and MoS_2), superconducting ($NbSe_2$ and FeSe) to topological insulators (Bi_2Se_3 and Sb_2Te_3). Through the combination of quantum confinement and enhanced electronic interactions, these atomically thin materials become strongly renormalized. Besides, different 2D materials can form various heterostructures by weak van der Waals forces. Figure 1.2 illustrates the available 2D materials and their electronic band structures covering a wide range of the electromagnetic spectra, from ultraviolet, visible, infrared to microwave, which would have different applications such as sensing, optical communication, and thermal imaging. We will introduce the properties of some typical individual 2D materials, followed by their van der Waals heterostructures.

1.1.2.1 Graphene

Graphene is a zero-gap semimetal with a single layer of 2D honeycomb carbon atoms. The resistivity will be reduced exponentially by adding electron/hole carriers. The interaction of the surrounding electrons around the honeycomb carbon atoms leads to the quasiparticles of massless Dirac fermions. Graphene has many

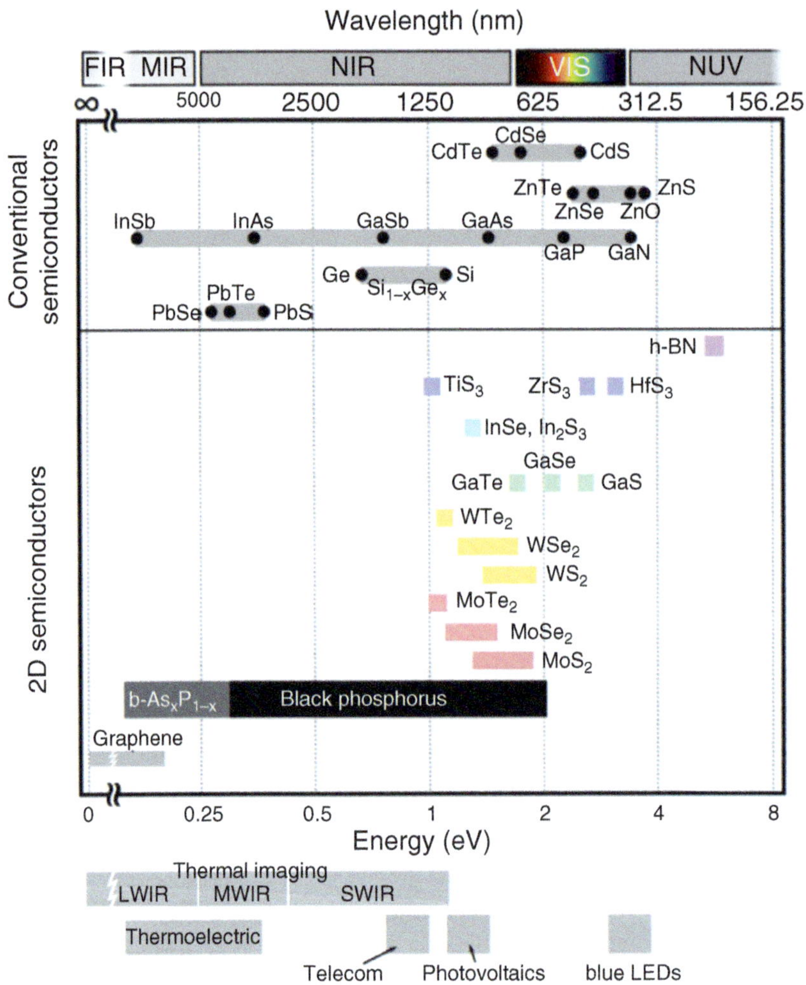

Figure 1.2 2D materials with potential applications in different wavelength regimes. Source: Reproduced with permission from Castellanos-Gomez [2].

attractive properties, such as high electron mobility (up to 2.5×10^5 cm^2 V^{-1} s^{-1}), high thermal conductivity (>3000 W mK^{-1}), high intrinsic strength of 130 GPa, and Young's modulus of 1 TPa.

1.1.2.2 Transition Metal Dichalcogenides

Transition metal dichalcogenides are atomically thin semiconductors with the formula MX2, with M, a transition metal element (Mo, W, or Nb), and X, a chalcogen element (S, Se, or Te). A pane of transition metal atoms is sandwiched by two layers of chalcogen atoms, as shown in Figure 1.3. The electrical and optical properties can be tuned by incorporating different transition metal elements. Transition metal dichalcogenides (TMDs) can vary from semiconducting to metallic characteristics, and some even demonstrate superconductivity characteristics (NbS_2 and $TaSe_2$).

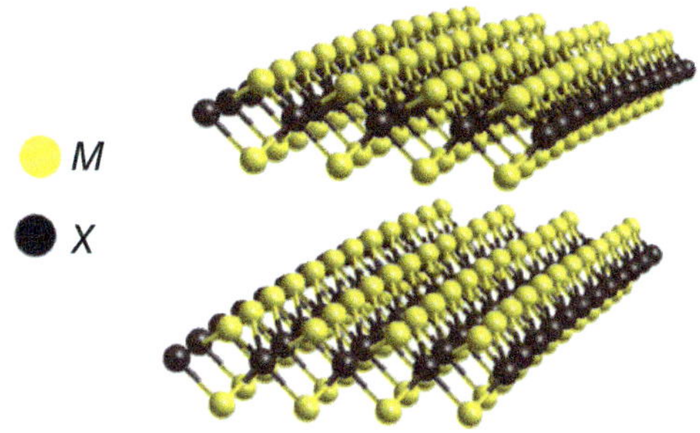

Figure 1.3 Structure of 2D molybdenum disulfide (MoS_2) layers. M atoms are in grey and X atoms are in yellow. Source: Reproduced with permission from Jariwala et al. [3].

Among various TMDs, MoS_2 is a semiconductor with an indirect bandgap of 1.3 eV. The MoS_2 will be converted into a direct bandgap with an energy of 1.8–1.9 eV when it is reduced to a monolayer, which is proved by photoluminescence (PL) measurements. Modified MoS_2 has demonstrated the ability to maintain permanent valley polarization, which is promising for valleytronics. Monolayer MoS_2 is also a candidate for spintronics due to its strong spin-orbit splitting. Monolayer TMDs have been applied in high-efficiency 2D optoelectronics devices such as light-emitting diodes (LEDs) and lasers.

1.1.2.3 2D Heterostructures

Within the wide range of materials available, the ways in which they can be combined are virtually limitless, each with its own unique properties and phenomena. For example, a tunneling device demonstration uses a thin layer of insulating hBN as a tunnel barrier between graphene sheets. 2D quantum wells can be formed by sequentially stacking different semiconducting transition metal dichalcogenides. Graphene/hBN heterostructure has been used to fabricate tunable LEDs. The development of nanoscale metamaterials has been achieved by coupling 2D materials with repeating nanostructures, revealing interesting plasmonic properties. The demonstrated applications of 2D heterostructures are in the fields of light detection, light generation, electronics, and some emerging directions such as single-photon emitters, quantum dot qubits, superconducting qubits, and topological quantum computing components.

1.2 Challenges and Prospect of Nano-optoelectronic Devices

Miniaturization of optoelectronic devices to the nanoscale offers enormous performance gains. As the volume of photoactive materials decreases, optoelectronic performance increases, including the operating speed, signal-to-noise ratio, and internal quantum efficiency. Over the past decades, the volume of photoactive materials in optoelectronic devices has been reduced by orders of magnitude. Relentless efforts have been made to overcome the limitations of further miniaturization. Among them, replacing Si with a new class of nanomaterials, such as 2D materials or heterostructures, is a promising candidate method. However, due to the lack of scalable fabrication methods, this direction is mostly limited to proof-of-concept research. Efforts are made toward the scalable fabrication of 2D heterostructures. CVD and metal organic chemical vapor deposition (MOCVD) are promising candidate methods. Nonetheless, the growth kinetics has not been fully understood. Large-scale and high-quality 2D HS with high controllability is also needed. Besides, the environmental stability of the device is a great challenge. To enable further development of nanodevices, new concepts regarding emerging materials and device configuration design need to be employed. Accordingly, the book will cover topics including (i) nanomaterials-based photodetector arrays for imaging, (ii) plasmonic photodetectors for sub-wavelength photodetection, (iii) optical resistance switch for optical sensing, (iv) optical interferometric sensing, (v) novel materials for clinical applications, (vi) computational imaging/sensing based on nanomaterials, and (vii) nanomaterial advances and on-chip integration.

1.2.1 III–V Compounds

Si is the most widely used semiconductor material due to its rich reserves, mature technology, and low cost, but it has an indirect band gap. Compared with Si, III–V compound semiconductors (such as GaAs, InP, GaN, and GaSb) and their alloys have direct band gaps. That is to say, the top of the valence band and the bottom of the conduction band are located at the same position in the wave vector space, and the electron–hole recombination does not need to exchange momentum, so it has high internal quantum efficiency and is predominantly used in optoelectronic devices [4].

GaAs is a typical III–V compound semiconductor material and is widely used in many optoelectronic fields, such as remote control, mobile phones, and lighting. Its band gap width is 1.42 eV (300 K), which can just absorb the high-energy part of the solar spectrum. Therefore, it is an ideal material for preparing solar cells. Researchers from the US Department of Energy's National Renewable Energy Laboratory (NREL) have developed a six-junction III–V solar cell with a 47.1% conversion efficiency rate under 143 suns of concentration. Its quaternary systems InGaAsP and InGaAlAs just cover the O band (1260—1360 nm, 1310 nm in the center) or C + L band (C band: 1530—1565 nm, 1550 nm in the center; L band: 1565—1625 nm) commonly used in optical communication. $In_xGa_{1-x}As$ photodetector can be used from 400 to 3600 nm, which is suitable for a wide range of applications, including optical communication, analysis, and measurement. Flexible InGaAs thin film materials have great potential in optoelectronic field. For practical fiber quantum key distribution (QKD), InGaAs/InP avalanche photodiodes are the NIR light detector of choice because they are compact and low cost, and allow cryogenic-free or even room-temperature operation [5].

Direct epitaxy on silicon is the most direct method to produce III–V compound semiconductors. However, the lattice mismatch between silicon and III–V compounds is very large (the mismatch between silicon and GaAs and InP is 4% and 8%, respectively), and this causes strain in the materials. In addition, silicon is a nonpolar crystal, whereas III–V compounds are polar. When the two are combined, an antiphase domain will exist at the interface. Therefore, obtaining high-quality epitaxial layers of III–V materials directly determines the performance of semiconductor optoelectronic devices. To solve these problems, buffer layers are usually introduced to reduce the defect density caused by the growth of III–V semiconductors on silicon. For example, the mismatch between Ge and GaAs is 0.08%, and the thermal expansion coefficient is relatively close. In order to suppress the antiphase domain, a silicon substrate with a certain deflection angle can be used, and the substrate can be pretreated at high temperature in the gas of V group elements. By this method, the diatomic steps can be formed on the silicon substrate and the surface reconstruction can be realized, which can suppress the antiphase domain.

1.2.2 Perovskites

Perovskite is a crystal material with the general molecular formula ABX3, which is octahedral in shape and has excellent structural properties. Perovskites have been intensively studied in many fields and applications owing to their excellent photoelectric efficiency and PL [6]. Especially as a promising photoelectric material, they have a wide range of applications in perovskite photovoltaic (PV) cells, LEDs [7, 8], low-threshold lasers [9, 10], electroluminescent devices [11–13], photodetectors [14], and photocatalysts [15]. Perovskite structure has strong designability and excellent photovoltaic performance, which has been a hot research direction in photovoltaics in recent years. Due to their high PL quantum efficiency, perovskites were reported as potential luminescence probes for cell imaging after overcoming the challenge of hypertoxicity by external encapsulation. Perovskite structure material is an important development path for next-generation LEDs for lighting or display. Perovskite is also the representative of the third-generation high-efficiency thin-film battery. In particular, organic perovskite is considered as a promising optoelectronic photoelectric semiconductor material. The explorations in photoelectric devices and bioscience broaden the application of perovskite. However, due to the difficulty in repetitively preparing high-quality perovskite thin films, low coupling efficiency of optical output, lead pollution, and other problems, there is still a certain distance from commercial application.

1.2.3 Organic Optoelectronic Materials

Organic optoelectronic materials are usually organic molecules rich in carbon atoms with large π-conjugated systems, which can be divided into four categories: conjugated conducting polymers, organic conjugated small molecules, phosphorescent heavy metal complexes, and aggregation-induced enhanced luminescence materials. Compared with inorganic optoelectronic materials, organic optoelectronic materials have lower costs, are thin in thickness, are light in material, and usually with higher light absorption coefficient. The material manufacturing process is simple and can be prepared in large areas and on flexible devices by solution method. In addition, organic materials have diversified structure compositions and wide property manipulation, which can obtain the required performance through molecular design. We can prepare nanodevices and molecular devices through bottom-up device assembly methods such as self-assembly. Organic optoelectronic materials are mainly used in organic solar cells, organic LEDs (OLEDs), and other display screens. Organic light-emitting materials are the key core materials of OLED panels, which directly determine the luminous characteristics of panels and are also one of the areas with the highest technical barriers. The application of organic optoelectronic materials in biomedicine has developed from early biosensors and detection to biological imaging and tumor treatment in recent years, especially high-resolution photoacoustic imaging and NIR imaging, providing a new method for the diagnosis of brain tumors and vascular tumors. Organic photovoltaics (OPV) will become an excellent candidate for energy supply on the Internet of Things (IoT) and intelligent wearable devices because of its unique mechanical flexibility, printability, and adjustable light absorption. However, organic photoelectric materials also have many disadvantages, such as low photoelectric conversion efficiency, short life, low mobility of carriers, high resistance, and poor durability. In addition, the synthesis steps of organic optoelectronic materials are cumbersome and not suitable for large-scale production. In organic semiconductors, the diffusion length of excitons is generally short, ranging between 10 and 20 nm [16]. This reduces efficiency, since the electrons accumulated as a result of solar energy conversion cannot easily flow through the photovoltaic semiconductor. Many of the advantages of organic solar cells are overridden by their relatively low efficiency, since this is the most important factor with respect to electricity generation by the cell.

References

1 Moutanabbir, O., Assali, S., Gong, X. et al. (2021). Monolithic infrared silicon photonics: the rise of (Si) GeSn semiconductors. *Appl. Phys. Lett.* 118 (11): 110502.

2 Castellanos-Gomez, A. (2015). Black phosphorus: narrow gap, wide applications. *J. Phys. Chem. Lett.* 6 (21): 4280–4291.

3 Jariwala, D., Sangwan, V.K., Lauhon, L.J. et al. (2014). Emerging device applications for semiconducting two-dimensional transition metal dichalcogenides. *ACS Nano* 8 (2): 1102–1120.

4 Sweeney, S.J. and Mukherjee, J. (2017). Optoelectronic materials and devices. In: *Springer Handbook of Electronic and Photonic Materials* (ed. S. Kasap and P. Capper) (Chapter 35), 897. Cham, Switzerland: Springer.

5 Comandar, L.C., Fröhlich, B., Dynes, J.F. et al. (2015). Gigahertz-gated InGaAs/InP single-photon detector with detection efficiency exceeding 55% at 1550 nm. *J. Appl. Phys.* 117: 083109.

6 Fu, Y., Zhu, H., Chen, J. et al. (2019). Metal halide perovskite nanostructures for optoelectronic applications and the study of physical properties. *Nat. Rev. Mater.* 4: 169–188.

7 Jana, A., Park, S., Cho, S. et al. (2022). Monodispersed perovskite quantum wells for efficient LEDs. *Matter* 5: 384–386.

8 Pacchioni, G. (2021). Highly efficient perovskite LEDs. *Nat. Rev. Mater.* 6: 108.

9 Yan, D., Shi, T., Zang, Z. et al. (2020). Stable and low-threshold whispering-gallery-mode lasing from modified $CsPbBr_3$ perovskite quantum dots@SiO_2 sphere. *Chem. Eng. J.* 401: 126066.

10 Wang, C., Dai, G., Wang, J. et al. (2022). Low-threshold blue quasi-2D perovskite laser through domain distribution control. *Nano Lett.* 22: 1338–1344.

11 Xie, P., Hang, H., Wang, S. et al. (2019). Perovskite bifunctional device with improved electroluminescent and photovoltaic performance through interfacial energy-band engineering. *Adv. Mater.* 31: 1902543.

12 Hang, J., Ren, B., Deng, S. et al. (2020). Voltage-dependent multicolor electroluminescent device based on halide perovskite and chalcogenide quantum-dots emitters. *Adv. Funct. Mater.* 30: 1907074.

13 Liu, D., Liu, X., Gan, Y. et al. (2022). Perovskite/organic hybrid white electroluminescent devices with stable spectrum and extended operating lifetime. *ACS Energy Lett.* 7: 523–532.

14 Ren, H., Chen, J.D., Li, Y.Q., and Tang, J.X. (2021). Recent progress in organic photodetectors and their applications. *Adv. Sci.* 8: 2002418.

15 Zhu, Y., Liu, Y., Miller, K.A. et al. (2020). Lead halide perovskite nanocrystals as photocatalysts for PET-RAFT polymerization under visible and near-infrared irradiation. *ACS Macro Lett.* 9: 725–730.

16 Sajjad, M.T., Ruseckas, A., and Samuel, I.D.W. (2020). Enhancing exciton diffusion length provides new opportunities for organic photovoltaics. *Matter* 3: 241–354.

2

2D Material-Based Photodetectors for Imaging

Wenshuo Xu[1,2,3], Zhuo Wang[1], and Andrew T. S. Wee[2,4]

[1]Shenzhen University, SZU-NUS Collaborative Innovation Center for Optoelectronic Science & Technology, International Collaborative Laboratory of 2D Materials for Optoelectronics Science and Technology of Ministry of Education, College of Optoelectronic Engineering, 3688 Nanhai Street, Shenzhen Guangdong 518060, China
[2]National University of Singapore, Department of Physics, 2 Science Drive 3, Singapore 117551, Singapore
[3]University of Cambridge, Department of Materials Science and Metallurgy, 27 Charles Babbage Road, Cambridge CB3 0FS, United Kingdom
[4]National University of Singapore, Centre for Advanced 2D Materials, 6 Science Drive 2, Singapore 117546, Singapore

2.1 Introduction

Advanced photodetectors play an important role in high-quality imaging. Figure 2.1 shows the key photodetection mechanisms. The photocurrent can be generated either by the excitation of charge carriers or by the heat produced as a result of exposure to light. In the former case, the optical transition is ascribed to photovoltaic, photoconductive, or photogating effect. The photovoltaic effect occurs when the energy of the absorbed photons is greater than the bandgap, so that the electron–hole pairs are excited and separated under the built-in electric field at the metal–semiconductor junction or p–n heterojunction (Figure 2.1a). The photoconductive effect refers to the situation where the photoinduced carriers lead to an increased density of charges and a decreased resistance of the semiconducting material. The excess carriers are drifted in the presence of applied bias, thereby forming photocurrent (Figure 2.1b). As a particular example, the photogating effect happens when the photocarriers are localized at the surface or defect sites of the semiconductor. Their trap states act as a gate, regulating the conductance of the channel. Light irradiation can also increase the temperature of photoactive materials. The photobolometric effect originates from uniform temperature change (ΔT) across the channel. The sign of the photocurrent is dependent on the induced variation in the resistivity, which is a function of temperature (Figure 2.1c). The photothermoelectric effect would happen if the light is illuminated at the junction of two materials with different thermoelectric properties or if the spot size of the incident light is smaller than the channel width (Figure 2.1d). A temperature gradient can be formed when the photon absorption is uneven across the channel,

Optical Imaging and Sensing: Materials, Devices, and Applications, First Edition.
Edited by Jiang Wu and Hao Xu.

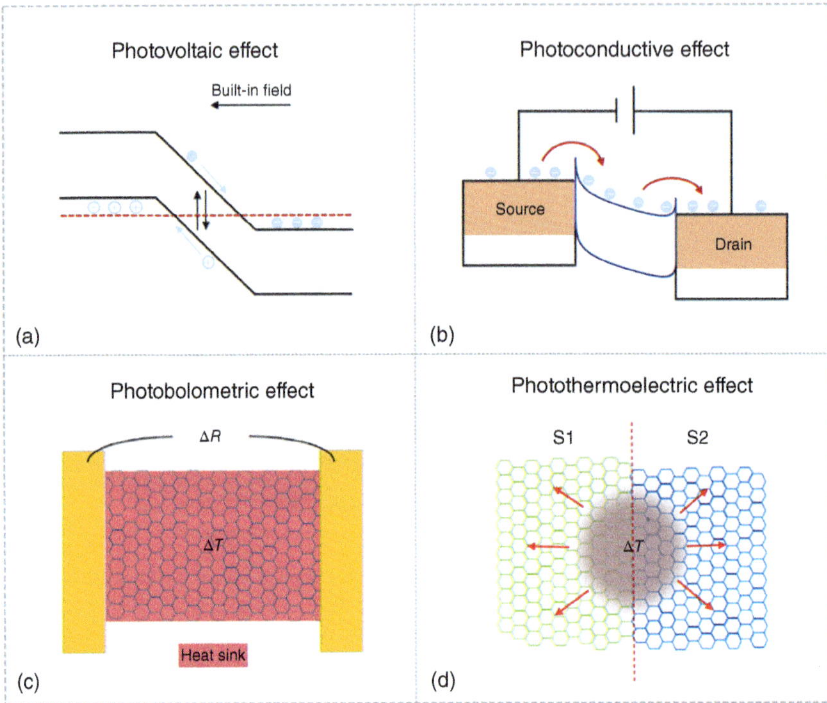

Figure 2.1 Photodetection mechanisms. (a) Photovoltaic effect. (b) Photoconductive effect. (c) Photobolometric effect. (d) Photothermoelectric effect. S1 and S2 represent two semiconductors with different thermoelectric properties.

giving rise to a photovoltage and driving a current through the detector. The sign of the photocurrent is related to the carrier polarity and the difference of Seebeck coefficients between the adjacent illuminated regions. The driving force to form a potential difference is built-in electric field for photovoltaic effect, while it is temperature gradient for the photothermoelectric effect. Both of them are based on the spontaneous transport of electrons and holes toward the two ends of the channel, offering access to self-driven photoresponse. In contrast, for photoconductive, photogating, and photobolometric effects, the photocurrent cannot be generated without an external bias voltage.

The performances of photodetectors are evaluated by several figure-of-merits, including photoresponsivity (R), external quantum efficiency (EQE), photoconductive gain, response speed, detectivity (D), or specific detectivity (D^*), and so on. R is the ratio of photogenerated voltage or photocurrent to the incident light power, expressed as $R = I_{ph}/P_{in}$, where I_{ph} represents the photocurrent and P_{in} is the power of the incident light. Its value depends on the light wavelength because the photon absorption is correlated with the bandgap of the semiconductor. EQE is the number of collected electron–hole pairs divided by the number of incident photons, i.e. $EQE = (I_{ph}/e)/(P_{in}/E_{ph})$, where e is the charge of an electron and E_{ph} is the energy of a photon. Photoconductive gain evaluates the capability of a photodetector in generating charge carriers in response to a single photon. Response speed is usually characterized by the rise and decay/fall times with the light pulse switched on and off, i.e. the time intervals needed for the output signal to increase from 10% to 90% and decrease from 90% to 10% of its on-state maximum, respectively. D is defined as the inverse of the noise-equivalent power. Its normalization result to the active detecting area and the detection bandwidth, known as D^*, has been widely used to quantify the sensitivity of a photodetector to signals against the noise. $D^* = R/(2eJ_d)^{1/2}$, where J_d is the dark current density, and the unit is Jones or cm $Hz^{1/2}$ W^{-1}.

The detection range is categorized according to the light wavelength. The absorption spectrum includes ultraviolet (UV) (10–400 nm), visible (400–750 nm), infrared (IR, 0.75–30 μm), and terahertz (THz) (30 μm to 3 mm) bands. Specifically, the UV range is further divided into UV-A (320–400 nm), UV-B (280–320 nm), and UV-C (100–280 nm), and the IR range is further divided into near-infrared (NIR, 750 nm to 3 μm), mid-infrared (MIR, 3–5 μm), long-wave infrared (LWIR, 8–12 μm), and very long-wave infrared (VLWIR, 12–30 μm). The full electromagnetic spectrum is shown in Figure 2.2, where the visible region is specified. For certain incident light, it is important to choose a photodetector with appropriate response wavelength. In Sections 2.2, 2.3 and 2.4, we will discuss about the visible, IR, and broadband photodetectors that are composed of two-dimensional (2D) materials.

Traditional thin-film materials fail to meet the increasing demands of modern device manufacture because of the limited detection wavelength, strict working conditions, high cost, and difficult fabrication. For instance, electronics and optoelectronics based on silicon play a leading role in the integrated circuit industry.

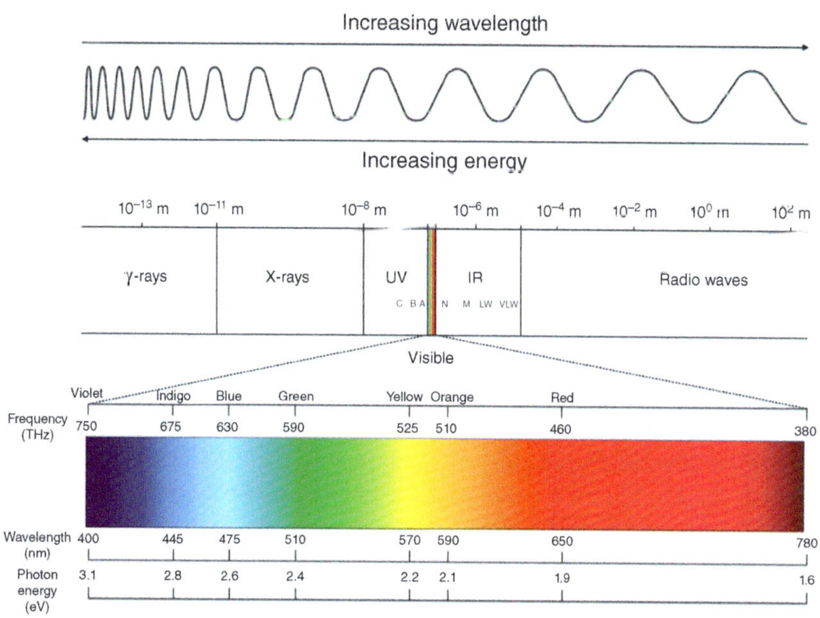

Figure 2.2 Electromagnetic spectrum. The wavelength for each region is indicated.

However, the indirect energy gap of silicon (1.12 eV) constricts its response to light within the visible–NIR regime [1]. GaAs and InSb have been employed for IR photodetectors, which often operate at cryogenic temperatures [2, 3], while the cooling requirement increases the expense and volume of the systems. Other examples include InGaAs and HgCdTe thin films, the growth procedures of which are complicated [4]. Therefore, it is necessary to develop some alternatives to overcome the above challenges.

A variety of 2D materials-based photodetectors have been reported in the past few years, including graphene, transition-metal chalcogenides (TMCs), transition-metal oxides (TMOs), black phosphorus (BP), arsenic phosphorus (AsP), and indium selenide (InSe). They can also form van der Waals heterostructures without the limitation of lattice matching. Many of these materials can interact with light and convert the photons to an electrical signal dictated by their unique band structures. Although graphene has large electron mobility ($\approx 10^5\ cm^2\ V^{-1}\ s^{-1}$), only $\approx$2.3% of the incident visible and IR light can be absorbed in monolayer thickness [5]. Furthermore, the zero-gap of graphene results in short lifetime of the photoinduced carriers, which are not favorable for efficient yield of photocurrents. As compared with graphene, the van Hove singularities in 2D transition metal dichalcogenides (TMDs) contribute to stronger light–matter interactions in visible and IR light ranges, and their tunable band gap by varying their layer number allows the detection of light with different wavelengths [6]. A promising method to exploit the properties of both graphene and TMD is to create hybrid devices where graphene is used as the electrodes and TMD acts as the photoactive channel. Moreover, owing to the atomic thickness and absence of dangling bonds, 2D materials often exhibit fast carrier transport and outstanding gate modulation ability. Previous works have demonstrated that 2D materials and their heterostructures are superior to conventional semiconducting materials in terms of the spectral coverage, mechanical flexibility, transparency, ease of processing, etc.

High photoresponsivity, fast response, low dark current, and large ON/OFF ratio are highly desired for advanced photodetectors. However, some pristine 2D materials cannot meet such demands. To enhance the device performance, several methods have been proposed [7]. Quick charge extraction is believed to assist the photodetection. For example, contacting the semiconductor with highly conductive 2D layers, usually graphene, accelerates the separation of electrons and holes, and reduces the contact resistance. Another strategy is building up vertical 2D TMD heterostructures to modify the band structure and match the cutoff wavelength of the incident light. In these heterostructures, adding a dielectric barrier layer can reduce the dark current and therefore improve the detectivity. In addition, plasmonic nanostructures, which support surface plasmons, can strengthen the light–matter interaction in materials, so they are used to improve the photoresponsivity of photodetectors based on 2D materials. Regarding the preparation methods, chemical vapor deposition (CVD) and molecular beam epitaxy are advantageous in producing clean interfaces. The strong interlayer coupling guarantees efficient charge transfer across the heterojunction. In addition to the structural design and performance improvement of 2D materials-based photodetectors, recent studies also focus on their large-scale manufacturing from single to arrayed devices and extension of their detection range from narrowband to broadband. All of these aspects are crucial for the development toward imaging applications and hence will be covered in this chapter.

2.2 Visible-Light Photodetectors

Silicon and germanium compounds with bandgaps of ≈1 eV have been used for visible-light photodetectors. Nevertheless, short-channel effects often take place when the channel length is approximately equal to the depletion layer widths of the source and drain junctions. Typical resultant phenomena may include barrier lowering, velocity saturation, quantum confinement, and hot carrier degradation [8]. In addition, the manufacturing costs of Si and Ge semiconductors are relatively high. These disadvantages have restricted the further miniaturization and large-scale integration of conventional photodetectors. The absorption wavelengths of many 2D TMDs are also in the visible-light spectrum, such as molybdenum dichalcogenides (MoX_2, X = S, Se, Te), tungsten dichalcogenides (WX_2, X = S, Se), tin dichalcogenides (SnX_2, X = S, Se), gallium chalcogenides (GaX, X = S, Se, Te), and indium selenides (InSe, In_2Se_3). They can thus be promising supplements to the traditional materials and have attracted much research interest.

The earliest demonstration of a 2D TMD-based phototransistor was reported by Yin et al. in 2012 [9]. Utilizing mechanically exfoliated monolayer MoS_2 as the channel, the device exhibits a photoresponsivity of 7.5 mA W^{-1} upon laser excitation at 488 nm with a response time of ≈50 ms. Follow-up studies were centered on the factors affecting the photoresponse [10].

First, the photoresponsivity is influenced by the external conditions, including the incident light wavelength, excitation power density, and applied electric field. The absorption of light is dependent on the intrinsic energy band structure and absorption coefficient of semiconducting materials. When the photon energy is larger than the bandgap, photocarriers, i.e. electron–hole pairs, are formed via the transition of electrons from the valence band to the conductance band. An increase in the concentration of photocarriers due to the enhanced absorption corresponds to a higher responsivity. Although more photocarriers can be produced as the excitation power increases, the reduced efficiency of carrier excitation and the enhanced Coulomb scattering may have a dominant effect on the generation of photocurrent. Therefore, the responsivity generally decreases with increasing light power density. The power dependence can be explained by Auger recombination, i.e. a nonradiative process in which electron–hole pairs recombine and give off energy to another electron in the same conduction band [11]. This effect is enhanced at high power density. It influences the lifetime of photocarriers and, thus, the efficiency of the device. Implementation of a gate bias can vary the carrier density in the semiconductor via modulation of the Fermi surface. This enables restriction of the dark current and improvement in the responsivity.

Second, the responsivity is dependent on the layer number of TMDs. Changing their layer number can modulate the bandgap and hence tune the response wavelength of the photodetector. For example, the bandgap of BP shows direct characteristics. Its value increases from ≈0.35 eV for the bulk to ≈2.0 eV for the monolayer [12]. Many TMDs also present increasing bandgaps as the layer number decreases, yet unlike BP, they undergo an indirect-to-direct transition when thinned down to the monolayer limit as a result of quantum confinement effects [13]. The layer-dependent bandgap dictates the cutoff wavelength in the absorption spectrum and influences the photodetection range. Lee et al. studied top-gate phototransistors based on mono-, bi-, and three-layer MoS_2 flakes. Mono- and bi-layer MoS_2 with optical energy gaps of 1.82 and 1.65 eV, respectively, exhibit maximal photoresponse in the red-light spectrum, whereas three-layer MoS_2 with an optical energy gap of 1.35 eV shows the best green-light detection capability [14]. Increasing thickness within a certain range often leads to enhanced optical absorption and responsivity. 2D ReS_2 has a direct band gap with a constant value (1.5 eV) irrespective of its layer number. The photodetector based on 30-nm ReS_2 can reach a high responsivity of 2.5×10^7 A W^{-1}, which is among the record-high values achieved by 2D materials [15]. This outstanding performance can be attributed to the combination of the relatively large thickness and the direct bandgap, which brings enhanced light absorption. As in other special cases, bulk PtS_2 and $PtSe_2$ are semimetals, whereas they become semiconductors with bandgap opening at reduced thicknesses below a few nanometers (≈3 nm for PtS_2 and ≈4 nm for $PtSe_2$). These enable photodetection in visible-light spectrum [16, 17].

Third, the photoresponse of TMDs is influenced by charge trapping in the TMD layers or at the TMD–substrate interface. Kufer and Konstantatos [18] encapsulated monolayer and bilayer MoS_2 with atomic layer-deposited HfO_2 to isolate them from environmental adsorbates. The increased carrier mobility due to the reduction in extrinsic impurities and the additional electrons induced by the positive charges in the top HfO_2 layer contribute to the observation of a high photoresponsivity (≈10^4 A W^{-1}). Jo et al. [19] enhanced the performance of a WSe_2 phototransistor via conversion of the device type from p- to n-channel with triphenylphosphine as the dopant. Further improvement was achieved by intercalating a hexagonal boron nitride (hBN) film underneath the WSe_2 to provide a charge-free environment and increase the carrier mobility. Wang et al. [20] reported a ferroelectric-driven MoS_2 photodetector with a poly(vinylidene fluoride-trifluoroethylene) film as the top gate. The remnant polarization of this polymer induces a strong electrostatic field, maintaining the MoS_2 in a fully depleted state and largely suppressing the generation of dark current. As a result, a responsivity of 2570 A W^{-1} along with a D^* of ≈2.2×10^{12} Jones was achieved at zero bias. The performance of the 2D TMD photodetector is also dependent on its surrounding atmosphere. This is ascribed to the charge transfer between the TMD and the absorbed gas molecules, which modulates the doping level and the carrier density in the TMD. Auger recombination can happen at increased carrier density, usually leading to higher response speed but lower photoresponsivity. Moreover, the adsorbates can serve as the recombination centers of photocarriers, thereby increasing the photocurrent relaxation rate and decreasing the response time. Lan et al. [21] demonstrated a sharp increase in the responsivity of multilayer WS_2 photodetector in vacuum

(18.8 mA W^{-1}) compared with that in air (0.2 μA W^{-1}) where the oxygen acts as an electron acceptor. The response times of these WS_2 photodetectors are on the order of sub-seconds in vacuum and <4.5 ms in air [21]. The preparation and processing methods of 2D materials also have crucial influences on their charge states and photodetecting properties. Solution-processed TMDs usually contain higher concentrations of defects and charge impurities, therefore displaying more charge-scattering centers and inferior performances to mechanically exfoliated and CVD-grown samples. Shim et al. studied the effect of oxygen plasma treatment on the optoelectronic properties of few-layer ReS_2 phototransistor. They showed decreases in the rise and decay times due to the formation of trap states within the bandgap of the surface ReS_2. Lu et al. [22] showed that the S vacancies in 2D GaS could lead to reduced optical responses and stability. Moreover, its photodetection range can be shifted from visible-to-UV (stoichiometric GaS) to UV-discriminating ($GaS_{0.87}$) via control of the defect concentration.

Fourth, the photoresponse is affected by the structure of the device. Kwon et al. [23] demonstrated that the multilayer MoS_2 phototransistor with a local bottom gate is superior to that with a global bottom gate. The former geometry, shown in Figure 2.3a, facilitates the trapping of the photoexcited holes in the channel region. Due to the accumulation of holes (Figure 2.3b), the potential barrier is reduced for the injection of electrons, delivering a larger thermionic current and a higher photoresponsivity (Figure 2.3c).

Fifth, the contact between the channel and the electrodes also strongly affects device performance. The work functions of the metal and the semiconductor are not identical to each other, so when they contact each other, their misaligned Fermi levels would result in band bending and formation of a Schottky barrier. The transport of carriers within the device is therefore restricted. The surface states in the bandgap of the semiconductor "pin" its Fermi-level position independent of the metal, known as Fermi-level pinning. Zhang et al. [25] investigated the effect of metal contacts on the photodetection performances of monolayer WSe_2 grown by CVD. The responsivity of the Pd-contacted device is as high as $\approx 1.8 \times 10^5$ A W^{-1}, while the value decreases by several orders of magnitude for the Ti-contacted device. However, the response time for the Ti contacts is <23 ms, much shorter than that for the Pd contacts (>5 seconds). These large variations result from the differences in the Schottky barriers between the metals and the WSe_2. Semimetallic graphene has also been utilized as the electrodes to provide good carrier mobility, and its high transparency can guarantee the optical transmission. The responsivity of the graphene-sandwiched WS_2 photodetector is about 4 orders of magnitude higher than that with Ti/Au contacts under 532-nm laser [26, 27]. Semimetallic and transparent graphene have also been used as the contact electrodes for other 2D semiconductors, such as perovskites, to improve their photodetection performances. Tu et al. [28] demonstrated a graphene/$(C_6H_5C_2H_4NH_3)_2PbI_4$/graphene photoconductor, showing narrowband responses at 517 nm with ultralow dark current (10^{-14} A), high D^* (10^{13} Jones), large ON/OFF ratio (10^6), and short rise time (10^2 μs). The photoresponse can be further enhanced by adopting direct CVD growth to construct graphene/semiconductor/graphene heterostructures. The epitaxial

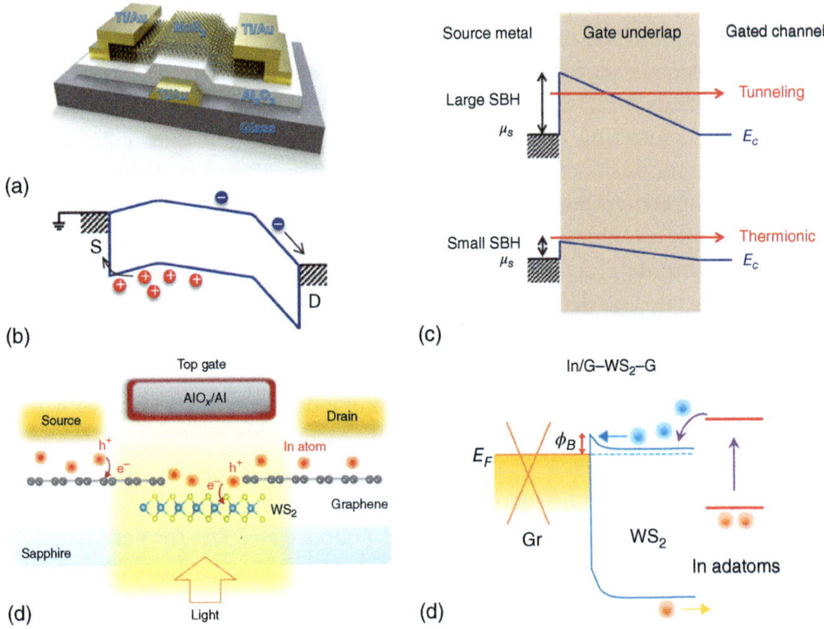

Figure 2.3 (a) Schematic 3D view of a MoS_2 photodetector with a local bottom-gate structure. (b) Energy band diagram in which the holes are accumulated and the low barrier for the generation of large photocurrent. (c) Tunnelling (upper) and near-thermionic emission (lower) through the gate underlap with a large and a small SBH, respectively, at the metal–MoS_2 interface. μ_s is the chemical potential at the source and E_c is the conduction band edge of MoS_2. (d) Cross-sectional view of a WS_2 photodetector with graphene contacts and chemisorbed In atoms. AlO_x/Al was used as the top gate. (e) Band diagram of the In/graphene–WS_2–graphene device, which illustrates the In-induced photogating effect. Source: (a–c) Reproduced with permission [23]. Copyright 2015, Wiley-VCH GmbH. (d, e) Reproduced with permission [24]. Copyright 2019, American Chemical Society.

graphene–semiconductor interfaces exhibit decreased Schottky barrier and contact resistance in comparison with that made via conventional layer-by-layer transfer, thus displaying higher electron transfer efficiency [29]. Another preparation method to obtain superior performances is introducing heteroatoms. Yeh et al. [24] enhanced the photogating effect in graphene–WS_2–graphene photodetector by covering the lateral junction with In atoms (Figure 2.3d). The photogenerated holes remain trapped in the In adatoms, during which the electrons are transferred from the In to the WS_2 channel and recirculate through the device (Figure 2.3e). The fast carrier dynamics results in a short response time of 40–65 μs, achieved simultaneously with a high responsivity of 2.6×10^3 A W^{-1}. Efficient photodetection was also achieved in TMD–TMD homojunctions. Wang et al. [30] employed two degenerately p-doped WSe_2 thin films as the drain and source to form an ohmic contact with the pristine WSe_2 channel. A fast photoresponse was realized as a result of the low RC time constant and high carrier transit speed across the interface. Since metallic TMDs have similar work functions to semiconducting TMDs, the

Schottky barrier height (SBH) is smaller at 2D TMD metal–2D TMD semiconductor junctions compared with pure metal–2D TMD semiconductor junctions, delivering lower contact resistance and larger gain. However, the response time could be relatively longer than that of a junction with a higher Schottky barrier. Zhang et al. [31] synthesized 2D TMD metal/2D TMD semiconductor vertical stacks of NbS_2/MoS_2 by CVD. Epitaxial growth was achieved due to the small lattice mismatch between NbS_2 and MoS_2. The responsivity of the NbS_2-contacted MoS_2 photodetector is 4 orders of magnitude higher than that for its Ti/Au-contacted counterpart, which was attributed to the Schottky barrier and the contact resistance of the NbS_2/MoS_2 heterostructure.

The responsivity of 2D materials-based photodetectors can also be improved by facilitating the carrier transport via formation of van der Waals interfaces. Lee et al. [32] utilized conducting NiO_x as the top gate on a few-layer MoS_2 channel (Figure 2.4a,b). This Schottky electrode can suppress interfacial electron scattering at the NiO_x–MoS_2 junction, allowing the MoS_2 to show near-intrinsic mobility of $1200\,cm^2\,V^{-1}\,s^{-1}$ (Figure 2.4c). Vertical p–n homojunctions can be formed by stacking TMDs with the same chemical composition but different doping levels. Such structures have limited densities of carrier trap sites due to the continuous band alignment and exhibit low contact resistance, giving high carrier transport efficiency through their interface. Li et al. [33] fabricated a multifunctional device based on p-MoS_2/n-MoS_2, where the p-type and the n-type MoS_2 were prepared using $AuCl_3$ and benzyl viologen as the dopants, respectively (Figure 2.4d). In contrast to unipolar MoS_2, this homojunction exhibits ambipolar transport behavior and can function as the channel of a phototransistor that shows photoresponse under biasing conditions (Figure 2.4e). Low Schottky barrier has been verified at graphene–TMD interfaces. Tan et al. [34] reported a hybrid photodetector based on a WS_2/MoS_2 heterobilayer with graphene contacts (Figure 2.4f). Type II band alignment is formed at the WS_2/MoS_2 junction, which leads to the separation of the photoexcited electrons and holes into different layers (Figure 2.4g). This can suppress the electron–hole recombination, and the photoinduced excitons tend to dissociate into spatially isolated carriers. The transfer curve indicates n-type behavior of this device (Figure 2.4h), suggesting that MoS_2 works as the main conductive channel. At forward biases, holes tend to accumulate near the graphene–TMD heterojunction at the source region due to the trapped states in the WS_2/MoS_2, leading to a positive photogating effect and a reduced Schottky barrier. The thermionic emission is increased as a consequence, thereby facilitating the electron transport from the graphene electrode to the WS_2/MoS_2 channel. Owing to the efficient circulation of carriers, this device exhibits a high responsivity of $2340\,A\,W^{-1}$ with a large gain over 3.7×10^4, outperforming most lateral TMD-based photodetectors investigated so far.

In addition to TMDs, groups IIIA, IV-A, IV-B, and ternary metal chalcogenides have also demonstrated potential in photodetection. GaS phototransistors exhibit high D^* in the range of $\approx 10^{12}$–10^{13} Jones, which surpasses that of commercial InGaAs devices [35]. The responsivity of GaTe photodetectors up to $10^4\,A\,W^{-1}$ is 3–4 orders of magnitude higher than the values of GaS and GaSe devices, owing to its larger photon absorption coefficient [36]. In_2Se_3 compounds exist in

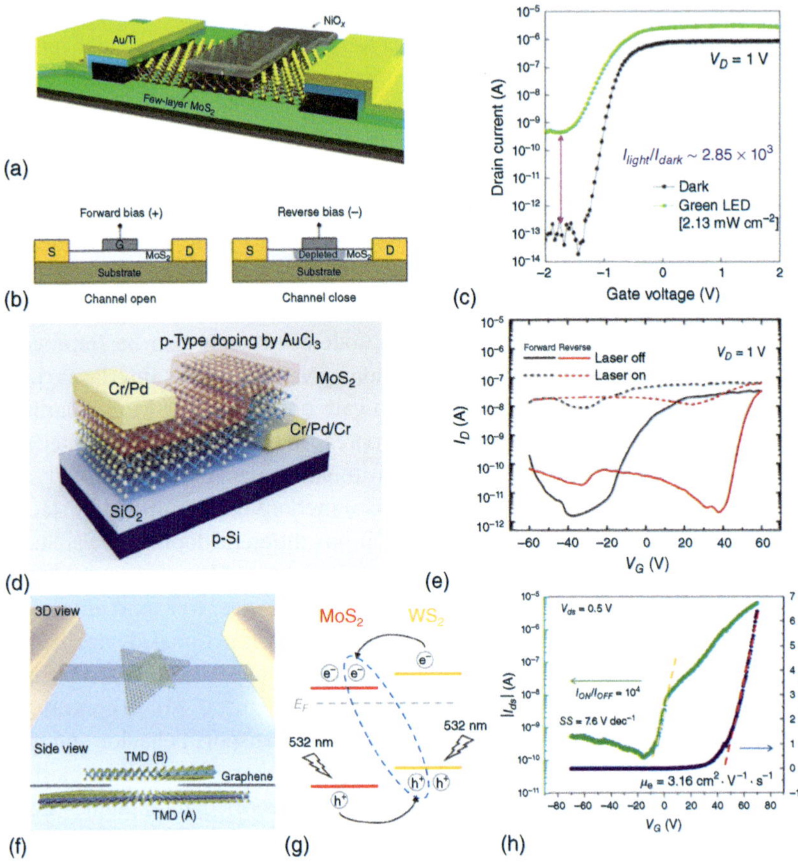

Figure 2.4 (a) Perspective and (b) cross-sectional views of a NiO_x-gated MoS_2 photodetector. (c) Transfer curves of this device under dark and illuminated conditions. Light source: 520-nm green light-emitting diode (LED). (d) 3D schematic of a photodetector based on p-MoS_2/n-MoS_2 vertical homojunction and (e) its transfer curves with applied forward and reverse biases. Inset: logarithmic dependences in the ON and the OFF states. Excitation source: 655-nm monochromator. (f) 3D view of a graphene-contacted photodetector based on WS_2/MoS_2 vertical heterojunction. (g) Schematic diagram of the type II band alignment in the WS_2/MoS_2 bilayer stack. (h) Transfer curves of the graphene/WS_2/MoS_2/graphene device under dark and illuminated conditions on logarithmic (green) and linear (blue) scales. Light source: 532-nm laser. Source: (a–c) Reproduced with permission [32]. Copyright 2015, American Chemical Society. (d, e) [33]/Springer Nature/Licensed under CC BY 4.0. (f–h) [34]/Reproduced with permission from John Wiley & Sons.

multiple structural phases. The most common form, α-In_2Se_3, has a direct bandgap of 1.3 eV and shows good visible-light-harvesting performance. Island et al. [37] demonstrated that the photodetection mechanism of multilayer α-In_2Se_3-based phototransistor is a combination of the gate-independent photoconductive effect and the photogating effect that is gate-modulable. The dominant mechanism responsible for the photocurrent generation can be tuned with bias voltages, giving rise to fast conduction in the OFF state and efficient gating in the ON state. Lei et al. [38] realized an avalanche effect in ultrathin InSe-based photodetectors by exploiting the large SBH between InSe and Al electrodes. This allows multiplication of the carriers and reduction in the dark current to pA scale as a function of the applied bias, leading to enhanced photoresponse. In contrast to most TMDs, group IVA chalcogenides with 1 : 1 stoichiometry are p-type semiconductors. They can form ohmic contacts with n-type metal dichalcogenides, offering promise for high-performance device applications of 2D heterostructures. Afzal et al. [39] studied the electronic and optoelectronic properties of vertically stacked p-GeSe/n-$PdSe_2$. The low Schottky barrier results in good rectification characteristics with a ratio up to 5.5×10^5. Photovoltaic measurements showed a high photoresponsivity of 1×10^3 A W^{-1} and ultrafast response within 2 μs. Ternary metal chalcogenides can provide multiple degrees of freedom to tailor their properties by altering the stoichiometric ratio. For instance, the bandgaps of $Sn_xGe_{1-x}S$, $Sn_xGe_{1-x}Se$, GeS_xSe_{1-x}, and SnS_xSe_{1-x} can be continuously tuned via control of their compositions [40]. Perumal et al. [41] optimized the performance of SnS_xSe_{1-x}-based phototransistors by changing the Se contents in SnS and showed the highest responsivity of 6×10^3 A W^{-1} achieved by $SnS_{0.8}Se_{1.2}$.

2.3 Infrared Photodetectors

IR photodetectors can be employed in many fields, such as military communication, remote monitoring, and thermal imaging. Their applications have been dominated by InAs, InSb, InGaAs, and HgCdTe alloys in the last few decades [42]. Nonetheless, the development of these devices is limited due to the high manufacturing costs, strict operation requirements, and environmental toxicity. Many traditional semiconductors also have severe short-channel effects. The scaling of Si-based transistors is supposed to fail when the gate length is below 5 nm. Such limit brings about challenging integration problems in achieving minimization while retaining good performances [43]. Considering these issues, 2D materials with bandgaps beyond silicon are desirable candidates for IR photodetection and can be fabricated using affordable techniques. Typical synthesis approaches include mechanical and liquid-phase exfoliation, molecular beam epitaxy (MBE) and CVD growth, pulsed laser deposition, and magnetron sputtering [44].

2D semiconductors with narrow bandgaps (<1 eV) are potential candidates for IR detection. BP and AsP are typical examples. Buscema et al. [12] conducted a pioneering study on the NIR photodetector based on few-layer BP, yet the responsivity is low (<1 mA W^{-1}). The following work demonstrated that the photocurrent could increase several times when the temperature is decreased from 300 to 20 K [45]. The enhancement is ascribed to the more efficient carrier generation and transport at low temperatures. However, the cryogenic operation is impractical for real BP devices. Other methods to improve the photoresponse include surface modification and chemical doping. Xiang et al. [46] demonstrated the improvement in the responsivity by functionalizing BP with MoO_3 and Cs_2CO_3, which gives rise to n- and p-type doping, respectively. Surface charge transfer was verified at the BP–dopants interfaces. The increased electron or hole mobility enhances the ambipolar transport behavior of BP. Xu et al. [47] synthesized centimeter-scale Se-doped BP nanosheets. With Se dopants, the Fermi level of BP shifts toward the conduction band, resulting in lowered Schottky barrier and decreased contact resistance. The Se doping also reduces the bandgap of BP. This may result in a larger absorption coefficient. As a consequence, the responsivity is increased to 15.33 $A W^{-1}$ at room temperature, 20 times higher than that of the pristine BP. Yu et al. [48] treated a selected region of the intrinsically p-type BP with electron dopants (benzyl viologen), forming an in-plane p–n junction. The large built-in potential at the p-BP/n-BP interface facilitates carrier transport and leads to improved responsivity.

Noble-metal dichalcogenides, such as PtX_2 and PdX_2 (X = S, Se, Te), can also be materials of choice for high-performance IR photodetection. The bandgap of atomically thin $PtSe_2$ is tunable via thickness control and defect engineering. Yu et al. [49] utilized different methods to prepare $PtSe_2$ nanosheets, including chemical vapor transport, CVD, and selenization of Pt films, and fabricated a bilayer $PtSe_2$ photodetector with the detection wavelength extended to the MIR region (10 μm). Its responsivity of ≈4.5 $A W^{-1}$ is on par with the commercial HgCdTe and quantum well-based devices. The role of defects in the photoresponse of 2D $PtSe_2$ is investigated through argon plasma treatment with controlled dosages. The X-ray photoelectron spectra and Fourier-transform IR spectra indicate the generation of Se vacancies, the concentration of which governs the bandgap evolvement of the $PtSe_2$. The absorption cut-off wavelength shows blue shifts with the defect concentration, corresponding to a decrease in the photon energy. Xu et al. [50] synthesized ultrathin $PdSe_2$ nanosheets (a few-nm-thick) by CVD. The as-grown $PdSe_2$ displays n-type behavior. Its electron mobility is on the order of $10^2\ cm^2\ V^{-1}\ s^{-1}$ and can remain stable over months. The $PdSe_2$ photodetector exhibits a photoresponsivity of 660 $A W^{-1}$ in the NIR regime (914 nm).

The ultranarrow bandgaps of semimetallic 2D materials allow them to exhibit absorption in the IR spectrum. As a typical example, 2D $PtSe_2$ displays a semiconductor-to-semimetal transition when its layer number increases. The bandgaps of monolayer and bilayer $PtSe_2$ are ≈1.2 and ≈0.3 eV, enabling them to show NIR and MIR photoresponses, respectively. The detection range can be extended toward the LWIR region by further increasing its layer number or introducing Se vacancies [49]. Moreover, the high electron mobility of $PtSe_2$ leads to fast IR photoresponse. Wang et al. [51] demonstrated that the rise and decay times of few-layer $PtSe_2$ photodetector are down to ≈29 and ≈156 ps, respectively, when the device is illuminated using 1560 nm light. Integrating these 2D semimetals with other materials can also increase the response speed of the photodetectors. Zhang and coworkers [52] integrated multilayer $PtSe_2$ into Cs-doped $FAPbI_3$ and obtained the fastest 2D perovskite photodetector. The rise/decay times of the self-powered device are 78/60 ns at the wavelength of 808 nm. The fast photoresponse can be attributed to the built-in electric field at the $PtSe_2/FA_{0.85}Cs_{0.15}PbI_3$ junction, which can efficiently separate the photogenerated electron–hole pairs, and to the high carrier mobilities of both constituting materials that allow short transit time of the photocarriers across the channel. Semimetallic graphene has also been utilized to achieve fast IR detection by forming heterointerfaces with TMD semiconductors. Tao et al. [53] constructed a graphene/MoS_2 vertical stack, the response time of which was estimated to be a few tens of nanoseconds under 905 nm light illumination. The atomically thin MoS_2 nanosheet between graphene and Si substrate not only acts as a tunneling layer for the photoexcited carriers but also passivates the interface trap states. However, the high dark current, short carrier lifetime, and low optical absorption efficiency often give rise to relatively low responsivity. The responsivity can be increased by sandwiching the MoS_2 with graphene. The top graphene exposed to air is more p-doped than the bottom graphene supported by SiO_2/Si. The different doping levels lead to the creation of an internal field. Asymmetric potential barriers are consequently formed between top graphene–MoS_2 and bottom graphene–MoS_2 junctions, which promote carrier tunneling and facilitate photocurrent generation. However, this effect is attenuated as the layer number of MoS_2 increases because of the reduced difference in the effective barrier heights [54].

IR photodetection has also been realized by fabricating 2D semiconductor/2D semiconductor or 2D semimetal/2D semiconductor heterojunctions with type II band alignment that gives rise to low-energy interlayer excitons (<1.65 eV) upon light excitation. For example, WS_2 and MoS_2 show responses to visible light because of their relatively wide bandgaps (>2 eV), while photoresponse at 1030 nm has been observed in WS_2/MoS_2 heterostructures [55]. The IR response originates from the shrunk energy interval of electrons transition induced by strong interlayer coupling, which enables the heterostructure to detect light with longer wavelength. Long et al. fabricated IR photodetectors based on AsP/MoS_2 [56] and $PdSe_2$/MoS_2 [57] vertical stacks (Figure 2.5a,b). The strong built-in electric fields at the interfaces efficiently reduce the dark noise currents. This enables the AsP/MoS_2 and $PdSe_2$/MoS_2 devices to show high sensitivities with the D^* on the order of 10^9 Jones at 3–5 μm (MIR) and beyond 10 μm (LWIR), respectively, surpassing many conventional IR materials. The comparisons are shown in Figure 2.5c. However, the indirect bandgaps are not favorable for optical generation. In contrast, the WSe_2/ReS_2 heterojunction shows a near-direct bandgap [59]. The photoexcitation is more efficient because almost no phonon-assisted processes are involved in the optical transition. Moreover, the large band offsets between WSe_2 and ReS_2 allow prompt dissociation of the photocarriers toward the contacts, resulting in a short response time (5 μs).

Mixed-dimensional heterostructures are alternatives to achieving extraordinary IR photoresponses. Jin et al. [60] reported the epitaxial growth of 0D PbS nanocrystals from 2D PbI_2 nanosheets through a liquid-phase sulfuration process. The PbS/PbI_2 photodetector exhibits a wide detection range up to 2 μm, a fast response within ≈400 μs, and a high D^* over 10^{12} Jones. The outstanding performances are mainly attributed to the role of the PbI_2 layer in passivating the defect states of the PbS crystals. Moreover, the photoresponse dynamics is improved via structural design of the device. When the PbS/PbI_2 heterostructure is placed on top of the Au electrodes, Au/PbI_2 Schottky contacts are formed, which can depress

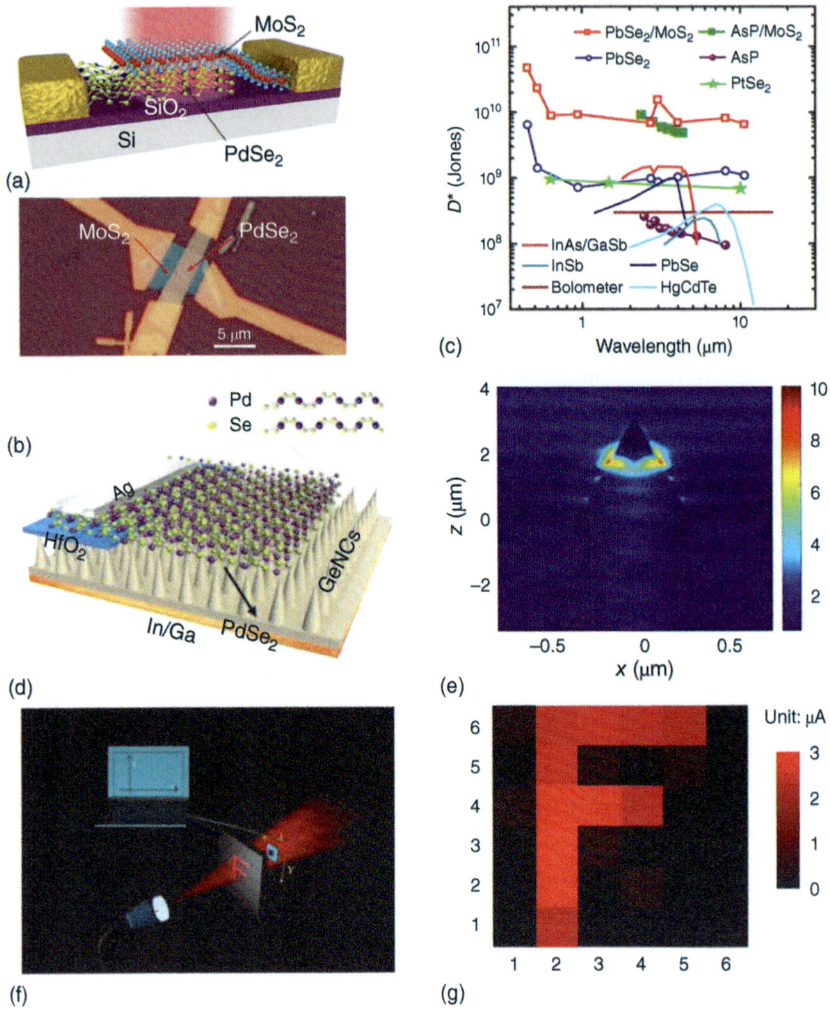

Figure 2.5 2D/2D and 2D/3D heterojunction photodetectors. (a–c) $PdSe_2/MoS_2$. (a) Schematic diagram and (b) optical microscopic image of the device. (c) Room-temperature D^* versus illumination wavelength of various 2D materials and conventional IR-detecting materials. (d–g) $PdSe_2$/Ge nanocone arrays. (d) Schematic diagram of the device. (e) Simulated electric-field density distribution under 1550 nm illumination. (f) Schematic diagram of the measurement setup for IR image sensing and the (g) resultant photocurrent map under 1550 nm illumination. Source: (a–c) Reproduced with permission [57]. Copyright 2019, American Chemical Society. (d–g) Reproduced with permission [58]. Copyright 2019, Wiley-VCH GmbH.

the dark current and expand the depletion region. Liang et al. [58, 61] proposed 2D/3D heterojunctions consisting of multilayer $PdSe_2$ and arrayed Ge nanocones (Figure 2.5d) or Si nanopyramids. The 3D geometric structures effectively confine the incident IR. The enhanced optical absorption is verified by the electric-field

Table 2.1 Key figures of merit for room-temperature IR photodetectors based on traditional semiconductors, 2D materials, and 2D/2D heterostructures.

Traditional material	Wavelength (μm)	R_{max}	Bias	Gain or *EQE*	Rise and fall times	D^* (Jones)
InAs [62]	4.1	4.6×10^{-2} A W^{-1}	2 V	35%		6.7×10^{7}
InGaAs	1.6	6.5×10^{3}	0.5 V	5.04×10^{5}%	70 ms, 280 ms	
InAsSb [42]	10.6	3×10^{-5} V cm^2 W^{-1}				1.5×10^{8}
InAs/GaSb [42]	11	2×10^{-3} A W^{-1}	1 V			1.3×10^{8}
HgCdTe [42]	10.6				<1 ns	2×10^{8}
PbSe [63]	4.4	6×10^{2} V W^{-1}			<2 μs	3.1×10^{9}
VO_x (bolometer) [64]	8–12	9.2×10^{3} V W^{-1}	28 μA			1.68×10^{8}
2D material	**Wavelength (μm)**	**R_{max} (A W^{-1})**	**Bias (V)**	**Gain or *EQE***	**Rise and fall times**	**D^* (Jones)**
BP [65]	7.7	2.2×10^{-3}	1.2			
AsP [56]	3.66	1.8×10^{-1}	0	6.1%		1.8×10^{8}
$MoS_{2.15}$ [66]	7.8	2.2×10^{-2}				
AsP/MoS_2 [56]	2.36	2.2×10^{-1}	0	11.36%		9.2×10^{9}
WSe_2/ReS_2 [59]	1.77	3	0	600%	5 μs, 5 μs	8.39×10^{10}
$PtSe_2$ [49]	10	4.5	0.1	17.5	1.1 ms, 1.2 ms	7×10^{8}
$PdSe_2$ [67]	1.064	708	1	8.27×10^{4}%	220 ms	1.31×10^{9}
$PdSe_2$/MoS_2 [57]	4.012	28.83	1			10^{10}

"G" represents graphene. For each material system, we only show the case of the highest *R* or *D* achieved to date.

energy density distribution, as shown in Figure 2.5e. Such light-trapping effect can facilitate the photoelectric conversion, thereby delivering improved photodetection capability.

These studies indicate that IR photodetection can be realized by employing 2D semimetals with narrow bandgaps or constructing type II heterojunctions with low interlayer exciton energy. The performance can be improved by fabricating mixed-dimensional heterostructures. As compared in Table 2.1, the D^* of 2D materials-based photodetectors surpasses the traditional counterparts. Further development would be desired in achieving ultrafast response at little expense of the responsivity. The advanced IR photodetectors bring new opportunities for biomedical and thermal imaging (Figure 2.5f,g).

2.4 Broadband Photodetectors

2D materials show variable bandgaps from 0 (graphene and metallic TMDs) to ≈6 eV (hBN) [68]. Broadband photodetection can be realized by employing materials that display strong absorption over a wide spectral range or constructing heterostructures with interlayer transitions.

BP has a direct bandgap, the value of which is strongly dependent on its layer number (0.35 eV for the bulk and 2.0 eV for a monolayer) and in-plane strain [69]. Its mobility is as high as $1 \times 10^3\ cm^2\ V^{-1}\ s^{-1}$ [70]. The disadvantages of BP mainly include the weak light absorption relative to traditional optoelectronic materials, large contact resistance with metals, and poor stability in air. To address these issues, Huang et al. [71] integrated Si photonics with BP photodetectors (Figure 2.6a,b), and exploited the optical confinement in the waveguide of Si to enhance the light–matter interactions (Figure 2.6c). Consequently, an increased responsivity of $23\ A\ W^{-1}$ with a noise-equivalent power of $<1\ nW\ Hz^{-1/2}$ was attained at 3.68 μm. Wang et al. [72] encapsulated the BP channel with a passivated layer of Al_2O_3 to inhibit the degradation, and the photoresponse can be retained for more than six months. Zhong et al. [75] improved the stability of BP via As doping. They investigated the optoelectronic properties of the As-doped BP ($As_{0.084}P_{0.916}$) and showed the photodetecting performances of this AsP-based device in the visible–NIR spectrum, with the highest responsivity of $37\ A\ W^{-1}$ at 532 nm. In addition, BP exhibits anisotropic transport behavior. This leads to different responsivities along zigzag and armchair directions (Figure 2.6d). The photodetection range can be extended by synthesizing AsP alloys. The cutoff wavelength is tunable (from 3.8 to 4.6 μm) by adjusting the As content (from 0% to 91%) [76]. High sensitivities of the AsP photodetectors have been obtained in the MIR and LWIR regions. The values of D^* are on the order of 10^{10} and 10^8 Jones, respectively, both higher than those of the commercially available counterparts [56, 76]. Liu et al. [77] covered the BP with graphene both to inhibit the degradation of BP and to improve the performance of the photodetector. The graphene not only serves as a protection layer but also promotes the carrier transport owing to its high mobility. As a result, the Schottky barrier is reduced between the BP channel and the metal electrodes, leading to efficient photocurrent extraction. The graphene/BP heterostructure device presents a long-term photoresponse at visible–NIR wavelengths (635–1600 nm), with a high photoresponsivity up to $3.3 \times 10^3\ A\ W^{-1}$ and a large photoconductive gain of 1.13×10^9.

The wide detectable range, however, poses a challenge to realize high wavelength selectivity. One feasible solution is hybridizing BP with other 2D materials that possess variable bandgaps. Yan and coworkers [73] fabricated a BP/MoS_2 photodiode, where the MoS_2 nanosheets act as an electrical contact and the top dielectric. Excellent wavelength selectivity with the full width at half-maximum (FWHM) down to 0.42 μm was realized in the range of 2–4 μm by adjusting the thickness of MoS_2 (Figure 2.6e). Xu et al. [78] created a p–n homojunction in a BP photodetector, where the n-type BP was formed by covering a Si_xN_y film on top of the p-type BP. The built-in electric field, which is gate-tunable, can facilitate the separation of the photogenerated electron–hole pairs. This device exhibits

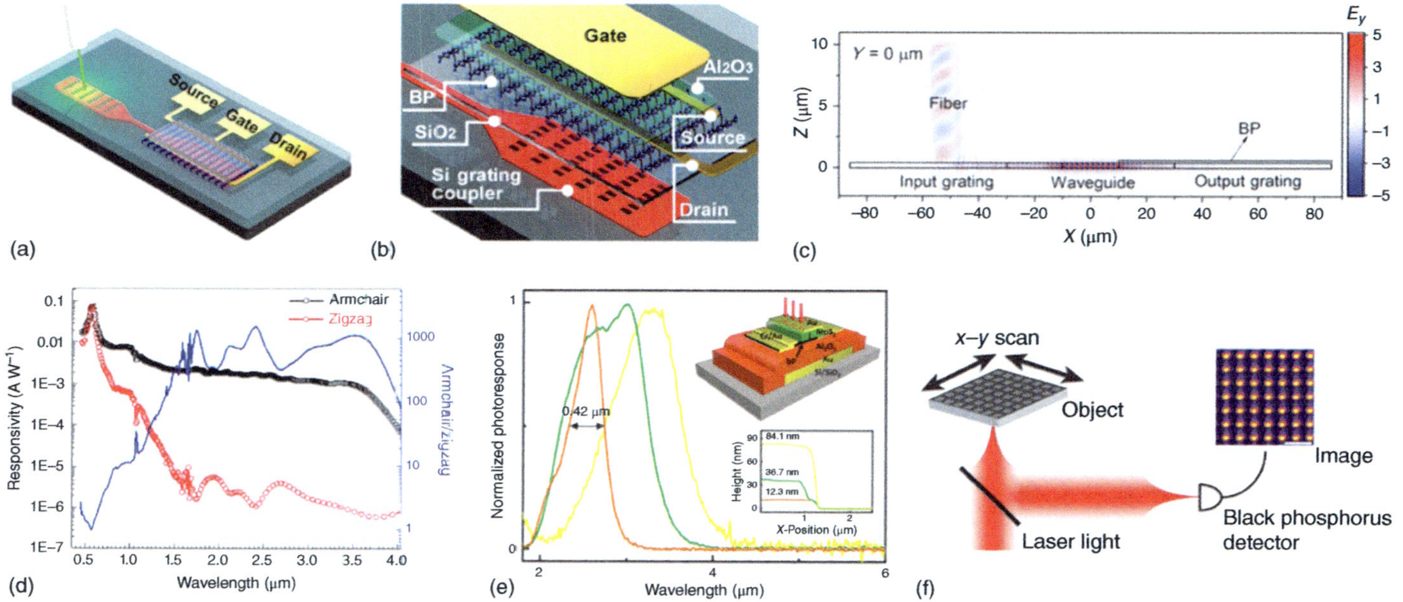

Figure 2.6 Ultrathin BP-based photodetectors. (a) Schematic of the Si waveguide-integrated device. (b) Zoomed-in view of (a) and side view of the (c) light distribution in this system. (d) Dependences of the responsivity on the direction and wavelength. (e) Spectral features of BP/MoS_2 diodes with different MoS_2 thicknesses. Inset: 3D structure of the device (upper) and atomic force microscopy height profiles of the MoS_2 layers (lower). (f) Simplified setup for imaging. Scale bar: 4 μm. Source: (a–c) Reproduced with permission [71]. Copyright 2019, American Chemical Society. (d) Reproduced with permission [72]. Copyright 2019, American Chemical Society. (e) Reproduced with permission [73]. Copyright 2020, American Chemical Society. (f) Reproduced with permission [74]. Copyright 2014, American Chemical Society.

higher performance compared to the pristine BP counterpart. Moreover, it shows dual-band photoresponse in both visible and IR regimes, providing access to high-resolution multispectral imaging (Figure 2.6f).

In terms of TMCs, broadband photodetection can be realized by layer number control, doping, and creation of van der Waals heterostructures. The bandgap of $PdSe_2$ is thickness-dependent and can be widened from 0 (bulk) to 1.3 eV (monolayer). Based on this exceptional characteristic, Liang et al. [67] demonstrated the ultrabroadband photoresponse of $PdSe_2$ across the visible–MIR spectrum. The highest responsivity of $\approx$700 A W^{-1} is obtained at the wavelength of 1064 nm. This superior performance is ascribed to the photogating effect, which causes trapping of the photogenerated electrons or holes into the localized states and therefore a prolonged recombination lifetime of the other type of photocarriers. In particular, the unique pentagonal structure of 2D $PdSe_2$ crystal endows it with anisotropy. As a result, the photoresponse of the device is highly sensitive to the polarization angle. Molecular doping can modify the structure, electronic and optical properties, as well as stability of diverse materials. This controllable method has been used to optimize the photodetection capabilities of 2D TMCs. Kang et al. showed performance improvements of MoS_2 and WSe_2 photodetectors via n- and p-doping with aminopropyltriethoxysilane (APTES) and octadecyltrichlorosilane (OTS) molecules, respectively. For the APTES-doped MoS_2, a larger population of negative trions are formed owing to the increased electron density. This restricts the production of neutral excitons that dominate radiative recombination. In contrast, the radiative recombination in WSe_2 mainly originates from negatively charged trions, the formation of which is suppressed due to the OTS-induced p-type doping. Quenched photoluminescence intensities confirm that the carrier recombination probabilities are decreased in both cases, leading to enhanced generation of photocurrents. As another example, the functionalization with didodecyldimethylammonium bromide (DDAB) can effectively dope InSe and then heal the Se vacancies (Figure 2.7a), leading to an ultrahigh photoresponsivity up to 10^6 A W^{-1} [79]. The defects at the InSe–SiO_2 interface may limit the electron transport by introducing localized states and participate in the Coulomb scattering. Such effects can be attenuated using a trap-free polymer as the gate dielectric, such as divinyltetramethyl disiloxanebis(benzocyclobutene) (BCB), thereby increasing the mobility and channel conductance (Figure 2.7b,c). As a result, the responsivity is 3–4 orders of magnitude higher than the InSe supported by SiO_2 (Figure 2.7d) [79]. Following this, a BP–InSe p–n junction was constructed to achieve faster response. While the InSe is n-doped by DDAB, the transport region for holes is protected by an insulating hBN film. The built-in electric field promotes the separation of photocarriers, giving rise to $\approx$5- and 10-fold decreases in the rise and fall times, respectively, at a small power density of 5 μW cm^{-2} (Figure 2.7e). To further extend the detection range, Ahmad et al. [80] fabricated a type-II InSe/$PdSe_2$ stack with strong interlayer transition. According to the density functional theory (DFT) calculations, the bandgap of the heterojunction is as narrow as 0.85 eV, which allows NIR excitation up to 1650 nm (Figure 2.7f). Alongside these studies, investigations have been

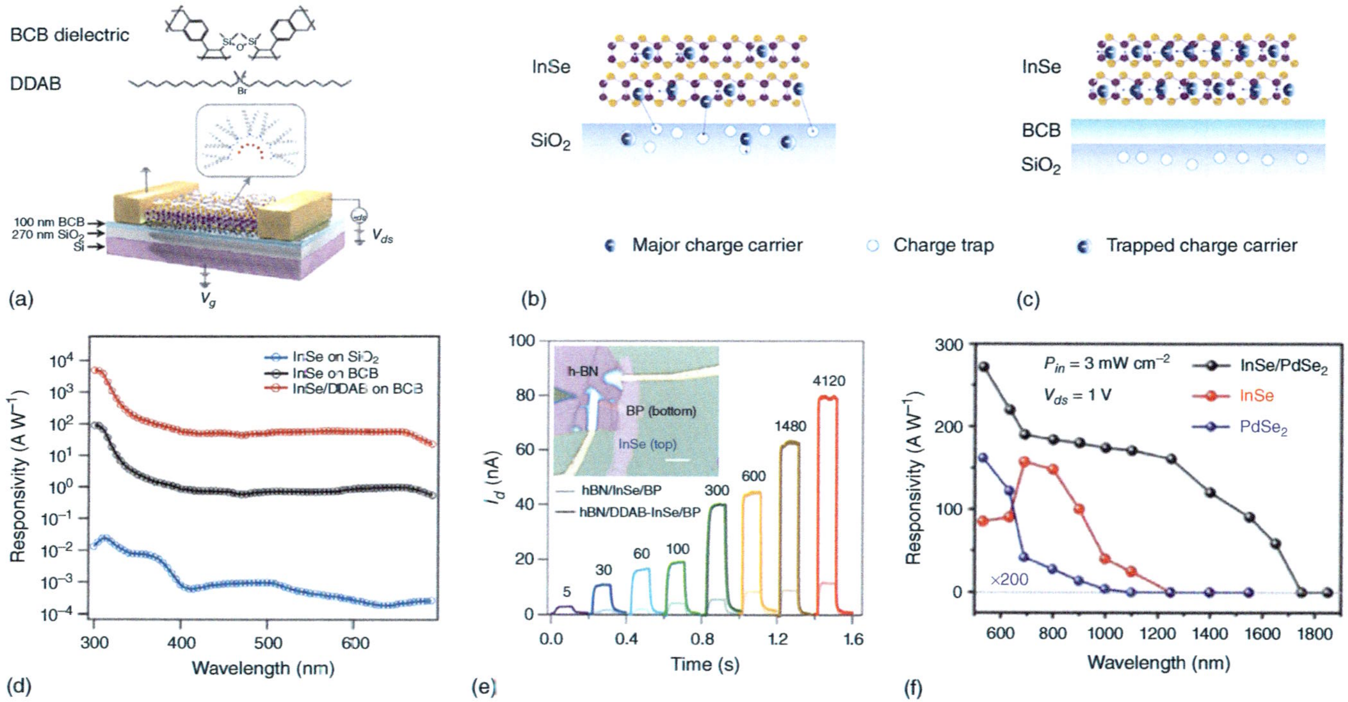

Figure 2.7 2D InSe-based broadband photodetectors. (a) Schematic illustration of the device along with the structures of a DDAB molecule and BCB polymer. (b, c) Schematic illustrations showing the carrier transport of InSe in contact with trap-containing SiO_2 (b) and trap-free BCB (c). (d) Spectral responses of InSe on SiO_2, InSe on BCB, and DDAB-functionalized InSe on BCB. (e) Time-resolved photoresponse at increasing power densities (5–4120 μW cm^{-2}) of hBN/InSe/BP and hBN/DDAB-InSe/BP heterostructures. Inset: optical microscopic image of InSe/BP partially covered by hBN. (f) Photoresponsivities of InSe/$PdSe_2$, InSe, and $PdSe_2$. Source: (a–e) [79]/Reproduced with permission from John Wiley & Sons. (f) Reproduced with permission [80]. Copyright 2021, Wiley-VCH GmbH.

conducted on diverse 2D materials and 2D/2D heterostructures for broadband photodetection. Those exhibiting superior responsivities are summarized in Table 2.2.

Broadband photodetection has also been achieved in mixed-dimensional structures. Based on the high carrier mobility and transparency of graphene, it can be employed in photodetectors via integration with other semiconductors. Previous studies have demonstrated the broad spectral response of 2D/3D hybrid photodetectors based on graphene/Si. For photodiodes, the electron–hole pairs are mainly generated in Si, while graphene functions as light-transmitting Schottky electrodes that extract the carriers. Nevertheless, the photoresponsivity is usually poor due to the absence of gain. For photogating devices, the photoexcited electrons or holes are transferred to graphene, while the oppositely charged carriers are trapped in the localized states. These lead to a large responsivity accompanied by a high dark current and a low response speed. To improve the overall performance, Guo et al. [107] optimized the structure of graphene/Si photodetector. As illustrated in Figure 2.8a, the integration of a wide Si ridge allows space between the signal and the ground electrodes, thus increasing the light absorption in the graphene layer. Figure 2.8b manifests the strongly localized and enhanced electric field near the metal strip. Moreover, the plasmonic waveguide in this hybrid device leads to an extended detection wavelength beyond 1.55 μm. The photocurrent derives from the photothermoelectric effect at 0 V, yet bolometric and photoconductive effects are dominant when a bias voltage is applied (Figure 2.8c).

The narrow light-sensitive spectral region of 2D TMDs restricts their applications in broadband photodetectors. For these materials, a broad range of detection can be achieved by integrating them with 3D semiconductors, forming 2D/3D p–n junctions. Wu et al. [108] demonstrated the utilization of WS_2/Si for a broadband self-driven photodetector (Figure 2.8d). The specific type II band structure dictates the interlayer optical excitation (Figure 2.8e), enabling the WS_2/Si device to show photodetection from deep UV (200 nm) up to mid-IR (3043 nm) (Figure 2.8f). The response speed increases with the light frequency. The rise and fall times are only 3.8 and 11 μs at 30 kHz, shorter than those of most 2D materials-based broadband photodetectors. In addition, the anisotropic absorption of the WS_2 film leads to high polarization sensitivity (Figure 2.8g). Yao et al. [110] proposed a heterojunction of SnS/Si, and demonstrated its broadband photoresponse (370–1064 nm, i.e. UV to NIR) with high responsivity (273 A W^{-1}), large *EQE* (4.2×10^4%), and outstanding D^* (7×10^{13} Jones). The excellent performance can be a result of the synergistic effects in the SnS ultrathin films due to their blended configuration, involving the efficient carrier separation across the dangling-bond-free interfaces of the lateral nanosheets, the fast carrier transport through the vertical components, and the carrier multiplication (i.e. multiple electron–hole pairs are generated with a single photon) induced by collision ionization.

Ultrathin TMCs can also function as an interfacial layer to improve the performances of graphene/Si-based photodetectors. He et al. [111] achieved a high responsivity of 8.96×10^4 A W^{-1} by intercalating a WS_2 film between Si and graphene. Since the WS_2 can provide additional trap states for holes, the

Table 2.2 Key figures of merit for advanced broadband photodetectors based on 2D materials and 2D/2D heterostructures.

Traditional material	Wavelength (nm)	R_{max}	Bias (V)	Gain or *EQE*	Rise and fall times	D^* (Jones)
Si[a]	350–1100 (970 nm)	0.725	5		65 ns, 65 ns (632 nm)	
Ge[a]	800–1800	0.88	3		600 ns, 600 ns	
InAs [81]	300–1100	4.4×10^3 A W^{-1}	15	1.0×10^6%		2.6×10^{11}
InSb [3] (77 K)	600–8000	1.0×10^3 V W^{-1}	−1.5	40%		2.8×10^{10}
InGaAs[b] [82]	400–1700	6.5×10^3 A W^{-1}	0.5	5×10^5%	70 ms, 280 ms	10^{12}
HgCdTe [83, 84] (200–212 K)	600–5000	8 A W^{-1}	−0.1		1.8 ns	3×10^{10}
2D material	**Wavelength (nm)**	**R_{max} (A W^{-1})**	**Bias (V)**	**Gain or *EQE***	**Rise and fall times**	**D^* (Jones)**
BP [45]	400–900	7×10^6 (20 K)	−1			
AsP [75]	450–2200	37	1		0.37 s, 0.40 s	1.24×10^9
MoS_2 (n-doped) [85]	520–850	5.75×10^3	5			
WSe_2 (CVD-grown) [25]	500–900	1.8×10^5	2	3.5×10^5	<23 ms	1×10^{14}
$PdSe_2$ [67]	532–4050	708	1	8.27×10^4%	220 ms	1.31×10^9
GaSe [86]	400–800	5×10^3	8		0.27 ms, 0.55 ms	
InSe (n-doped, on BCB dielectric) [79]	300–690	10^6	1	5.7×10^8%	4.9 ms, 28.6 ms	10^{13}
In_2Se_3 [37]	405–940	9.8×10^4	0.05		9 s	3.3×10^{13}
hBN/InSe (n-doped)/BP (on BCB dielectric) [79]	300–690	10^3	1	3.4×10^5%	0.96 ms, 2.97 ms	10^{11}
Fluorine-G/G [87]	255–4290	6.2×10^3	1	2×10^5	80 ms, 200 ms	5×10^{10}
Fluorine-G/MoS_2 [88]	350–850	8×10^3	0.3		1.35 s, 1.41 s	10^{13}
BP/G [77]	635–1600	3.3×10^3	1	1.13×10^9	3 ms	
BP/MoS_2 [89]	532–1550	24	3	1.2×10^3%		
G/MoS_2 [53]	400–1120	3×10^4	5	5.8×10^4	17 ns, 1.1 μs	

(Continued)

Table 2.2 (Continued)

2D material	Wavelength (nm)	R_{max} (A W^{-1})	Bias (V)	Gain or *EQE*	Rise and fall times	D^* (Jones)
G/$MoSe_2$ [90]	550–950	1.3×10^4	1			
G/$MoTe_2$ [91]	532–1064	970.82	1	4.69×10^8	78 ms, 375 ms	1.55×10^{11}
MoS_2/$MoTe_2$ [92]	550–1550	0.046	0		60 μs, 25 μs	1.06×10^8
MoS_2/ReS_2 [93]	400–1100	42.61	1	6.6×10^3%	20 ms, 19 ms	2.81×10^{13}
ReS_2/$ReSe_2$ [94]	405–1310	6.78×10^6	−5		0.28 s, 1.14 s	
MoS_2/G/WSe_2 [95]	400–2400	10^4	1		53.6 μs, 30.3 μs	10^{15}
G/SnS_2 [96]	365–2240	7.76×10^3	2		6 s	8.9×10^{13}
WSe_2/SnS_2 [97]	400–900	244	−1		13 ms, 24 ms	
WSe_2/$SnSe_2$ [98]	532–1550	588	−3	1367%	16 ms, 45 ms	4.4×10^{10}
G/Bi_2Te_3 [99]	400–1550	35	1	83	8.7 ms, 14.8 ms	
G/MoS_2/WS_2 [100]	400–2000	6.64×10^7	−3	2.06×10^8	21.86 ms, 7.01 ms	
G/hBN/MoS_2 [101]	400–886	180	3	1.67×10^3	0.23 s, 0.25 s	2.6×10^{13}
G/MoS_2/G [102]	405–2000	414	−1	9.67×10^4%	692.5 ms	3.2×10^{10}
G/WS_2/GaSe/G [103]	270–740	149	2		37 μs, 43 μs	4.3×10^{12}
G/InSe/GaSe/G [104]	270–920	350	2			
BP/$PdSe_2$ [105]	532–1310	9.6×10^5	−4	9.4×10^6	6 s, 20 s	5.8×10^{13}
InSe/$PdSe_2$ [80]	532–1650	270	1	6.4×10^4%	53 ms, 72 ms	4.6×10^{10}
hBN/$MoTe_2$/G/SnS_2/hBN [106]	405–1550	10^5	1	10^6%	17.6 ms, 72.3 ms	10^{13}

Traditional material-based broadband photodetectors are shown as benchmarks. For each material system, we only show the case of the highest responsivity achieved to date. These photodetectors work at room temperature, unless indicated otherwise.

a) https://www.thorlabs.com/.

b) https://www.sensorsinc.com/.

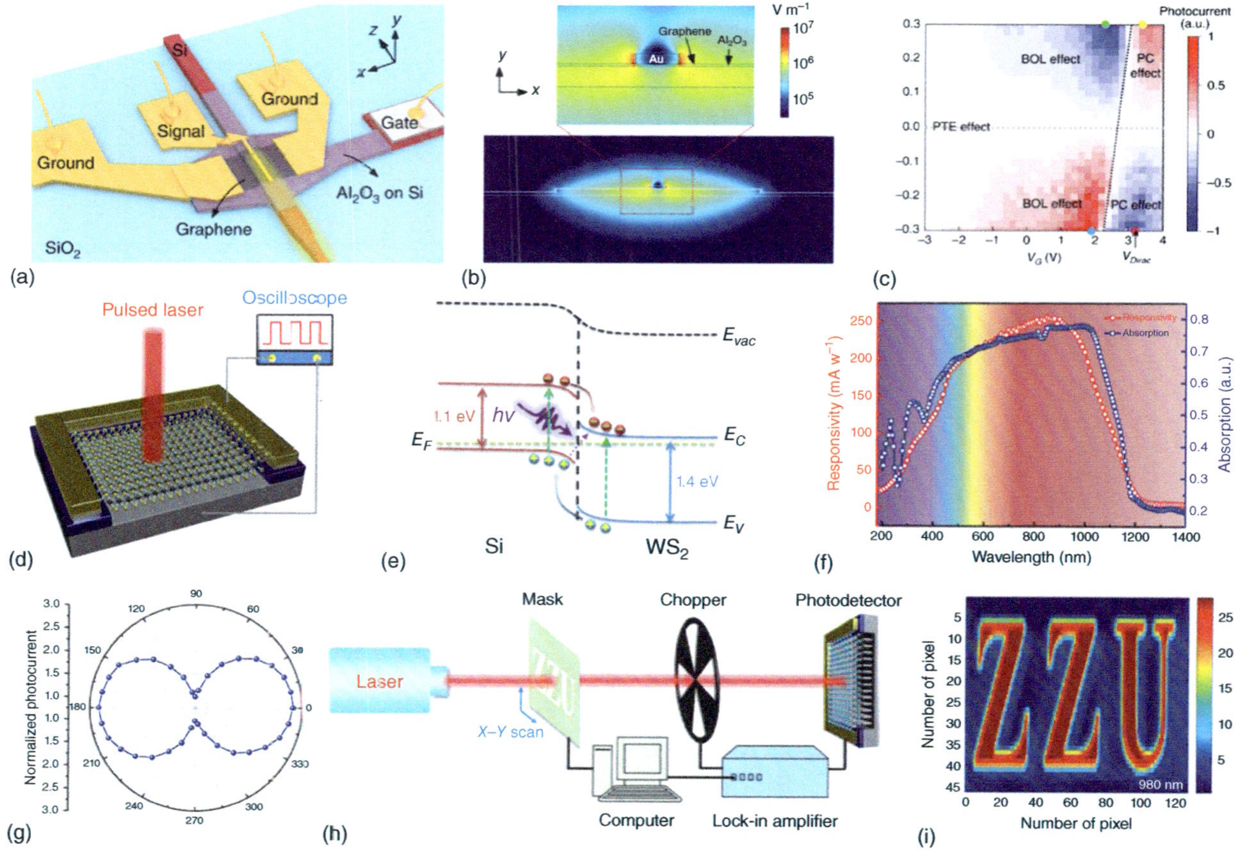

Figure 2.8 (a–c) Graphene/Si photodetector. (a) Schematic illustration of the hybrid device. (b) Side and top views of the electric-field distribution. (c) Photocurrent map as a function of the gate voltage (V_G) and the bias voltage (V_b). PTE, photothermoelectric; PC, photoconductive; BOL, bolometric. (d–g) WS_2/Si photodetector. (d) Schematic illustration with a pulsed laser. (e) Energy band diagram of the heterojunction upon light excitation. (f) Wavelength dependences of the responsivity and absorption. (g) Normalized photocurrent versus the polarization angle. (h, i) $PdSe_2$/Si photodetector. (h) Schematic of the setup for imaging and the (i) result under NIR illumination. Source: (a–c) [107]/Springer Nature/Licensed under CC BY 4.0. (d–g) Reproduced with permission [108]. Copyright 2019, American Chemical Society. (h, i) Reproduced with permission [109]. Copyright 2020, Royal Society of Chemistry.

responsivity and D^* first increase with the thickness of WS_2, while decreasing at large thicknesses because of the much lower light transmittance. The rise and fall times monotonically increase with the WS_2 thickness, which is attributed to the longer electron transport paths and the slower recombination of photogenerated carriers. This graphene/WS_2/Si device exhibits broad spectral photoresponse in the range of 400–1100 nm. Lu et al. [112] coupled graphene with 2D In_2S_3 and constructed a double heterojunction of In_2S_3/graphene/Si. The In_2S_3 serves to enhance the light absorption of graphene/Si, reduces the dark current, facilitates the separation of photoexcited carriers, and delivers a photoconductive gain. Based on combined photogating and photovoltaic effects, In_2S_3/graphene/Si photodetector exhibits both ultrahigh photoresponsivity (4.53×10^4 A W^{-1}) and fast response speed (rise and fall times: ≈33 and 40 μs, respectively). The photogain exceeds 1.4×10^5, which is 1 order of magnitude larger than that of the graphene/Si device.

In particular, semimetallic 2D TMDs may exhibit distinct photodetection behavior due to their ultranarrow bandgaps. Zeng et al. [113] synthesized wafer-scale $PtTe_2$ nanosheets via epitaxial growth on Si. This was achieved by tellurization of a Pt thin film that had been predeposited onto the substrate. The single-crystalline $PtTe_2$ domains with a preferential orientation along the [001] zone axis show reduced interfacial defects and efficient out-of-plane carrier transportation. Transparent bilayer graphene was used as the top electrode to enhance the carrier collection, and In/Ga alloy was employed as an ohmic contact to the Si substrate. 2D $PtTe_2$ possesses a highly tilted Dirac cone. This unique band characteristic is favorable for the separation of photogenerated electron–hole pairs and can suppress their recombination. These render the $PtTe_2$/Si Schottky heterojunction capable of ultrabroadband photodetection from UV to MIR regions. At wavelengths in the range of 200–1100 nm (UV–visible–NIR), the incident light is absorbed in both the $PtTe_2$ layer and the Si substrate, after which the photoexcited carriers are separated, driven by the in-built potential. The electrons drift from $PtTe_2$ to Si, whereas the holes drift in the opposite direction and are collected by the graphene electrode, forming a photocurrent. At wavelengths beyond 1100 nm, Si is not able to absorb the light due to its limited bandgap, and the photoresponse solely originates from the $PtTe_2$. Benefiting from the ultrathin depletion layer at $PtTe_2$–Si heterointerface, the electrons in $PtTe_2$ can tunnel directly to the conduction band of Si, giving pronounced photocurrent in a wide spectral range up to 10.6 μm.

Apart from Si, Ga-based semiconductors can also couple with 2D TMDs as a 3D support to extend their detection spectra. Jain et al. [114] reported the broadband photoresponse of 2D MoS_2/3D GaN. Such a van der Waals heterostructure presents complementary light absorption profiles for MoS_2 (visible) and GaN (UV). Moreover, the type II band alignment enables photodetection in the NIR regime via the generation of interlayer excitons. High responsivities were obtained across the UV–visible–NIR region. Geum et al. [115] showed the fast responses and good detectivities of $MoS_2/In_{0.53}Ga_{0.47}As$ photodetectors at wavelengths ranging from 400 (visible) to 1550 (NIR) nm. They also demonstrated the imaging capability of the device array with a scanning laser, suggesting that the $MoS_2/In_{0.53}Ga_{0.47}As$ heterostructure can be used as a broadband imager in focal plane apparatus.

Pristine 2D perovskites usually display superior photoelectric properties within the visible spectrum. Their detection wavelength can be extended through compositional variation or chemical doping. The optical bandgap of $(CH_3NH_3)_{0.5}$ $(NH_2CHNH_2)_{0.5}Pb_{1-x}Sn_xI_3$ is narrowed from 1.55 to 1.17 eV as the Sn content increases from 0 to $x = 0.5$ [116], rendering it capable of NIR photodetection [117]. Such bandgap narrowing and visible-to-NIR detection range extension was also been realized for $(C_6H_5C_2H_4NH_3)_2PbI_4$ via incorporation of Sn ions, whereby the photocurrent is derived from the dissociation of excitons in the self-trapped states that are introduced by the Sn dopants [118]. However, these photodetectors that consist only of 2D perovskite(s) exhibit low responsivities in the IR regime, inspiring the fabrication of hybrid photodetectors.

The broadband photoresponse of 2D materials-based devices can also be improved by incorporating 1D nanostructures. Graphene exhibits ultrafast dynamics in photodetecting processes due to the short carrier lifetime, high intrinsic carrier mobility, and Auger-induced charge multiplication effects. However, its single-atom thickness limits the absorption of photons. Single-wall carbon nanotubes (SWCNTs) are effective light absorbers, while SWCNTs-based devices show either poor photoresponsivity or slow response. Given their complementary properties and photoresponses in a wide spectral range, Liu et al. [119] hybridized graphene with SWCNTs to achieve high-performance broadband photodetection. The large inbuilt potential, along with the efficient charge transfer at the graphene/SWCNT junction, enables fast separation of the electron–hole pairs and suppresses the recombination of dissociated carriers. As a result, the hybrid photodetector exhibits a large photogain of $\approx 10^5$ along with a fast response time of $\approx 100\,\mu s$ across visible to NIR region. Wu et al. [109] fabricated a 2D/1D heterostructure constituting $PdSe_2$ nanosheets and vertically aligned Si nanowires. The $PdSe_2$/Si device exhibits outstanding photovoltaic behavior with a large D^* (3.19×10^{14} Jones) and a wide detection wavelength range (0.2–4.6 μm), ascribed to the broadband light absorption of $PdSe_2$ and the strong light confinement induced by Si nanowire arrays. In addition, the device shows high polarization sensitivity owing to the unique asymmetric crystal structure of $PdSe_2$. The fast response speed allows monitoring of nanosecond pulsed signals. Moreover, this device can be utilized for high-resolution IR imaging (Figure 2.8h,i). The contrast ratios are over 10^3 for 980 nm (NIR) and over 10^2 for 4.6 μm (MIR). Apart from such metal dichalcogenides, elemental Te is a new member of layered materials and also an intrinsic p-type semiconductor with a tunable bandgap (0.35–1.2 eV). It shows high mobility, strong light absorption, and good stability in air. The photoresponse of Te can be enhanced by forming a p–n or Schottky junction. Yao et al. [120] integrated 2D Te flakes with 1D CdS nanobelts and utilized the heterostructure for a self-powered photodiode. The response spectrum ranges from 520 (visible) to 980 (NIR) nm. The recombination of carriers is suppressed due to the type II band alignment, giving a high photo-to-dark current ratio of 9.2×10^3.

Enhanced charge transport has also been realized at 2D/0D junctions. Selamneni et al. [121] constructed a 2D/0D homojunction consisting of MoS_2 ultrathin flakes and MoS_2 quantum dots. Localization of the electric field was also observed in this hybrid structure due to the misaligned Fermi levels of 2D MoS_2 and 0D MoS_2

at their unipolar interface. The coupling with 0D Pb-based chalcogenides (e.g. PbS, $Pb_{1-x}Sn_xSe$, and $Pb_{1-x}Sn_xTe$) can enable various 2D materials to show broadband photodetection capabilities. Kufer et al. [122] presented a 2D MoS_2/0D PbS phototransistor. The light absorption is strong throughout the visible–NIR region, with the peak value tailorable by controlling the size of the colloidal PbS quantum dots. In addition, due to the efficient carrier separation at the p–n (PbS–MoS_2) interface and fast carrier transport through the MoS_2 channel, the device exhibits a high photoresponsivity of $106\,A\,W^{-1}$ along with backgate-dependent detectivity. Luo et al. [123] demonstrated a 2D Bi_2O_2Se photodetector sensitized with 0D PbSe quantum dots. Type II band alignment was formed at the Bi_2O_2Se–PbSe interface. The incorporation of PbSe quantum dots into Bi_2O_2Se nanosheets extends the detection spectrum from 660 nm (visible) to >2 μm (NIR). In spite of this, Bi_2O_2Se photodetectors usually exhibit slow response time (≈4 ms) as a result of the bolometric effects. Hossain et al. [124] resolved these issues by combining 2D nonlayered Bi_2O_2Se nanosheets with 0D $CsPbBr_3$ nanocrystals. This device shows a comparable responsivity ($\approx 10^3\,A\,W^{-1}$) but much shorter rise/fall times (≈12/24 μs) in the visible–NIR range. The excellent performance is attributed to the strong and broadband absorption of $CsPbBr_3$, along with the high carrier mobility of Bi_2O_2Se. Experimental and theoretical investigations verified the efficient charge transport across the type I heterojunction. This work implies the potential of 0D halide perovskites as a light-harvesting component in hybrid photodetectors to achieve high responsivity.

2.5 Plasmon-Enhanced Photodetectors

2D materials are promising candidates for next-generation ultrathin photodetectors in different spectral regions. However, being atomically thin results in much shorter light–matter interaction length in 2D materials than that in bulk materials, which hinders their optoelectronic performance. To improve the light–matter interactions, plasmonic nanostructures which support surface plasmon resonance have been introduced into 2D material photodetectors [125–127]. Over the past 10 years, significant developments, including device performance and physical mechanisms in plasmon-enhanced photodetectors, have been achieved. In this section, we present basic concepts in plasmonic nanostructures, an introduction of light–matter interaction between plasmonic nanostructures and 2D materials, and a discussion of current advances in plasmon-enhanced photodetectors.

Plasmonic nanostructures (e.g. metal nanospheres, nanorods, nanocubes, arrays of nanopatterns, slits, and gratings) have attracted considerable attention because of the significantly boosted light–matter interaction by a localized surface plasmon resonance (LSPR) effect in optoelectronic devices. The principle of LSPR can be understood in a simple model of a harmonic oscillator. As shown in Figure 2.9a, an oscillating optical electric field illuminating the metal nanostructures acts as an applied force and drives the electrons of the metal to move along the direction of the electric field [128]. The delocalized electron cloud and the positive nuclei of the

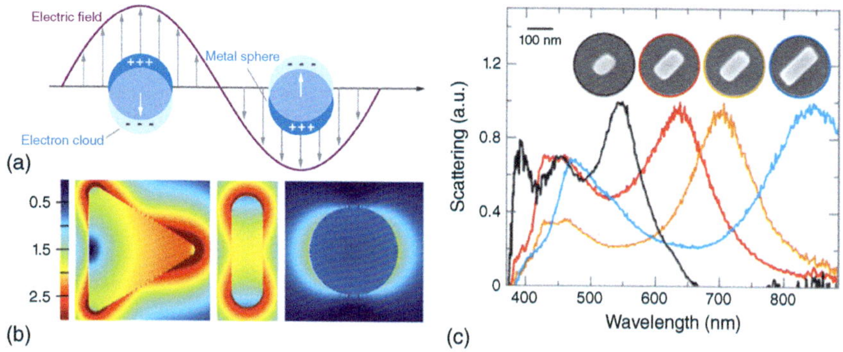

Figure 2.9 (a) Schematic of localized surface plasmons in metal spheres. (b) Logarithmic-scale electric-field intensity enhancement contours of differently shaped Au nanocrystals. The nanocrystals have different shapes but the same volume. The nanoplate (left) in an equilateral triangle has an edge length of 87 nm and a thickness of 10 nm. The length and diameter of the nanorod (middle) are 102 and 30 nm, respectively. The nanosphere (right) has a diameter of 50 nm. (c) Scanning electron micrographs of individual Ag nanobars with different aspect ratios and their corresponding dark-field scattering spectra. Source: (a) Reproduced with permission [128]. Copyright 2007, Annual Reviews. (b) Reproduced with permission [129]. Copyright 2014, Wiley-VCH GmbH. (c) Wiley, Benjamin J. et al. 2007/with permission of American Chemical Society.

metal have strong Coulombic attraction, which acts as a restoring force. When the oscillating electric field becomes in phase with the Coulombic attraction at a certain frequency, the incident light is largely absorbed by the metal, and a very strong electric field is concentrated at the edges or gaps of the metal nanostructure [131].

The resonant properties such as field distribution and resonant wavelength of the LSPR effect can be efficiently modified by changing material composition, shape, size, and spacing of the metal nanostructures. Typically, noble metals such as Au, Ag, and Al with high free electron concentration ($>10^{21}$ cm^{-3}) are used for fabricating plasmonic nanostructures in the visible and NIR spectral range [132]. Au nanostructures with different shapes show different extent of increase in field intensity (Figure 2.9b) [129]. Compared to Au nanospheres, nanoprisms, and nanorods show stronger field enhancements, especially around the tip regions. The maximum field enhancement for plasmonic metal nanostructures reported from theoretical calculations ranges from c. 50 to c. 10 000. The resonant wavelength is also affected by the geometry of the metal nanostructures, which can be calculated by Mie theory analytically solving Maxwell's equations or simulated by finite-difference time-domain (FDTD) and COMSOL software. It can also be experimentally characterized by measuring the absorption, scattering, or reflectance spectra. As shown in the scattering spectra of Ag nanobars with different aspect ratios (Figure 2.9c), there are two resonance peaks, with one transverse plasmon resonance peak in the blue region at ≈460 nm and one longitudinal plasmon resonance peak in the visible or NIR region [130]. Both resonance peaks undergo a red shift as the size of Ag nanobars increases, indicating that the resonant response can be easily tuned by geometrical tailoring.

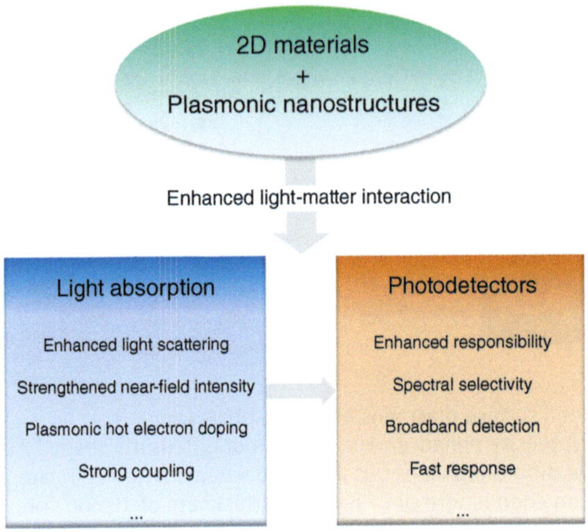

Figure 2.10 A list of the mechanisms that contribute to the enhanced light absorption in the hybrid system of plasmonic nanostructures and 2D materials and the enhanced parameters of photodetectors.

Plasmonic nanostructures can significantly enhance the light–matter interaction of nearby materials that have matched absorption window with plasmon resonance. 2D materials are a promising candidate to make full use of this plasmonic effect, as their thin structure enables them to efficiently utilize the highly localized electromagnetic field, which exponentially decays along the perpendicular direction with a decay length on the order of 10 nm [133]. The interplay of plasmonic nanostructures and 2D materials has been extensively reported in plasmon-enhanced light absorption and photodetectors. Figure 2.10 summarizes the mechanisms that contribute to the enhanced light absorption in the hybrid system and the enhanced parameters of photodetectors based on them. The enhanced light–matter interaction in the hybrid structure features in enhanced light absorption, owing to the enhanced light scattering, strengthened near-field intensity, plasmonic hot electron doping, and strong coupling. These effects will bring in enhanced photoresponsivity, spectral selectivity, broadband detection, and fast response in photodetectors.

With advances in preparation techniques for plasmonic nanostructures by chemical methods, electron beam lithography, and focused ion beam, various plasmonic nanostructures have been used to improve the photodetection performance of 2D materials. The following paragraphs will discuss the mechanism and performance of plasmon-enhanced photodetectors based on graphene, transition-metal dichalcogenides, InSe, and so on. The hybrid system of surface plasmon and 2D materials was first reported in graphene. Graphene has intrinsically weak light absorption ($\approx$2.3% for pristine graphene). Its short photocarrier lifetime leads to poor photoresponsivity (5×10^{-4} A W^{-1}) and internal quantum efficiency (6–16%) [131]. An efficient method to enhance light absorption is to utilize the near-field enhancement from the excitation of the plasmonic effect induced by the metal nanostructures. By placing the plasmonic nanostructures close to the graphene, the photoresponse of graphene-based photodetectors can be efficiently improved. Figure 2.11a shows the schematic illustration of the process to couple Au nanoparticles with graphene devices [125]. After the Au nanoparticles are introduced, the photocurrent exhibits

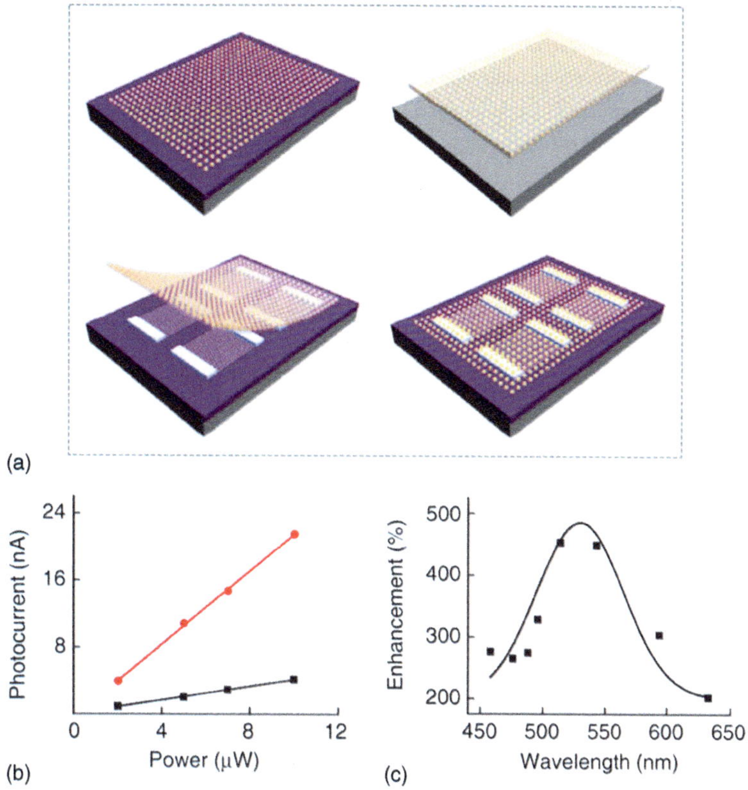

Figure 2.11 (a) Schematic illustration of the fabrication processes to integrate plasmonic nanoparticles with graphene. Gold nanoparticles are obtained by thermally annealing 4 nm-thick gold film on the SiO_2 substrate. PMMA is spin coated on the substrate, followed by a lifting-off process using buffered oxide etching to obtain free-standing PMMA/Au-nanoparticle thin film. PMMA/Au-nanoparticle thin film is fished by premade back-gate CVD graphene transistors. PMMA is dissolved away to leave gold nanoparticles on the surface of the graphene devices. Gray, blue, yellow, and white represent silicon, silicon dioxide, gold nanoparticles, and gold electrode, respectively. (b) Photocurrent generated as a function of laser power. The red and black lines indicate the responses of a typical device with and without Au nanoparticles, respectively. (c) Photocurrent enhancement as a function of laser wavelength. The black squares represent the experimental data points, and the solid line is the guide to the eye. Source: Reproduced with permission [125]. Copyright 2011, Nature Publishing Group.

a 15-fold enhancement compared to the pristine graphene photodetectors at laser power of 10 μW (Figure 2.11b). The increase in the photocurrent is attributed to plasmon-enhanced carrier excitation of graphene electrons. The spectral-dependent photocurrent enhancement shows an obvious peak around 530 nm, which is close to the extinction peak of graphene–nanoparticle conjugate and strongly suggests that the enhancement is indeed originating from the plasmon resonance-enhanced absorption in graphene devices (Figure 2.11c).

In addition to the localized surface plasmon-enhanced photodetectors of graphene in visible light range, other types of plasmon-enhanced graphene photodetectors have been reported. One is utilizing the propagating surface plasmon. Ferrari and coworkers coupled graphene with a plasmonic grating and exploited the resulting surface plasmon polaritons to deliver the collected photons to the junction region of a metal–graphene–metal photodetector [134]. This gives a 400% enhancement of responsivity and a 1000% increase in photoactive length, combined with tunable spectral selectivity. The interference between surface plasmon polaritons and the incident wave introduces new functionalities, such as light flux attraction or repulsion from the contact edges, enabling the tailored design of the photodetector's spectral response. The other is using the intrinsic plasmon resonance of graphene itself. Graphene's interband absorption is a frequency-independent constant $\pi\alpha \approx 2.3\%$ in the visible and MIR ranges, where α is the fine structure constant [135]. Cai et al. [136] showed a room-temperature THz detector based on large-area arrays of epitaxial graphene microribbons contacted by metal electrodes, whose responsivity is significantly improved by the plasmon-enhanced absorption via graphene microribbons. In recent years, a Ge nanodot array was also reported to be integrated with graphene to construct a plasmon-enhanced photodetector [137]. Owing to the LSPR, light scattered by an individual Ge nanodot is collected by the adjacent nanodots to increase the photocurrent of the graphene photodetector and shorten the response time (18.7 μs).

Transition-metal dichalcogenides such as MoS_2, $MoSe_2$, WS_2, and WSe_2 have enhanced light absorption in the visible light range as compared with graphene, due to the opening of electronic band gap. Their absorption in light and performance in photodetectors are further improved by incorporating plasmonic nanostructures. Miao et al. [138] demonstrated a large enhancement of photocurrent response obtained by coupling few-layer MoS_2 with Au plasmonic nanostructure arrays (Figure 2.12a). At the plasmon resonance wavelength (620 nm) of the Ag arrays, a sevenfold enhancement in the photoresponsivity is observed (Figure 2.12b,c). Lan et al. [139] reported gate-tunable plasmonic photodetectors that consist of monolayer MoS_2 photodetectors with Ag nanoarrays (Figure 2.12d–f). The plasmonic photodetector has an ultrahigh photoresponsivity of 2.7×10^4 A W^{-1}, achieving a sevenfold enhancement in the photocurrent compared to pristine photodetector. This predominately benefits from the combination of the enhancement of the photon absorption rate via the strongly localized electromagnetic field, and the gate-tunable plasmon-induced photocarrier-generation rate in the monolayer MoS_2. Plasmonic nanostructures are also useful in expanding the response wavelength range of photodetectors. For instance, WS_2 and MoS_2 individually have no response to IR light. Wang et al. [55] reported that heterostructures formed by these two materials can detect longer wavelengths compared to individual components due to the interlayer coupling, but this IR detection ability is limited (Figure 2.12g,h). The heterostructures with Au nanoparticles can enhance the IR detection at 1030 nm by ≈25-fold due to the plasmon resonance (Figure 2.12i).

As the rising star of 2D III–VI compounds semiconductor, InSe attracts a lot of attention due to its smaller electron effective mass ($m^* = 0.143m0$) compared with MoS_2 ($m^* = 0.45m0$), and high mobility over 10^3 cm^2 V^{-1} s^{-1} at room temperature [140]. One challenge for InSe photodetectors is to increase their quantum efficiency

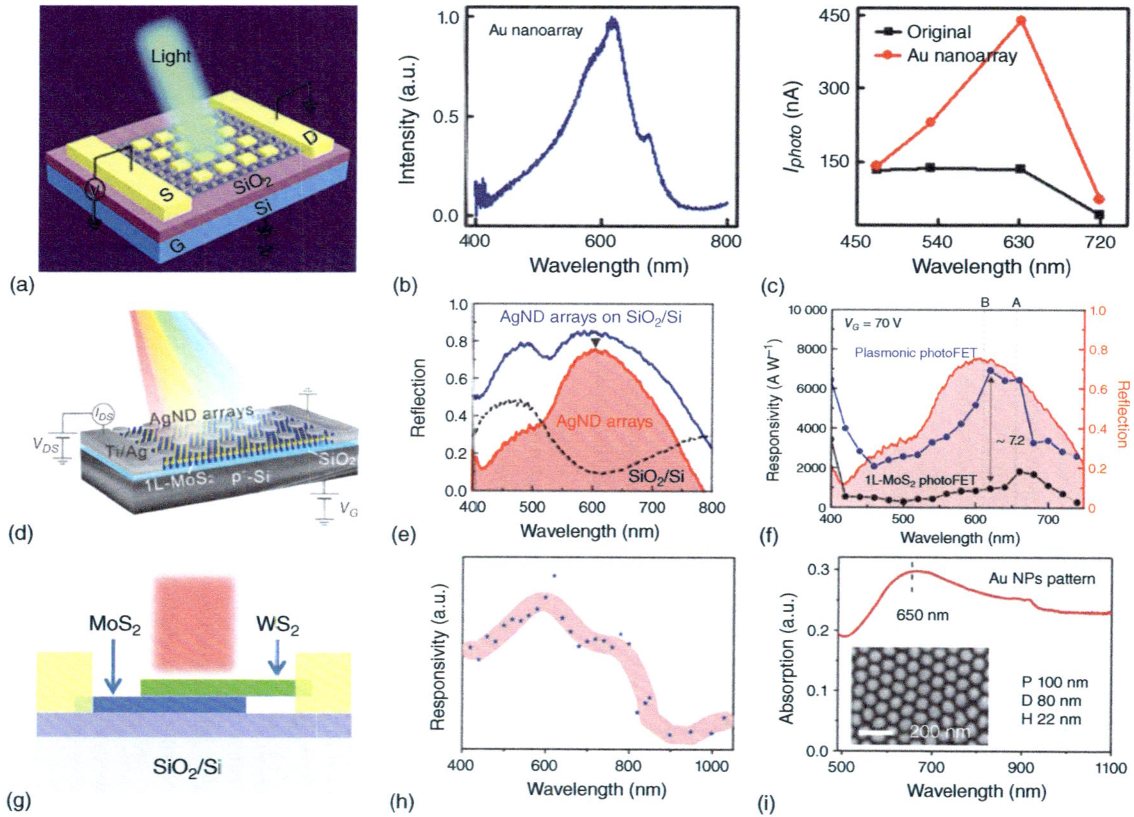

Figure 2.12 (a) Schematic of few-layer plasmon-enhanced MoS_2 phototransistors. (b) Plasmonic resonance absorption of periodic Au nanoarrays on few-layer MoS_2. (c) Photocurrent of MoS_2 phototransistors with and without Au nanoarrays as a function of illumination wavelength. (d) Schematic of monolayer MoS_2 plasmonic photodetector that consists of a monolayer MoS_2 and square arrays of Ag nanodisks. (e) Reflection spectra of the designed 2D plasmonic crystals on a SiO_2/Si substrate (blue line) and the SiO_2/Si substrate (black dashed line). (f) Photoresponsivity of the plasmonic photodetector and pristine photodetector as a function of illumination wavelength. (g) Schematic of photodetector fabricated on heterostructures of WS_2 and MoS_2. (h) Wavelength response curve of photodetector in visible and infrared range. (i) Absorption spectrum of Au nanoparticles pattern, inset is the scanning electron microscopy image of Au nanoparticles. Source: (a–c) Reproduced with permission [138]. Copyright 2015, Wiley-VCH GmbH. (d–f) Reproduced with permission [139]. Copyright 2021, American Chemical Society. (g–i) Reproduced with permission [55], © 2018, John Wiley & Sons.

and generate as many photocarriers as possible. Halas et al. realized an avalanche effect in an InSe photodetector by exploiting the large Schottky barrier formed between InSe and Al electrodes, and they used Al disk nanoantennas to enhance the light absorption of the device (Figure 2.13a) [38]. The scattering peak of Al nanodisks overlaps with the peak of the photoresponse spectrum of layered InSe (≈510 nm), so the photoresponse in this wavelength region is enhanced (Figure 2.13b,c). With 543 nm incident laser light, the quantum efficiency is increased from 108% (without Al disks) to 866% (with Al disks). This is attributed to the local electromagnetic field enhancement that contributes to carrier generation in the InSe, and hot electrons generated by the decay of the surface plasmons of the Al nanodisks provide additional photogenerated carriers. Another challenge for InSe photodetectors is that most of the devices can only operate with a high responsivity in the short wavelength region, limiting their ability of photodetection in visible or NIR region. Dai et al. [141] reported visible to NIR photodetection in an InSe photodetector using Au plasmonic nanoparticles (Figure 2.13d). This intriguing phenomenon is attributed to the wavelength-selective enhancement of pristine responsivities by hybridized quadrupole plasmons resonance of Au nanoparticles. In addition, the response time can be decreased by depositing Au nanoparticles on the surface of InSe nanosheets (Figure 2.13e). This may arise from the effective surface plasmon–exciton coupling process. The maximum enhancement ratio in responsivity reaches up to ≈1200% at a wavelength of 685 nm (Figure 2.13f).

In the plasmon-enhanced photodetectors discussed above, the plasmonic nanostructure and 2D materials are generally in direct contact. In this case, the 2D materials can be placed in a position that makes full use of the enhanced local field and results in strong light–matter interactions. Furthermore, the interfacial charge-transfer transition by directly exciting hot electrons of plasmonic nanostructures into the coupled semiconductor acceptor benefits the enhancement of photocurrent. However, the metal–semiconductor junctions in such systems have strong thermionic emission, which may elevate undesirable dark current. Sun et al. [142] fabricated novel SnSe photodetectors by noncontact plasmonic nanoparticles, consisting of a hybrid structure of few-layer $SnSe_2$ transferred onto a fused silica (SiO_2) plate with embedded Ag nanoparticles (Figure 2.13g). The dark current of the 2D photodetector does not increase obviously because of the extra surface barrier and decrease of interface states. Moreover, after incorporation of Ag nanoparticles (NPs), the response time of photodetectors decreases considerably at excitation of 405 nm, i.e. only 11% of that of the pristine $SnSe_2$ photodetectors (Figure 2.13h). The responsivity of the photodetector is enhanced by about 900 times with the bias voltage of 0.1 V, due to the well-preserved intrinsic features of $SnSe_2$ and LSPR effect (Figure 2.13i).

In summary, plasmon-enhanced photodetectors typically combine metal nanostructures, for example, nanoparticles, nanoarrays, or grating with a light-active 2D material. The properties of surface plasmons are exploited to impart specific characteristics to a photodetector, such as spectral selectivity, or enhance the performance of the photodetector by increasing its absorption at certain wavelengths. We summarize the enhancement factors of plasmon-enhanced photodetectors based on

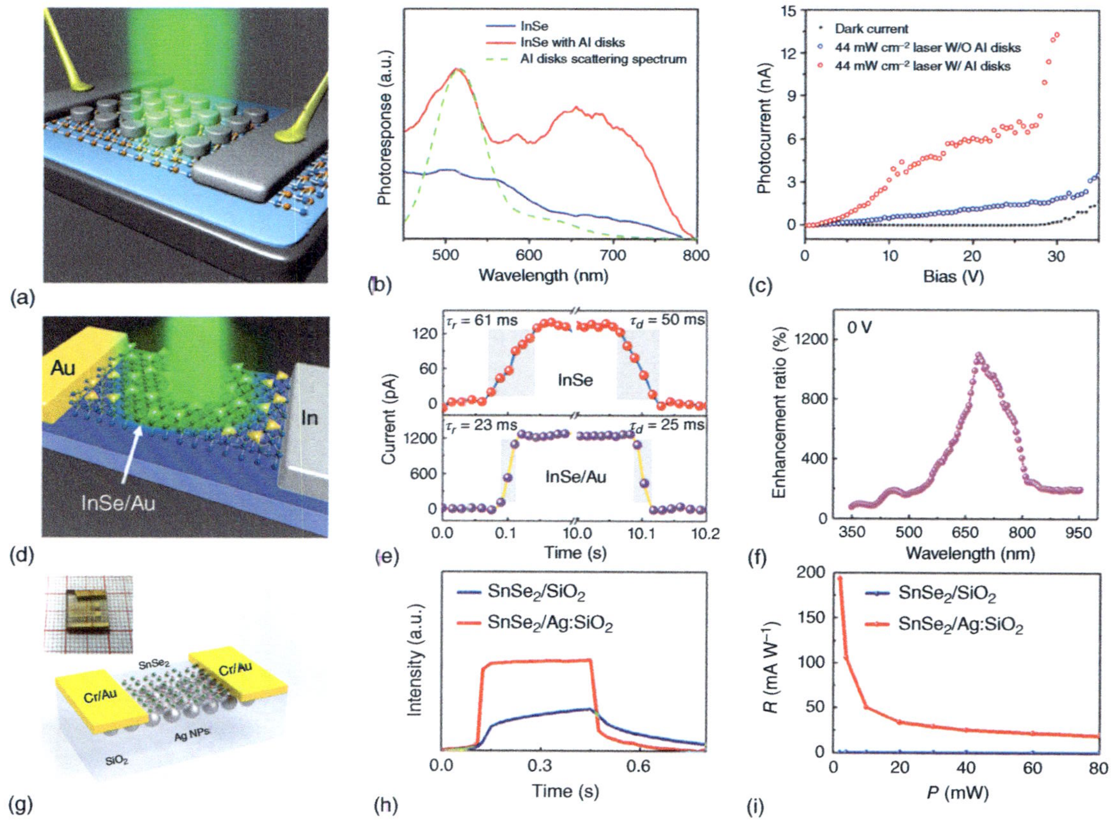

Figure 2.13 (a) Schematic of InSe avalanche photodetector patterned with an array of Al nanodisks. (b) Photocurrent spectra of InSe (blue), InSe decorated with Al nanodisks (red), and the dark-field scattering spectrum of the Al nanodisk arrays (green). (c) Photocurrent response of InSe device with (red hollow circle) and without (blue hollow circle) Al nanodisks at different laser intensities. The solid black square is the measured dark current. (d) Schematic diagram of the device structure of InSe/Au surface plasmon photodetector. (e) Time-resolved photoresponse of InSe photodetector with and without Au surface plasmon upon 685 nm light illumination at 0 V. (f) Responsivity enhancement of InSe surface plasmon photodetector via surface plasmon resonance with illumination wavelengths ranging from 350 to 950 nm. (g) Schematic of the photodetector structure based on $SnSe_2$/AgNP:SiO_2. (h) Response time of $SnSe_2$ nanosheets with and without embedded Ag ions at wavelength of 405 nm. (i) Photoresponsivity of $SnSe_2$/AgNP:SiO_2 and $SnSe_2$:SiO_2 versus incident power at 405 nm. Source: (a–c) Reproduced with permission [38]. Copyright 2015, American Chemical Society. (d–f) Reproduced with permission [141]. Copyright 2018, American Chemical Society. (g–i) Reproduced with permission [142]. Copyright 2021, Wiley-VCH GmbH.

Table 2.3 Comparison of plasmon enhancement factors of 2D materials-based photodetectors.

Hybrid structure	Wavelength (nm)	Voltage (V)	Enhancement factor (fold)	R_{max} (A W^{-1})
Graphene/Au NP [125]	530	/	4–15	2.2×10^{-3}
Au NP/MoS_2 [138]	632	1	3	
Ag nanowire/MoS_2 [126]	532	5	1192	59.6
Au NP/MoS_2 [143]	980	15	≈30	6.4×10^{-2}
Ag NP/MoS_2/Au array [144]	532	10	8.8	287.5
Au NP/WS_2 [145]	590	2	30	1050
Al NP/InSe [38]	543	30	8	
Au NP/InSe [141]	685	0	12	0.244
$SnSe_2$/Ag:SiO_2 [142]	405	0.1	881	

graphene, MoS_2, WS_2, InSe, and SnSe in Table 2.3. The enhancement factor ranges from several-fold to over 1000-fold, but the total photoresponsivity for these photodetectors is still not satisfactory. To further improve the performance, more judiciously designed plasmonic nanostructures with high enhancement in electromagnetic field and low loss are required.

2.6 Large-Scale and Flexible Photodetectors

Scalable integration of advanced photodetectors holds much prospect for industry adoption. Efforts have been made to realize low-cost, controllable production of the active materials. In parallel, high flexibility of light-sensitive 2D nanosheets is desired for the development of wearable optoelectronics. Given the large optical absorption coefficient, high mobility, and good flexibility of Sn chalcogenides (e.g. SnS_2, SnS, SnSe, and SnTe), these materials are proper selections to construct high-performance flexible devices. In terms of the synthesis approach, Su et al. [146] demonstrated the controlled growth of SnS_2 crystal arrays via CVD, before which thin metal pads (Pd/Cr) were patterned as seeds at designated positions (Figure 2.14a). Zheng et al. [148] combined microetching technique with van der Waals epitaxial growth to pattern diverse 2D metal chalcogenides on mica substrates. Organic solvents (e.g. acetone or ethanol) containing polydimethylsiloxanes (PDMS) oligomers can be used as the ink, which is pipetted onto the surface of a PDMS stamp with microwell arrays. The mica substrate is then placed on top of the stamp and is removed after the solvent is dried. Precise control of the layer thickness and crystal orientation has been achieved. The patterned In_2Se_3 arrays were integrated into flexible photodetectors, reaching a high responsivity of ≈1650 A W^{-1}. These methods offer access to fabricating a batch of devices on one chip with predesigned layout.

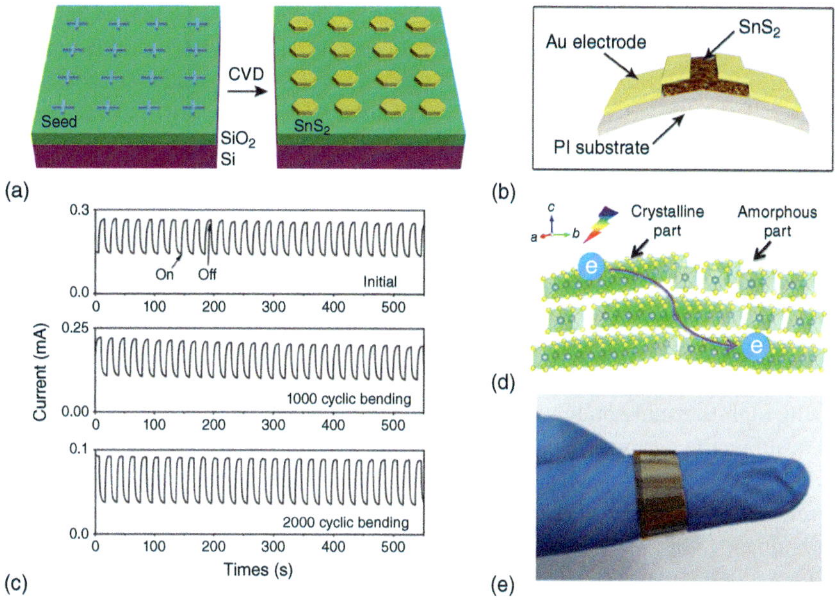

Figure 2.14 (a) Schematic diagram showing the growth of SnS_2 2D crystals with the locations predefined by the patterned seed arrays. (b) Schematic of the flexible SnS_2 photodetector on a polyimide (PI) substrate. (c) Photocurrents at 0, 1000, and 2000 bending cycles. (d) Mixed structure of crystalline/amorphous SnS_2 thin films and its maintained photoresponse after bending. (e) Photograph of the wearable device. Source: (a) Reproduced with permission [146]. Copyright 2015, American Chemical Society. (b–e) Reproduced with permission [147], © 2020, American Chemical Society.

Another challenge is to obtain atomically thin 2D nanosheets with strong interlayer coupling, such as SnS, PtS_2, $PtSe_2$, and so on. This could be achieved by using a molten-phase precursor that has weaker van der Waals interactions. Krishnamurthi et al. [149] developed a novel method to prepare SnS films with the thickness down to 0.8 nm (a single-unit cell) and a lateral dimension of several millimeters by touch printing the naturally sulfurized surface of Sn liquid onto arbitrary substrates. Further increase in the sample area is likely to be achieved by MBE and CVD techniques [150]. Such materials can find applications in flexible photodetectors by using polymer substrates (Figure 2.14b). The SnS_2 [147], SnSe [151], and SnTe [152] photodetectors exhibit high stability over cyclic bending tests. The SnS_2 thin films prepared by thermal evaporation and postannealing can retain at least 91.7% of their initial photoelectric responsivity after 1000 bending cycles with a small radius of 1 mm (Figure 2.14c) [147]. The excellent durability was attributed to the mixed phases consisting of both crystalline and amorphous SnS_2 (Figure 2.14d), showing potential for application-wearable devices (Figure 2.14e).

Flexible devices are typically made on top of polymers with high bendability. The choice of substrate may dictate the photodetecting performances. For example, the responsivity of few-layer InSe photodetector on a rigid SiO_2/Si chip can reach 12.3 A W^{-1}, while the value decreases to 3.9 A W^{-1} when the device is supported by a polyethylene terephthalate (PET) film [153]. Dielectric polymer substrates that are free from traps can suppress the hysteresis of 2D materials. In this regard, BCB is an ideal candidate because of its solution processability, facile preparation, excellent thermal and chemical resistance, and adaptability to electron beam microfabrication. It is advantageous compared with other trap-free dielectrics, including polyimide (PI, complex precursor-involved synthesis), poly(methyl methacrylate) (PMMA, soluble in organic solvents), and hBN (costly for large area growth). Negative photoconductivity was demonstrated in PI-supported BP devices [154], which results from the strong photothermal effect induced by the PI. Due to the low thermal conductivity of PI, the heat generated upon IR illumination efficiently enhances the phonon scattering and decreases the carrier mobility in the BP, giving rise to a negative photocurrent. This phenomenon cannot be observed if utilizing high-thermal-conductivity substrates such as SiO_2/Si.

To evaluate the performances of flexible photodetectors, it is important to understand the effects of strain on the properties of the photoactive materials. The spectral coverage of many TMD-based devices is restricted within the visible range owing to their bandgap (1.65–3.10 eV). In-plane strain can modulate the electronic structures of 2D semiconductors. The shrinkage of the bandgap can enhance the photoresponsivity in the visible region and extend their sensing capabilities beyond fundamental limits. Thai et al. [155] presented the biaxial strain-dependent photoresponse of graphene-contacted MoS_2 devices up to the NIR regime, in which the MoS_2 film was subject to a controllable pneumatic bulging pressure (Figure 2.15a,b). Linear array of the strained MoS_2 photodetectors shows visible–NIR imaging functionality in a line-scanning system (Figure 2.15c,d). However, the NIR photoresponsivity achieved using this method is relatively low (e.g. 407 and 23.5 mA W^{-1} under 785 and 904 nm illumination, respectively, with 1.19% tensile strain). Kim et al. [156] demonstrated the strain-induced continuous and reversible variation of the operating wavelengths of BP from 0.22 to 0.53 eV. Compressive and tensile strains were applied in sequence by thermally quenching the preannealed polyethylene terephthalate glycol (PETG) substrate and mechanically bending it in a two-point apparatus. The peak responsivity is 2.5 A W^{-1} at 2.5 μm with 0.7% tensile strain.

External strain can also interact with the built-in electric field and regulate the behavior of photoinduced carriers. Wu et al. [157] first revealed the piezophototronic effect in 2D crystals, i.e. monolayer MoS_2. Under tensile strain, positive polarization charges are generated in the MoS_2 near the Pd/Au–MoS_2 interface. The resultant piezoelectric field decreases the conduction band energy of MoS_2. Hence, the Schottky barrier is reduced, and the depletion region is shrunk. Such band tilting hinders the separation of the photogenerated electron–hole pairs while facilitating the production of dark current, resulting in lower sensitivity of the photodetector. In contrast, increased photoresponse could be expected when the device is subject to small compressive strain. Yet, large compressive strain causes significant upward bending of the valence band maximum and formation of a new energy barrier at the drain contact. The holes are therefore trapped in the vicinity of the Pd/Au–MoS_2 junction, making it unfavorable for the circulation of free carriers.

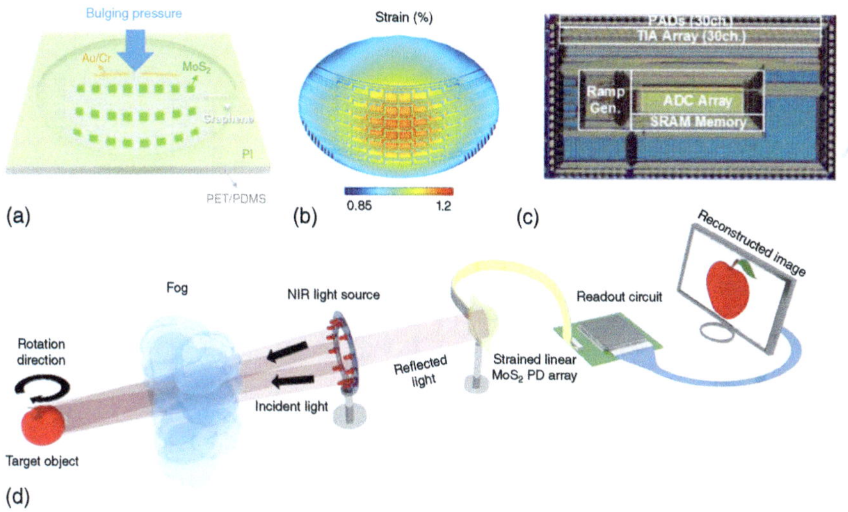

Figure 2.15 Graphene/MoS_2 photodetector arrays with strain-induced spectral extension to the NIR regime and its imaging application. (a) Schematic illustration of device under bulging pressure. (b) Finite element analysis-simulated strain distribution in a bulged PI film. The geometric deformation is magnified by 10 times. (c) Microphotograph of a line-scanning apparatus. (d) Schematic illustration of the NIR-imaging system that can operate even in foggy environment. Source: Reproduced with permission Thai et al. [155], © 2021, American Chemical Society.

Based on the piezotronic properties of ZnO, Du et al. [158] investigated the strain effects on the performance of WSe_2/ZnO photodetector (Figure 2.16a,b). The discrepancy is duc to the type II band alignment and the role of WSe_2 in light absorption. When tensile strain is applied, positive piezopolarization charges are generated in the ZnO, and negative charges are thus accumulated in the WSe_2 near the heterointerface (Figure 2.16c,e). The increased band slope provides extra driving force for the transport of the photoexcited carriers, similar to the voltage-gating effects. Therefore, a larger photocurrent can be obtained in a self-powered mode (Figure 2.11g). When compressive strain is applied, negative piezo charges are generated in the ZnO, and the band of WSe_2 is up-tilted (Figure 2.16d,f). This reconstruction suppresses the carrier transport, thereby decreasing the photoresponse (Figure 2.16g). Maiti et al. [159] wrapped few-layer $MoTe_2$ with a nonplanarized Si waveguide structure to introduce uniaxial tensile strain. The bandgap of $MoTe_2$ is modified from 1.04 (unstrained) to 0.80 eV (4% strained) due to the coupling with the waveguide's optical mode at 1550 nm, which enables wavelength-selective light absorption.

Dai et al. [160] investigated the stacking symmetry-determined piezoelectricities in 2D InSe. The direction of the in-plane polarizations is identical in the adjacent layers for γ-phase InSe with ABC-stacking sequence, whereas reverse direction was observed for β-InSe that presents AB-stacking sequence. The noncentrosymmetry of γ-InSe was confirmed by evident second-harmonic generation characteristics, and this structural feature is responsible for the different strain dependences of the

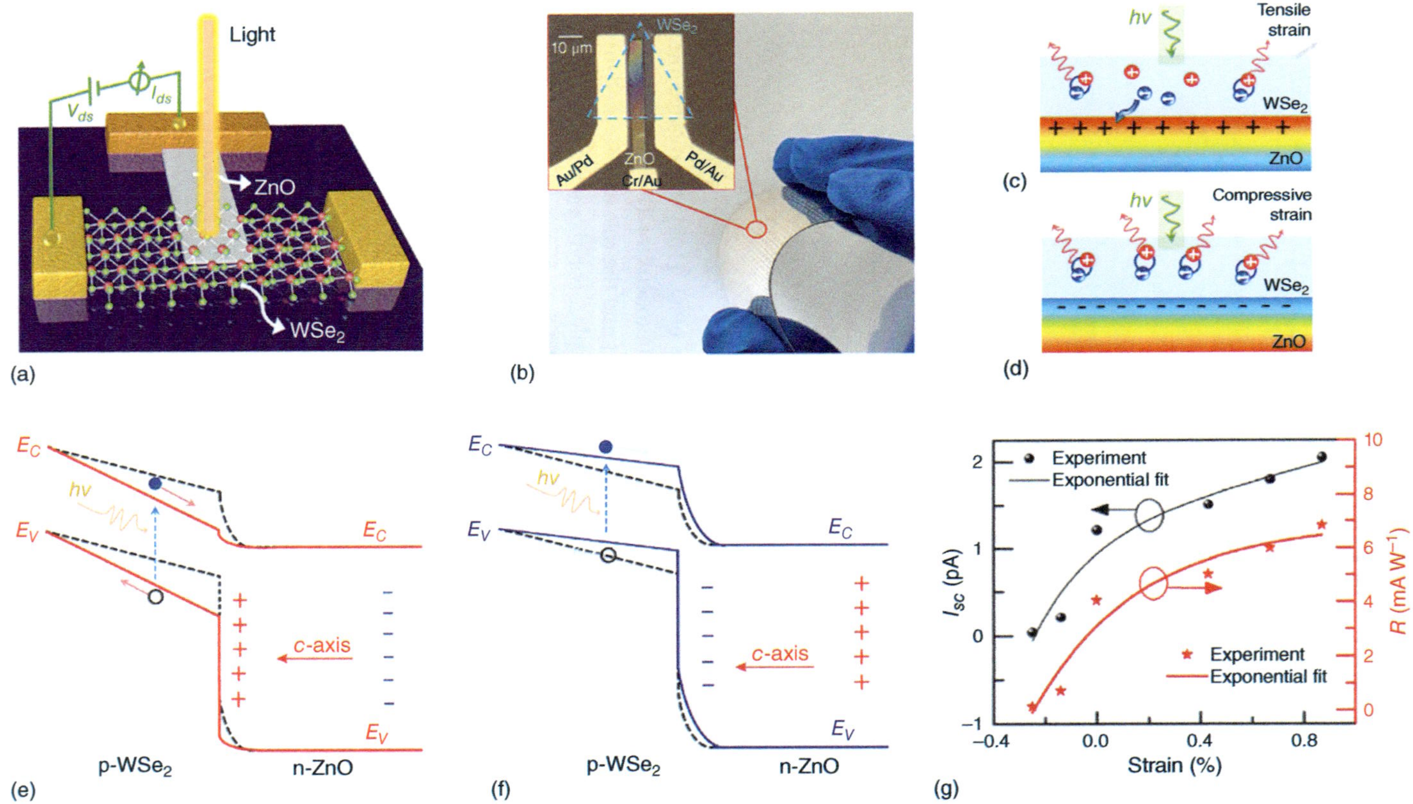

Figure 2.16 Strain effects on the performance of the photodetector based on few-layer WSe_2 nanosheet and ZnO nanobelt. (a) Operation schematic with an external circuit and light illumination. (b) Optical microscopic image of the flexible integrated device under tensile strain. Inset: magnified view. (c, d) Schematics of photogenerated carriers and (e, f) energy-band diagrams under compressive (c, e) and tensile (d, f) strains. Black circles: electrons; dark blue dots: holes; red and blue marks: positive and negative piezopolarization charges, respectively. Black dotted line: initial band structure; red and blue lines: band structures under tensile and compressive strains, respectively. (g) Strain dependence of short-circuit current (left axis) and photoresponsivity (right axis). Source: Reproduced with permission Du et al. [158], © 2019, Elsevier.

photoresponse. Piezoresistive effect is induced by tensile strain in β-InSe photodetector. The carrier density is increased owing to the red shift of the bandgap, and the barrier heights of both Schottky junctions are decreased at the same time. For γ-InSe photodetector, the piezo charges stem from coupled piezoresistive/piezoelectric effect. The piezopotential further increases the energy difference between the two Schottky junctions, whereby facilitating the production of photocurrent. As a result, the applied strain can more effectively modulate the performance of the γ-InSe device. A small tensile strain of 0.62% brings 7- and 10-fold increments in the responsivity and response speed, respectively. These studies open up new opportunities for the scalable integration of 2D materials-based photodetectors and indicate great promise for next-generation wearable (opto)electronics.

2.7 Summary

In summary, 2D materials feature unique characteristics, including atomic thickness, high room-temperature carrier mobility, and excellent flexibility. These enable them to outperform the conventionally used semiconductor thin films in photodetection and can promote the minimization and integration of photodetectors. Broadband photoresponse from IR to UV regions has been achieved due to their tunable bandgaps over a wide spectral range. Various strategies have been developed to enhance the photodetecting performances in terms of the responsivity, response speed, detectivity, etc. As regards the photoactive materials, the layer number control, surface modification, chemical doping, strain and defect engineering, and ferroelectric modulation have crucial impacts on their light-detection capabilities. For the device, the channel–electrode conductance, structural design, and environment are of importance for achieving high performances. In addition, the good mechanical flexibility renders 2D materials promising in wearable and biointegrated devices. These open up prospects for versatile applications such as multicolor imaging and optical communications.

There remain some issues for the commercialization of 2D materials-based photodetectors. The light absorption of 2D nanosheets (e.g. graphene and TMDs) with atomic thicknesses is relatively low compared to conventional photodetecting materials. In addition, Auger scattering could be significant due to the strong Coulomb interaction, and the attenuated electrostatic shielding effect would result in large exciton-binding energy. These pose difficulties in achieving high quantum efficiencies and are likely to be circumvented by exploring new 2D materials (e.g. 2D perovskites) that display both low exciton-binding energy and long carrier-diffusion pathlengths. 2D III–V and II–VI semiconductors exhibit high charge-carrier mobilities and large optical absorption coefficients. However, cost-effective techniques are under investigation for the wafer-scale synthesis of these materials with high uniformity and crystallinity, which could meet industry demands. In another aspect, many 2D semiconductors have wide bandgaps, while those with narrow bandgaps (e.g. BP and AsP) are prone to oxidation under ambient environment. Noble-metal dichalcogenides present excellent air stability, but their large-scale synthesis is still challenging and expensive. The fabrication and optoelectronic properties of stable narrow-bandgap 2D materials are therefore worth of further studies to achieve long-wavelength photodetection. More research is also needed to combine high quantum efficiency with fast response in a single photodetector. Improvement in the overall performances may be realized by constructing hybrid devices based on novel 2D heterostructures. These will inspire future work toward the development of 2D functional materials-based photodetectors for advanced imaging applications.

References

1 Dyck, R.H. and Weckler, G.P. (1968). *IEEE Trans. Electron Devices* 15: 196.
2 Levine, B.F., Bethea, C.G., Hasnain, G. et al. (1990). *Appl. Phys. Lett.* 56: 851.
3 Michel, E., Xu, J., Kim, J.D. et al. (1996). *IEEE Photonics Technol. Lett.* 8: 673.
4 Seo, S.W., Jokerst, N.M., Cho, S.Y. et al. (2004). *IEEE J. Sel. Top. Quantum Electron.* 10: 686.
5 Nair, R.R., Blake, P., Grigorenko, A.N. et al. (2008). *Science* 320: 1308.
6 Britnell, L., Ribeiro, R.M., Eckmann, A. et al. (2013). *Science* 340: 1311.
7 Yan, F., Wei, Z., Wei, X. et al. (2018). *Small Methods* 2: 1700349.
8 Veeraraghavan, S. and Fossum, J.G. (1989). *IEEE Trans. Electron Devices* 36: 522.
9 Yin, Z., Li, H., Li, H. et al. (2012). *ACS Nano* 6: 74.
10 Omkaram, I., Hong, Y.K., and Kim, S. (2018). *Two-Dimensional Materials for Photodetector*. London: IntechOpen.
11 Steiauf, D., Kioupakis, E., and Van De Walle, C.G. (2014). *ACS Photonics* 1: 643.
12 Buscema, M., Groenendijk, D.J., Blanter, S.I. et al. (2014). *Nano Lett.* 14: 3347.
13 Zhang, L. and Zunger, A. (2015). *Nano Lett.* 15: 949.
14 Lee, H.S., Min, S.W., Chang, Y.G. et al. (2012). *Nano Lett.* 12: 3695.
15 Shim, J., Oh, A., Kang, D.H. et al. (2016). *Adv. Mater.* 28: 6985.
16 Das, T., Yang, E., Seo, J.E. et al. (2021). *ACS Appl. Mater. Interfaces* 13: 1861.
17 Zhang, L., Yang, T., Sahdan, M.F. et al. (2021). *Adv. Electron. Mater.* 7: 2100559.
18 Kufer, D. and Konstantatos, G. (2015). *Nano Lett.* 15: 7307.
19 Jo, S.H., Kang, D.H., Shim, J. et al. (2016). *Adv. Mater.* 28: 4824.
20 Wang, X., Wang, P., Wang, J. et al. (2015). *Adv. Mater.* 27: 6575.
21 Lan, C., Li, C., Yin, Y., and Liu, Y. (2015). *Nanoscale* 7: 5974.
22 Lu, Y., Chen, J., Chen, T. et al. (2020). *Adv. Mater.* 32: 1906958.
23 Kwon, J., Hong, Y.K., Han, G. et al. (2015). *Adv. Mater.* 27: 2224.
24 Yeh, C.H., Chen, H.C., Lin, H.C. et al. (2019). *ACS Nano* 13: 3269.
25 Zhang, W., Chiu, M.H., Chen, C.H. et al. (2014). *ACS Nano* 8: 8653.
26 Tan, H., Fan, Y., Zhou, Y. et al. (2016). *ACS Nano* 10: 7866.
27 Lan, C., Zhou, Z., Zhou, Z. et al. (2018). *Nano Res.* 11: 3371.
28 Tu, Y., Xu, Y., Li, J. et al. (2020). *Small* 16: 2005626.
29 Chen, T., Sheng, Y., Zhou, Y. et al. (2019, 6430). *ACS Appl. Mater. Interfaces* 11: 6421.

30 Wang, T., Andrews, K., Bowman, A. et al. (2018). *Nano Lett.* 18: 2766.
31 Zhang, P., Bian, C., Ye, J. et al. (2020). *Sci. China Mater.* 63: 1548.
32 Lee, H.S., Baik, S.S., Lee, K. et al. (2015). *ACS Nano* 9: 8312.
33 Li, H.M., Lee, D., Qu, D. et al. (2015). *Nat. Commun.* 6: 6564.
34 Tan, H., Xu, W., Sheng, Y. et al. (2017). *Adv. Mater.* 29: 1702917.
35 Hu, P., Wang, L., Yoon, M. et al. (2013). *Nano Lett.* 13: 1649.
36 Liu, F., Shimotani, H., Shang, H. et al. (2014). *ACS Nano* 8: 752.
37 Island, J.O., Blanter, S.I., Buscema, M. et al. (2015). *Nano Lett.* 15: 7853.
38 Lei, S., Wen, F., Ge, L. et al. (2015). *Nano Lett.* 15: 3048.
39 Afzal, A.M., Iqbal, M.Z., Mumtaz, S., and Akhtar, I. (2020). *J. Mater. Chem. C* 8: 4743.
40 Feng, F., Wu, J., Wu, C., and Xie, Y. (2015). *Small* 11: 654.
41 Perumal, P., Ulaganathan, R.K., Sankar, R. et al. (2016). *Adv. Funct. Mater.* 26: 3630.
42 Piotrowski, J. and Rogalski, A. (2004). *Infrared Phys. Technol.* 46: 115.
43 Desai, S.B., Madhvapathy, S.R., Sachid, A.B. et al. (2016). *Science (80-.)* 354: 99.
44 Yao, J. and Yang, G. (2020). *Nanoscale* 12: 454.
45 Huang, M., Wang, M., Chen, C. et al. (2016). *Adv. Mater.* 28: 3481.
46 Xiang, D., Han, C., Wu, J. et al. (2015). *Nat. Commun.* 6: 6485.
47 Xu, Y., Yuan, J., Fei, L. et al. (2016). *Small* 12: 5000.
48 Yu, X., Zhang, S., Zeng, H., and Jie, Q. (2016). *Nano Energy* 25: 34.
49 Yu, X., Yu, P., Wu, D. et al. (2018). *Nat. Commun.* 9: 1545.
50 Xu, W., Jiang, J., Ma, H. et al. (2020). *Nano Res.* 13: 2091.
51 Wang, Y., Yu, Z., Tong, Y. et al. (2020). *Appl. Phys. Lett.* 116: 211101.
52 Zeng, L.H., Chen, Q.M., Zhang, Z.X. et al. (2019). *Adv. Sci.* 6: 1901134.
53 Tao, L., Chen, Z., Li, X. et al. (2017). *npj 2D Mater. Appl.* 1: 19.
54 Yu, W.J., Vu, Q.A., Oh, H. et al. (2016). *Nat. Commun.* 7: 13278.
55 Wang, G., Li, L., Fan, W. et al. (2018). *Adv. Funct. Mater.* 28: 1800339.
56 Long, M., Gao, A., Wang, P. et al. (2017). *Sci. Adv.* 3: e1700589.
57 Long, M., Wang, Y., Wang, P. et al. (2019). *ACS Nano* 13: 2511.
58 Luo, L.B., Wang, D., Xie, C. et al. (2019). *Adv. Funct. Mater.* 29: 1900849.
59 Varghese, A., Saha, D., Thakar, K. et al. (2020). *Nano Lett.* 20: 1707.
60 Jin, Z., Hu, C., Lan, Y. et al. (2020). *Adv. Opt. Mater.* 8: 2001319.
61 Liang, F.X., Zhao, X.Y., Jiang, J.J. et al. (2019). *Small* 15: 1903831.
62 Lim, H., Tsao, S., Zhang, W., and Razeghi, M. (2007). *Appl. Phys. Lett.* 90: 131112.
63 Johnson, T.H., Cozine, H.T., and McLean, B.N. (1965). *Appl. Opt.* 4: 693.
64 Chen, C., Yi, X., Zhang, J., and Zhao, X. (2001). *Infrared Phys. Technol.* 42: 87.
65 Chen, X., Lu, X., Deng, B. et al. (2017). *Nat. Commun.* 8: 1672.
66 Xie, Y., Liang, F., Wang, D. et al. (2018). *Adv. Mater.* 30: 1804858.
67 Liang, Q., Wang, Q., Zhang, Q. et al. (2019). *Adv. Mater.* 31: 1807609.
68 Xia, F., Wang, H., Xiao, D. et al. (2014). *Nat. Photonics* 8: 899.
69 Liu, H., Neal, A.T., Zhu, Z. et al. (2014). *ACS Nano* 8: 4033.
70 Li, L., Yu, Y., Ye, G.J. et al. (2014). *Nat. Nanotechnol.* 9: 372.
71 Huang, L., Dong, B., Guo, X. et al. (2019). *ACS Nano* 13: 913.

72 Wang, J., Rousseau, A., Eizner, E. et al. (2019). *ACS Photonics* 6: 3092.
73 Bullock, J., Yan, W., Shresha, V.R. et al. (2020). *ACS Nano* 14: 13645.
74 Engel, M., Steiner, M., and Avouris, P. (2014). *Nano Lett.* 14: 6414.
75 Zhong, M., Meng, H., Ren, Z. et al. (2021). *Nanoscale* 13: 10579.
76 Amani, M., Regan, E., Bullock, J. et al. (2017). *ACS Nano* 11: 11724.
77 Liu, Y., Shivananju, B.N., Wang, Y. et al. (2017). *ACS Appl. Mater. Interfaces* 9: 36137.
78 Xu, Y., Liu, C., Guo, C. et al. (2020). *Nano Energy* 70: 104518.
79 Wang, Y., Wang, H., Gali, S.M. et al. (2021). *Adv. Funct. Mater.* 31: 2103353.
80 Ahmad, W., Liu, J., Jiang, J. et al. (2021). *Adv. Funct. Mater.* 31: 2104143.
81 Liu, Z., Luo, T., Liang, B. et al. (2013). *Nano Res.* 6: 775.
82 Tan, H., Fan, C., Ma, L. et al. (2016). *Nano Micro Lett.* 8: 29.
83 Rogalski, A. (2020). *HgCdTe Photodetectors*. Elsevier Ltd.
84 Madejczyk, P., Gawron, W., Piotrowski, A. et al. (2011). *Infrared Phys. Technol.* 54: 310.
85 Kang, D.H., Kim, M.S., Shim, J. et al. (2015). *Adv. Funct. Mater.* 25: 4219.
86 Cao, Y., Cai, K., Hu, P. et al. (2015). *Sci. Rep.* 5: 8130.
87 Du, S., Lu, W., Ali, A. et al. (2017). *Adv. Mater.* 29: 1700463.
88 Sharma, R., Sharma, P., Sahoo, K.R. et al. (2021). *Mater. Today*. https://doi.org/10.1016/j.mattod.2021.06.012.
89 Ye, L., Li, H., Chen, Z., and Xu, J. (2016). *ACS Photonics* 3: 692.
90 Liu, B., You, C., Zhao, C. et al. (2019). *Chin. Opt. Lett.* 17: 020002.
91 Yu, W., Li, S., Zhang, Y. et al. (2017). *Small* 13: 1700268.
92 Chen, Y., Wang, X., Wu, G. et al. (2018). *Small* 14: 1703293.
93 Wadhwa, R., Agrawal, A.V., Kushavah, D. et al. (2021). *Appl. Surf. Sci.* 569: 150949.
94 Jo, S.H., Lee, H.W., Shim, J. et al. (2018). *Adv. Sci.* 5: 1700423.
95 Long, M., Liu, E., Wang, P. et al. (2016). *Nano Lett.* 16: 2254–2259.
96 Zhao, Y., Tsai, T.-Y., Wu, G. et al. (2021). *ACS Appl. Mater. Interfaces* 13: 47198.
97 Zhou, X., Hu, X., Zhou, S. et al. (2018). *Adv. Mater.* 30: 1703286.
98 Xue, H., Dai, Y., Kim, W. et al. (2019). *Nanoscale* 11: 3173.
99 Qiao, H., Yuan, J., Xu, Z. et al. (2015). *ACS Nano* 9: 1886.
100 Xiong, Y.F., Chen, J.H., Lu, Y.Q., and Xu, F. (2019). *Adv. Electron. Mater.* 5: 1800562.
101 Vu, Q.A., Lee, J.H., Nguyen, V.L. et al. (2017). *Nano Lett.* 17: 453.
102 Gao, S., Wang, Z., Wang, H. et al. (2021). *Adv. Mater. Interfaces* 8: 2001730.
103 Lv, Q., Yan, F., Wei, X., and Wang, K. (2018). *Adv. Opt. Mater.* 6: 1700490.
104 Yan, F., Zhao, L., Patanè, A. et al. (2017). *Nanotechnology* 28: 27LT01.
105 Afzal, A.M., Dastgeer, G., Iqbal, M.Z. et al. (2020). *ACS Appl. Mater. Interfaces* 12: 19625.
106 Li, A., Chen, Q., Wang, P. et al. (2019). *Adv. Mater.* 31: 1805656.
107 Guo, J., Li, J., Liu, C. et al. (2020). *Light Sci. Appl.* 9: 29.
108 Wu, E., Wu, D., Jia, C. et al. (2019). *ACS Photonics* 6: 565.
109 Wu, D., Jia, C., Shi, F. et al. (2020). *J. Mater. Chem. A* 8: 3632.
110 Yao, J., Zheng, Z., and Yang, G. (2018). *ACS Appl. Mater. Interfaces* 10: 38166.

111 He, T., Lan, C., Zhou, S. et al. (2021). *J. Mater. Chem. C* 9: 3846.
112 Lu, J., Zheng, Z., Yao, J. et al. (2019). *Small* 15: 1904912.
113 Zeng, L., Wu, D., Jie, J. et al. (2020). *Adv. Mater.* 32: 2004412.
114 Jain, S.K., Low, M.X., Taylor, P.D. et al. (2021). *ACS Appl. Electron. Mater.* 3: 2407.
115 Geum, D.M., Kim, S., Khym, J.H. et al. (2021). *Small* 17: 2007357.
116 Hao, F., Stoumpos, C.C., Chang, R.P.H., and Kanatzidis, M.G. (2014). *J. Am. Chem. Soc.* 136: 8094.
117 Xu, X., Chueh, C.C., Jing, P. et al. (2017). *Adv. Funct. Mater.* 27: 1701053.
118 Pan, Y., Wang, H., Li, X. et al. (2020). *J. Mater. Chem. C* 8: 3359.
119 Liu, Y., Wang, F., Wang, X. et al. (2015). *Nat. Commun.* 6: 8589.
120 Yao, J., Chen, F., Li, J. et al. (2021). *Nanotechnology* 32: 415201.
121 Selamneni, V., Ganeshan, S.K., and Sahatiya, P. (2020). *J. Mater. Chem. C* 8: 11593.
122 Kufer, D., Nikitskiy, I., Lasanta, T. et al. (2015). *Adv. Mater.* 27: 176.
123 Luo, P., Zhuge, F., Wang, F. et al. (2019). *ACS Nano* 13: 9028.
124 Hossain, M.T., Das, M., Ghosh, J. et al. (2021). *Nanoscale* 13: 14945.
125 Liu, Y., Cheng, R., Liao, L. et al. (2011). *Nat. Commun.* 2: 579.
126 Bang, S., Duong, N.T., Lee, J. et al. (2018). *Nano Lett.* 18: 2316.
127 Sun, B., Wang, Z., Liu, Z. et al. (2019). *Adv. Funct. Mater.* 29: 1900541.
128 Willets, K.A. and Van Duyne, R.P. (2007). *Annu. Rev. Phys. Chem.* 58: 267.
129 Jiang, R., Li, B., Fang, C., and Wang, J. (2014). *Adv. Mater.* 26: 5274.
130 Singh, J., Kaurav, N., Choudhary, K.K., and Okram, G.S. (2007). *Nano Lett.* 7: 1032.
131 Li, X., Zhu, J., and Wei, B. (2016). *Chem. Soc. Rev.* 45: 3145.
132 Tao, L., Chen, Z., Li, Z. et al. (2021). *InfoMat* 3: 36.
133 Yang, A., Wang, D., Wang, W., and Odom, T.W. (2017). *Annu. Rev. Phys. Chem.* 68: 83.
134 Echtermeyer, T.J., Milana, S., Sassi, U. et al. (2016). *Nano Lett.* 16: 8.
135 Mak, K.F., Sfeir, M.Y., Wu, Y. et al. (2008). *Phys. Rev. Lett.* 101: 2.
136 Cai, X., Sushkov, A.B., Jadidi, M.M. et al. (2015). *Nano Lett.* 15: 4295.
137 Gao, M., Tian, Z., Tang, S. et al. (2020). *Adv. Mater. Interfaces* 7: 2001122.
138 Miao, J., Hu, W., Jing, Y. et al. (2015). *Small* 11: 2392.
139 Lan, H.Y., Hsieh, Y.H., Chiao, Z.Y. et al. (2021). *Nano Lett.* 21: 3083.
140 Feng, W., Zheng, W., Cao, W., and Hu, P. (2014). *Adv. Mater.* 26: 6587.
141 Dai, M., Chen, H., Feng, R. et al. (2018). *ACS Nano* 12: 8739–8747.
142 Sun, X., Sun, J., Xu, J. et al. (2021). *Small* 17: 2102351.
143 Guo, J., Li, S., He, Z. et al. (2019). *Appl. Surf. Sci.* 483: 1037.
144 Wu, Z.Q., Yang, J.L., Manjunath, N.K. et al. (2018). *Adv. Mater.* 30: 1706527.
145 Liu, Y., Huang, W., Chen, W. et al. (2019). *Appl. Surf. Sci.* 481: 1127.
146 Su, G., Hadjiev, V.G., Loya, P.E. et al. (2015). *Nano Lett.* 15: 506.
147 Lei, Y., Luo, J., Yang, X. et al. (2020). *ACS Appl. Mater. Interfaces* 12: 24940.
148 Zheng, W., Xie, T., Zhou, Y. et al. (2015). *Nat. Commun.* 6: 6972.
149 Krishnamurthi, V., Khan, H., Ahmed, T. et al. (2020). *Adv. Mater.* 32: 2004247.
150 Li, S., Wang, Y., Cheng, P. et al. (2021). *Small* 17: 2101154.

151 Xu, H., Hao, L., Liu, H. et al. (2020). *ACS Appl. Mater. Interfaces* 12: 35250.

152 Liu, J., Li, X., Wang, H. et al. (2020). *ACS Appl. Mater. Interfaces* 12: 31810.

153 Tamalampudi, S.R., Lu, Y.-Y., Kumar, U.R. et al. (2014). *Nano Lett.* 14: 2800.

154 Miao, J., Song, B., Li, Q. et al. (2017). *ACS Nano* 11: 6048.

155 Thai, K.Y., Park, I.J., Kim, B.J. et al. (2021). *ACS Nano* 15: 12836–12846.

156 Kim, H., Uddin, S.Z., Lien, D.-H. et al. (2021). *Nature* 596: 232.

157 Wu, W., Wang, L., Yu, R. et al. (2016). *Adv. Mater.* 28: 8463.

158 Du, J., Liao, Q., Hong, M. et al. (2019). *Nano Energy* 58: 85.

159 Maiti, R., Patil, C., Saadi, M.A.S.R. et al. (2020). *Nat. Photonics* 14: 578.

160 Dai, M., Chen, H., Wang, F. et al. (2019). *ACS Nano* 13: 7291.

3

Surface Plasmonic Resonance-Enhanced Infrared Photodetectors

Boyang Xiang[1], *Guiru Gu*[2], *and Xuejun Lu*[1]

[1] *University of Massachusetts Lowell, Department of Electrical and Computer Engineering, One University Avenue, Lowell, MA 01854, USA*
[2] *Stonehill College, Department of Physics, 320 Washington Street, Easton, MA 02357, USA*

3.1 Introduction

Infrared (IR) photodetectors and focal plane arrays (FPA) are key components for numerous sensing and imaging applications, including standoff chemical and biological sensing, IR target detection and tracking, night vision, multispectral imaging, light detection and ranging (Lidar), and computer vision for autonomous driving. Tremendous improvement has been achieved in both IR photodetector materials and optics systems as well as device structures (e.g. resonant cavities [1–3] and surface gratings [4]) to improve the performance of IR photodetectors.

The advancement of nanofabrication technologies such as electron-beam (E-beam) lithography and deep ultraviolet (DUV) stepper photolithography enables the fabrication of high-precision nanostructures with low surface roughness-induced scattering loss. This allows the fabrication of high-quality surface plasmonic resonance (SPR) [5–12] structures and their integration onto photonics devices [10, 13–20].

SPR [7, 21] (also referred to as surface plasmon polaritons, SPPs) had been utilized for a long time in stained glass, such as the rose windows of Notre Dame Cathedral, before its principle was discovered by Ritchie in 1957 [21]. Engineering the SPR resonance and the electric field (E-field) offers a promising emerging technology to provide strong surface confinement of electromagnetic (EM) energy. Such strong surface confinement leads to significant near-field enhancement [5, 9] for sensor and photodetector sensitivity improvement, such as SPR-enhanced Raman spectroscopy [22–24]. It also enables high absorption using a thin absorption layer and thus leads to significant enhancement in IR photodetectors [5, 6, 8, 11, 25–27]. This chapter focuses on the SPR enhancement of IR photodetectors.

Optical Imaging and Sensing: Materials, Devices, and Applications, First Edition.
Edited by Jiang Wu and Hao Xu.

3.2 Brief Review of Basic Concepts of SPR and SPR Structures

3.2.1 Plasma Oscillations in Metals

Metals contain free-electron gases. The electron-density fluctuation and the Coulomb-restoring force result in plasma oscillations. The frequency of the plasma oscillations is the plasma frequency ω_p, which is related to the electron density n_0 and the permittivity of the metal ε by:

$$\omega_p = \sqrt{\frac{e^2 n_0}{m_0 \varepsilon}} \tag{3.1}$$

where n_0 is the electron density, ε is the permittivity of the material, $e = -1.6 \times 10^{-19}$(C) is the charge of an electron, and m_0 is the static mass of a free electron, which will be replaced by the effective mass of an electron if it is in metal or semiconductor materials.

3.2.2 Complex Permittivity and the Drude Model

The complex permittivity ε_c of isotropic conductive EM materials can be written as:

$$\varepsilon_c = \varepsilon\left(1 + \frac{\sigma}{j\omega\varepsilon}\right) = \varepsilon\left(1 - \frac{e^2 n_0}{\omega^2 m \varepsilon}\right) = \varepsilon\left(1 - \frac{\omega_p^2}{\omega^2}\right) \tag{3.2}$$

where σ is the conductivity of the EM material. The detailed derivation of Eq. (3.2) is presented by Gu et al. in Ref. [10, 28]. Equation (3.2) is derived under the assumption of frictionless electron motion (i.e. the lossless condition). The electron motions with the friction force can be expressed as:

$$m_0 \ddot{x} = -m_0 \gamma \dot{x} + e\vec{E} \tag{3.3}$$

where x is the position of the electrons and the $-m_0\gamma\dot{x}$ term represents the friction force, and the $e\vec{E}$ term is the Coulomb electric force. The current density $\vec{J}$ can be written as:

$$\vec{J} = ne\dot{x} \tag{3.4}$$

Combining Eqs. (3.3) and (3.4) and changing it to the phasor domain, one gets:

$$j\omega \frac{m_0}{ne}\vec{J} = \frac{m_0\gamma}{ne}\vec{J} + e\vec{E} \tag{3.5}$$

Plugging in $\vec{J} = \sigma\vec{E}$, one gets:

$$\sigma = \frac{ne^2}{m_0}\left(\frac{j\omega}{-\omega^2 - j\omega\gamma}\right) \tag{3.6}$$

From Eq. (3.6), one can rewrite Eq. (3.2) as:

$$\varepsilon_c = \varepsilon\left(1 + \frac{\sigma}{j\omega\varepsilon}\right) = \varepsilon\left(1 - \frac{\omega_p^2}{\omega^2 + j\omega\gamma}\right) \tag{3.7}$$

Equation (3.7) is the well-known Drude model [29]. It shows that permittivity depends on frequency. When the frequency ω is much smaller than the plasmonic

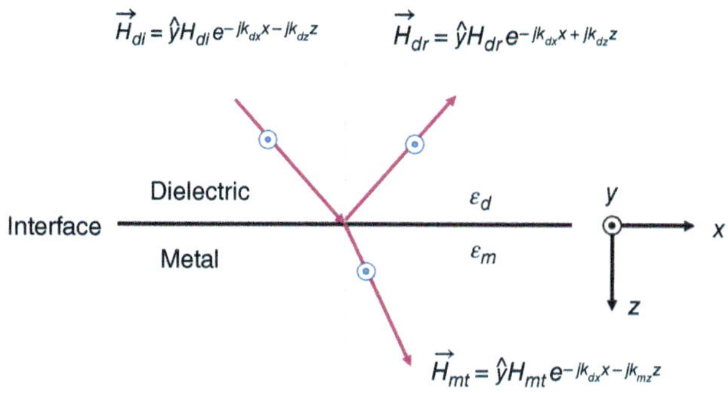

Figure 3.1 Scheme of TM wave incidence on the dielectric and metal interface from the dielectric material.

frequency ω_0, the metal has a large negative permittivity ε_c, whereas when the frequency ω is much larger than the plasmonic frequency ω_0, the metal behaves like regular dielectric materials. At certain frequencies, the permittivity can reach close to zero, which is referred to as epsilon near-zero materials [30–32].

3.2.3 Surface Plasmonic Waves at the Semi-infinite Dielectric and Metal Interface

The analysis of plasmonic waves at a dielectric/metal interface was given by Raether [33]. A brief summary of the surface plasmonic waves at the semi-infinite dielectric and metal interface is presented as follows.

Figure 3.1 shows the incidence of a transmagnetic (TM) plane wave from the dielectric material on a dielectric/metal interface with the H-field in the y-direction. The dielectric and the metal materials are semi-infinite with permittivities of ε_d and ε_m, respectively. The permittivity of metal ε_m is the complex permittivity. The H fields of the incident wave (i.e. $\vec{H}_{di}$), the reflection wave (i.e. $\vec{H}_{dr}$), and the transmission wave (i.e. $\vec{H}_{mt}$) are marked on the figure.

The E-fields are related to the magnetic fields by:

$$\vec{E} = \frac{1}{j\omega\varepsilon}\nabla \times \vec{H} \tag{3.8}$$

From Eq. (3.8), the E_x can be expressed as:

$$\vec{E}_{di} = \hat{x}\frac{jk_{dz}}{j\omega\varepsilon_d}H_{di}e^{-jk_{dx}x-jk_{dz}z} \tag{3.9}$$

$$\vec{E}_{dr} = \hat{x}\frac{-jk_{dz}}{j\omega\varepsilon_d}H_{dr}e^{-jk_{dx}x+jk_{dz}z} \tag{3.10}$$

$$\vec{E}_{mt} = \hat{x}\frac{jk_{2z}}{j\omega\varepsilon_m}H_{mt}e^{-jk_{mx}x-jk_{mz}z} \tag{3.11}$$

where k_{dx} and k_{dz} are the x and z components of the propagation constant k_d, respectively. k_{mx} and k_{mz} are the x and z components of the propagation constant k_m, respectively.

$$\mathbf{k_{dx}^2 + k_{dz}^2 = k_d^2 = \omega^2 \varepsilon_d \mu_d} \tag{3.12}$$

$$\mathbf{k_{mx}^2 + k_{mz}^2 = k_m^2 = \omega^2 \varepsilon_m \mu_m} \tag{3.13}$$

where $\mu_d \approx \mu_m \approx \mu_0$ are the permeability of the dielectric, metal, and vacuum, respectively. The boundary conditions are: H_y and E_x continuous at the dielectric/metal interface (i.e. $z = 0$).

Applying the boundary conditions and defining $r = H_{dr}/H_{di}$ and $t = H_{mt}/H_{di}$, one gets:

$$1 + r = t \tag{3.14}$$

$$k_{dx} = k_{mx} \tag{3.15}$$

$$\frac{k_{dz}}{\varepsilon_d}(1 - r) = \frac{k_{mz}}{\varepsilon_m} t \tag{3.16}$$

From Eqs. (3.14) and (3.16), the coefficients t and r can be obtained in terms of k_{dz}, k_{mz}, and ε_m and ε_d:

$$t = \frac{2k_{dz}\varepsilon_m}{k_{dz}\varepsilon_m + k_{mz}\varepsilon_d} \tag{3.17}$$

$$r = \frac{k_{dz}\varepsilon_m - k_{mz}\varepsilon_d}{k_{mz}\varepsilon_d + k_{mz}\varepsilon_d} \tag{3.18}$$

Under certain excitation conditions to be discussed in Section 3.2.4, the reflection can be zero if $k_{dz}\varepsilon_m - k_{mz}\varepsilon_d = 0$, i.e.

$$\frac{k_{dz}}{\varepsilon_d} = \frac{k_{mz}}{\varepsilon_m} \tag{3.19}$$

Combining Eqs. (3.12), (3.13), and (3.19), one gets:

$$\mathbf{k_{dx}^2 + k_{mz}^2 = k_{dx}^2 + \frac{\varepsilon_m^2}{\varepsilon_d^2} k_{dz}^2 = \omega^2 \varepsilon_m \mu_0} \tag{3.20}$$

$$\mathbf{k_{dx}^2 + k_{dz}^2 = \omega^2 \varepsilon_d \mu_0} \tag{3.21}$$

$$\mathbf{k_{dz}^2 = \omega^2 \mu_0 \frac{\varepsilon_d^2}{\varepsilon_d + \varepsilon_m}} \tag{3.22}$$

$$\mathbf{k_{mz}^2 = \omega^2 \mu_0 \frac{\varepsilon_m^2}{\varepsilon_d + \varepsilon_m}} \tag{3.23}$$

$$\mathbf{k_{dx} = k_{mx} = k_0 Re\left[\sqrt{\frac{\varepsilon_{mr}\varepsilon_{dr}}{(\varepsilon_{dr} + \varepsilon_{mr})}}\right] + jk_0 Im\left[\sqrt{\frac{\varepsilon_{mr}\varepsilon_{dr}}{(\varepsilon_{dr} + \varepsilon_{mr})}}\right]} \tag{3.24}$$

$$\mathbf{k_{mz} = k_0 \sqrt{\frac{\varepsilon_{mr}^2}{(\varepsilon_{dr} + \varepsilon_{mr})}}} \tag{3.25}$$

where $\varepsilon_{dr} = \frac{\varepsilon_d}{\varepsilon_0}$ and $\varepsilon_{mr} = \frac{\varepsilon_m}{\varepsilon_0}$ are the relative permittivity, ε_0 is the permittivity of vacuum, and Re(.) and Im(.) are taking the real and the imaginary parts, respectively.

Table 3.1 Complex permittivity ε_{mr} of gold at different wavelengths.

λ (μm)	$\varepsilon_{mr,\,re}$	$\varepsilon_{mr,\,im}$	$k_{x,\,re}$	$k_{x,\,im}$	$k_{mz,\,re}$	$k_{mz,\,im}$
10.0	−5.05E+03	1.09E+03	k_0	0	$7.6k_0$	$71.5k_0$
5.0	−1.31E+03	1.41E+02	k_0	0	$1.9k_0$	$36.3k_0$
3.0	−4.75E+02	3.07E+01	k_0	0	$0.7k_0$	$21.8k_0$
0.95	−37.2	2.32	$1.01k_0$	0	$0.19k_0$	$6.2k_0$
0.80	−23.4	1.55	$1.02k_0$	$0.002k_0$	$0.16k_0$	$4.9k_0$
0.5	−2.68	3.09	$1.07k_0$	$0.12k_0$	$0.69k_0$	$2.1k_0$

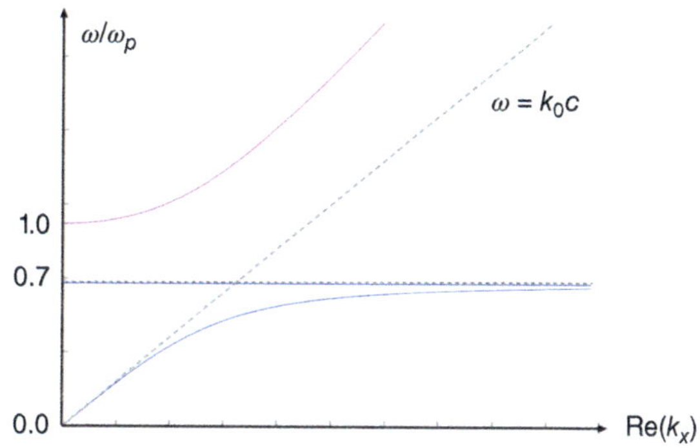

Figure 3.2 Dispersion relationship of the surface plasmonic waves.

The complex permittivity ε_{mr} of gold at different wavelengths is listed in Table 3.1 [34]. The real ($k_{x,\,re}$) and imaginary ($k_{x,\,im}$) parts of k_{spp} and $k_{mz,\,re}$ and $k_{mz,\,im}$ are also included in Table 3.1 for $\varepsilon_{dr} = 1$ (i.e. vacuum or air).

From Table 3.1, since the $k_{x,\,re} \geq k_0$, the surface plasmonic wave cannot be directly excited from air with zero reflection field. Figure 3.2 shows the dispersion relationship of the surface plasmonic waves. The dashed line is the k_0 and ω relation in free space.

3.2.4 Prism-Coupled Surface Plasmonic Wave Excitation

Surface plasmonic waves can be excited by the prism-coupling techniques with the Otto or Kretschmann configurations [35, 36] or through grating-coupling. Figure 3.3a,b shows the coupling schemes of the Otto and the Kretschmann configurations, respectively. The Otto configuration has a small air gap between the prism and the metal film, whereas the Kretschmann configuration makes the thin metal film in close contact with the prism. The Otto configuration is analyzed below.

For the Otto configuration, when the incidence angle is θ larger than the critical angle $\theta_0 = \sin^{-1}\left(\frac{1}{n_d}\right)$ with $n_d = \sqrt{\varepsilon_{dr}}$ as the refractive index of the prism, total internal reflection occurs and generates an evanescence wave traveling along the

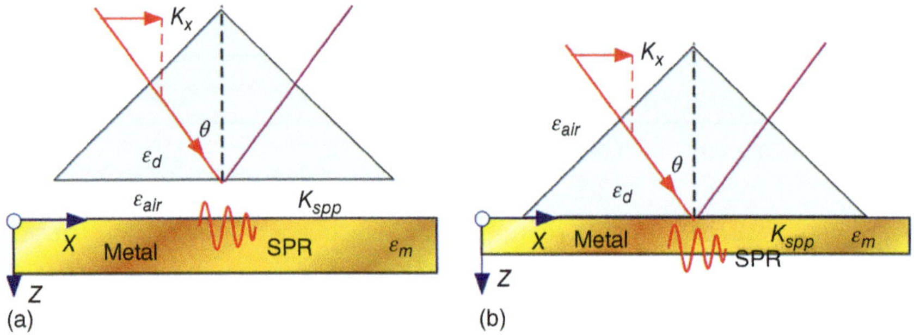

Figure 3.3 Surface plasmonic excitation by prism-coupling with the Otto or Kretschmann configurations. Source: Otto [35] and Shalabney and Abdulhalim [36]. (a) Otto configuration; (b) Kretschmann configuration. The Otto configuration has a small air gap between the prism and the metal film, whereas the Kretschmann configuration makes the thin metal film in close contact with the prism.

bottom surface. The x component k_{ax} and the z component k_{az} of the evanescence wave are:

$$k_{ax} = k_{dx} = k_0\sqrt{\varepsilon_{dr}} \sin \theta \tag{3.26}$$

$$k_{az} = k_0[(\varepsilon_0 - \varepsilon_{dr} \sin^2 \theta)]^{1/2} \tag{3.27}$$

To excite surface plasmonic waves on the metal surface under the air gap (Figure 3.3a), the x-component the wave vector in the air k_{ax} must equal to that of the x wave vector of the surface plasmonic wave k_{mx}, i.e.

$$\mathbf{k}_{ax} = \mathbf{k}_0\sqrt{\boldsymbol{\varepsilon}_{dr}} \sin\boldsymbol{\theta} = \mathbf{k}_{mx} = \mathbf{k}_0\mathbf{Re}\left[\sqrt{\frac{\boldsymbol{\varepsilon}_{mr}}{(1+\boldsymbol{\varepsilon}_{mr})}}\right] \tag{3.28}$$

The incident angle of the light to excite the surface plasmonic wave is thus:

$$\theta = \sin^{-1}\left(\mathrm{Re}\left[\sqrt{\frac{\varepsilon_{mr}}{\varepsilon_{dr}(1+\varepsilon_{mr})}}\right]\right) \tag{3.29}$$

3.2.5 Surface Grating-Coupled Surface Plasmonic Wave Excitation

In addition to the prism coupling, surface plasmonic waves can also be excited by providing an additional wave vector through a grating coupler. Figure 3.4a,b shows the surface plasmonic wave excitation through a grating coupler and the matching of the wave vectors, respectively.

For the TM wave, the incident $\vec{H}$ field can be expressed as:

$$\vec{H}_{1i} = \hat{y}H_{1i}e^{-jk_{1x}x-jk_{1z}z} \tag{3.30}$$

where $k_{1x} = k_0 \sin \theta$, $k_{1z} = k_0 s \cos \theta$, and θ is the incidence angle. The reflection and the transmission $\vec{H}$ fields are:

$$\vec{H}_{1r} = \hat{y}H_{1r}e^{-jk_{1x}x+jk_{1z}z} \tag{3.31}$$

$$\vec{H}_{2t} = \hat{y}H_{2t}e^{-jk_{2x}x-jk_{2z}z} \tag{3.32}$$

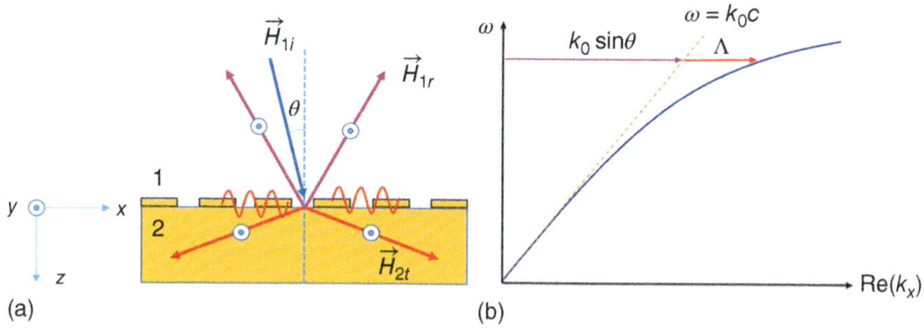

Figure 3.4 Surface plasmonic excitation by a grating coupler: (a) Schematic illustration of the surface plasmonic excitation and the *H* vectors; (b) diagram for the wave vector-matching.

The corresponding *x*-components of the $\vec{E}$ fields are:

$$\vec{E}_{1i} = \hat{x}\frac{jk_{1z}}{j\omega\varepsilon_1}H_{1i}e^{-jk_{1x}x-jk_{1z}z} \tag{3.33}$$

$$\vec{E}_{1r} = \hat{x}\frac{-jk_{1z}}{j\omega\varepsilon_1}H_{1r}e^{-jk_{1x}x+jk_{1z}z} \tag{3.34}$$

$$\vec{E}_{2t} = \hat{x}\frac{jk_{2z}}{j\omega\varepsilon_2(x)}H_{2t}e^{-jk_{2x}x-jk_{2z}z} \tag{3.35}$$

where $\varepsilon_2(x)$ is the periodical grating-induced permittivity variation, which can be expressed using Fourier series expansion, i.e.

$$\varepsilon_2(x) = \sum_{m=-\infty}^{\infty} \varepsilon_m e^{i\frac{2m\pi}{d}x} \tag{3.36}$$

Taking the first-order approximation of Eq. (3.35) and applying the $\vec{E}$ field boundary condition at the interface, one gets the wave vector-matching condition shown in Figure 3.4b, i.e.:

$$k_{2x} = k_{spp} = k_{1x} + \Lambda \tag{3.37}$$

where $\Lambda = \frac{2\pi}{d}$ is the grating vector and d is the grating period. The surface plasmonic wave can be excited when Eq. (3.37) is met, i.e. the wave vector of the surface plasmonic wave matches that of the incident wave.

3.3 Surface Plasmonic Wave-Enhanced QDIPs

3.3.1 Two-Dimensional Metallic Hole Array (2DSHA)-Induced Surface Plasmonic Waves

Another well-known structure to excite surface plasmonic waves is metal film perforated with periodical holes with the diameter of the holes smaller than the wavelength of the incidence light [7], also known as the two-dimensional metallic hole

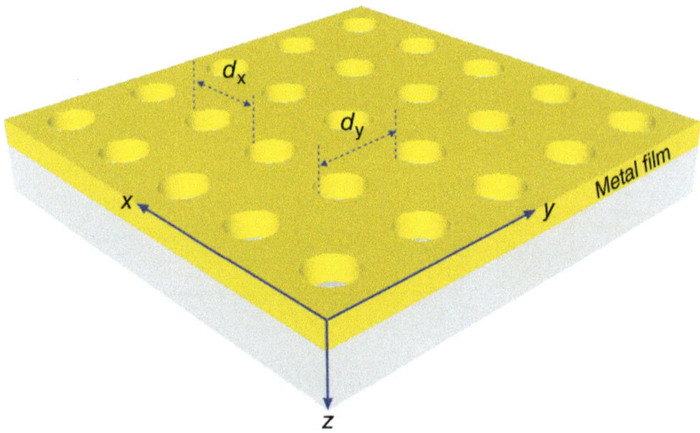

Figure 3.5 2DSHA structure for surface plasmonic wave excitation. The 2DSHA structure is a rectangular lattice with the grating periods of d_x and d_y in the x- and y-directions, respectively.

array (2DSHA) structure [7]. Figure 3.5 shows the schematic view of the 2DSHA structure of a rectangular lattice with the grating periods of d_x and d_y in the x- and y-directions, respectively.

From Eq. (3.37), for the normal incidence light, the fundamental wavelengths of the excited surface plasmonic waves for the polarization of x- and y-directions are [11]:

$$\lambda_x = d_x \mathrm{Re}\left[\sqrt{\frac{\varepsilon_{mr}\varepsilon_{dr}}{(\varepsilon_{dr} + \varepsilon_{mr})}}\right] \tag{3.38}$$

$$\lambda_y = d_y \mathrm{Re}\left[\sqrt{\frac{\varepsilon_{mr}\varepsilon_{dr}}{(\varepsilon_{dr} + \varepsilon_{mr})}}\right] \tag{3.39}$$

From Eqs. (3.38) and (3.39), by choosing different periods in the x- and y-directions, one can achieve polarimetric detection where the polarizations are differentiated by the plasmonic resonant wavelengths [11, 26].

The transmission profiles of the 2DSHA structure are the Fano-type transmission profile [37], which can be generally expressed as [8, 38]:

$$T = |t_B|^2 \frac{(\xi + q)^2}{\xi^2 + 1} \tag{3.40}$$

where t_B is the transmission coefficient of direct nonresonant background transition and ξ is a dimensionless ratio of the energy offset to the energy loss, which can be expressed as [8]:

$$\xi = \frac{E - (E_{spp} + \hbar\Delta)}{\hbar\Gamma/2} \tag{3.41}$$

where Γ is the linewidth of the Fano-type line profile, which also represents the energy loss rate, $\hbar\Delta$ corresponds to the resonant energy shift [8], and q^2 defines the resonant transition to the nonresonant transition probability ratio [37]. Fano-type transmission profiles can be found in many photonic devices [39]. The mechanism of the Fano-type transmission lies in the interference of nonresonant and resonant transmissions [37]. The Fano-type profile formula was originally developed for transitions among atomic energy levels [37]. To our best knowledge, there is no derivation of the Fano transmission formula for the 2DSHA structure. In this chapter, we present a simplified derivation of the Fano transmission formula for the 2DSHA structure. For simplicity, we only consider the modes in the x-direction under the linearly x-polarized light excitation. Similar approach can be applied to the y-direction as well. The fundamental mode of the surface plasmonic wave is:

$$\vec{E}_{spp,0} = \vec{x}E_0 e^{ik_{spp}x} \tag{3.42}$$

where E_0 is the magnitude of the fundamental mode. The resonant conditions for the SPR modes are:

$$k_{spp,m}d = 2m\pi \tag{3.43}$$

Due to hole-induced scattering, the surface plasmonic waves experience multiple reflections in the multipath traveling. The total surface plasmonic wave is thus:

$$\vec{E}_{spp} = \vec{x}\sum_m E_0(r_s)^m e^{imk_{spp}x} \tag{3.44}$$

where r_s is the scattering coefficient. Equation (3.44) can be rewritten as:

$$\vec{E}_{spp} = \vec{x}E_0 \frac{e^{ik_{spp}x}}{1 - r_s e^{ik_{spp}x}} \tag{3.45}$$

Equation (3.45) is the total surface plasmonic waves on the top surface. For a thin metal, the E-M waves can transmit to the back surface with an attenuation factor $e^{-\alpha L}$, where L is the thickness of the metal film. The nonresonant transmission wave $\vec{E}_n$ is thus:

$$\vec{E}_n = \vec{x}E_0 e^{-\alpha L} \tag{3.46}$$

The surface current $\vec{J} = \sigma\vec{E}_{spp}$ driven by the surface plasmonic waves will induce an electric dipole $\vec{D}$ across the holes in the x-direction, which will generate surface plasmonic waves on the back surface with a certain phase delay determined by the conductivity σ and the phase shift in the hole. Assuming a 90° phase shift, the back-side surface plasmonic waves generated through the holes can be written as:

$$\vec{E}_{spp,r} = \vec{x}E_0 \frac{it_h e^{ik_{spp}d}}{1 - r_s e^{ik_{spp}d}} \tag{3.47}$$

where t_h is the transmission ratio through the hole. Combining Eqs. (3.46) and (3.47), the E-M waves $\vec{E}_t$ at the backside hole, which is a sum of the non-resonant transmission (E.q. 3.46) and the resonant transmission (E.q. 3.47). It can be expressed as:

$$\vec{E}_t = \vec{x}E_0 e^{-\alpha L}\left(\frac{1 - r_s\cos(k_{spp}d) - e^{\alpha L}t_h\sin(k_{spp}d) - ir_s\sin(k_{spp}d) + ie^{\alpha L}t_h\cos(k_{spp}d)}{1 - r_s e^{ik_{spp}d}}\right) \tag{3.48}$$

Assuming large scattering coefficient, i.e. $r_s \approx 1$, $\alpha \approx 0$, and taking the first-order approximation, the intensity $I_t = |\vec{E}_t|^2$ can be simplified as:

$$I_t = |\vec{E}_t|^2 \approx e^{-2\alpha L}E_0^2 t_h^2 \frac{[\sin(k_{spp}d)]^2 + [r_s/t_h\sin(k_{spp}d) + 1]^2}{1 - 2r_s\cos(k_{spp}d) + r_s^2} \tag{3.49}$$

Equation (3.49) can be expressed for an off-resonant k by replacing the k_{spp} with k:

$$I_t = |\vec{E}_t|^2 \approx e^{-2\alpha L} E_0^2 t_h^2 \frac{[\sin(kd)]^2 + [r_s/t_h \sin(kd) + 1]^2}{1 - 2r_s \cos(kd) + r_s^2} \tag{3.50}$$

One can evaluate a k vector that is slightly off-resonance, i.e. $k = k_{spp} + \Delta k$. At near resonance, i.e. $k_{spp}d = 2\pi$, $\cos(kd) \approx 1 - (\Delta kd)^2 = 1 - \frac{(\hbar\omega - \hbar\omega_0)^2}{\hbar^2 c^2/d^2}$, and $\sin(kd) \approx \Delta kd = \frac{(\hbar\omega - \hbar\omega_0)}{\hbar c/d}$.

Equation (3.50) can be expressed as:

$$I_t = |\vec{E}_t|^2 \approx e^{-2\alpha L} E_0^2 \frac{(r_s)^2 \left[\frac{(\hbar\omega - \hbar\omega_0)}{(\hbar c/d)(1+r_s)/\sqrt{2r_s}} + \frac{t_h/r_s}{(1+r_s)/\sqrt{2r_s}}\right]^2}{1 + \frac{(\hbar\omega - \hbar\omega_0)^2}{[\hbar^2 c^2/d^2][(1+r_s)^2/2r_s]}} = I_0 |t_B|^2 \frac{[q + \xi]^2}{1 + \xi^2} \tag{3.51}$$

where $\xi = \frac{(\hbar\omega - \hbar\omega_0)}{(\hbar c/d)(1+r_s)/\sqrt{2r_s}}$, $q = \frac{t_h/r_s}{(1+r_s)/\sqrt{2r_s}}$, and $|t_B|^2 = e^{-2\alpha L} E_0^2 (r_s)^2$.

3.3.2 2DSHA Surface Plasmonic Structure-Enhanced QDIP

Metallic 2DSHA structures show extraordinary optical transmission (EOT) of over 100% due to the excitation of resonant surface plasmonic waves [7]. The surface plasmonic structure can also function as a band spectral filter with properly designed surface plasmonic structures. The resonant conditions in Eq. (3.43) can be expressed in terms of the wavelengths:

$$\lambda_{sp} = \frac{d}{\sqrt{i^2 + j^2}} \mathrm{Re}\left\{ \left[\frac{\varepsilon_m \varepsilon_d}{\varepsilon_m + \varepsilon_d} \right]^{1/2} \right\} \tag{3.52}$$

where d is the lattice constant, (i, j) are the orders of the grating vectors, ε_m and ε_d are the dielectric constants of the metal and GaAs, respectively. The peak transmission wavelength varies linearly with the period d of the plasmonic structure. Figure 3.6 shows the simulated transmission spectra with different periods d using CST Microwave Studio®.

Figure 3.7 shows the simulated near-field E-field profile at a 2DSHA structure with an incident plane wave illuminated from the top and traveling in the z-direction. The polarization of the E-field is in the x-direction, and the E-field amplitude is set to $1\,\mathrm{V\,m^{-1}}$. The E-field is much stronger near the hole regions, i.e. the majority of the energy is confined near the 2DSHA/dielectric interference [12]. The extended region of the E-field depends on the diameters of the holes [12]. One can exploit such surface energy confinement effect to achieve photodetector enhancement with a thin active absorption layer such as quantum dots (QD) [8, 11, 12, 26, 40]. By engineering the diameters of the holes, one can control the overlap of the E-field with the active light absorption below the 2DSHA/dielectric interference and thus achieve the highest light absorption [12].

Figure 3.8a,b shows the schematic structure of the 2DSHA plasmonic structure-enhanced quantum dot IR photodetector and an scanning electron

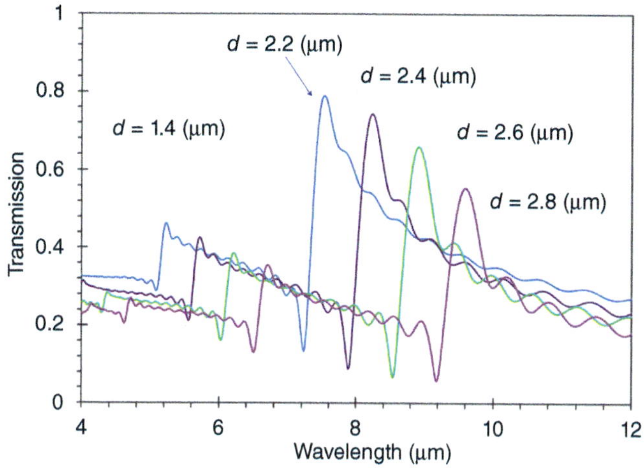

Figure 3.6 Simulated transmission of the 2DSHA structures with different periods d. The transmission peak wavelengths increase linearly with the periods d.

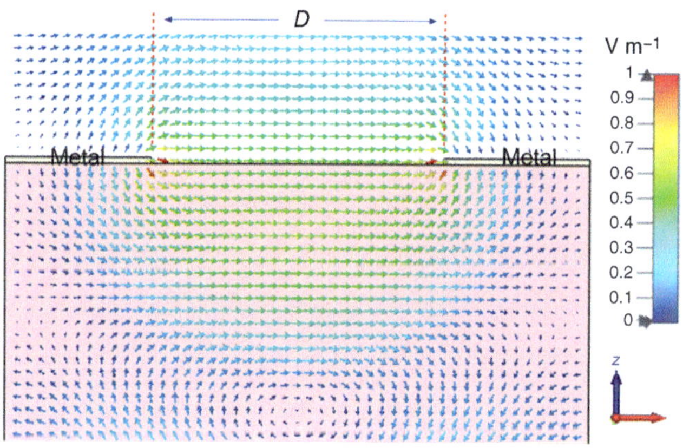

Figure 3.7 Simulated E-field profile of the 2DSHA structure at a top plane-wave illumination with the x-polarization and the amplitude of $1\,\mathrm{V\,m^{-1}}$.

microscopy (SEM) image of the 2DSHA structure, respectively. As shown in Figure 3.8a, the 2DSHA is fabricated on the 10 layers of QDs. The light illumination is from the top 2DSHA side, i.e. top illumination. Backside-configured 2DSHA structures have also been developed and show higher performance enhancement [5, 6].

Figure 3.9 shows the measured photocurrent of the 2DSHA-enhanced quantum dot infrared photodetector (QDIP) (red curve) compared with the reference photodetector (referred to as REF, hence force) without the 2DSHA plasmonic structure (blue curve). The period of the 2DSHA plasmonic structure is 1700 nm, corresponding to $i = 1$ and $j = 1$ in Eq. (3.52). The measured enhancement peak wavelength is at 5.4 μm. The peak wavelength is consistent with the peak wavelength calculation from Eq. (3.52).

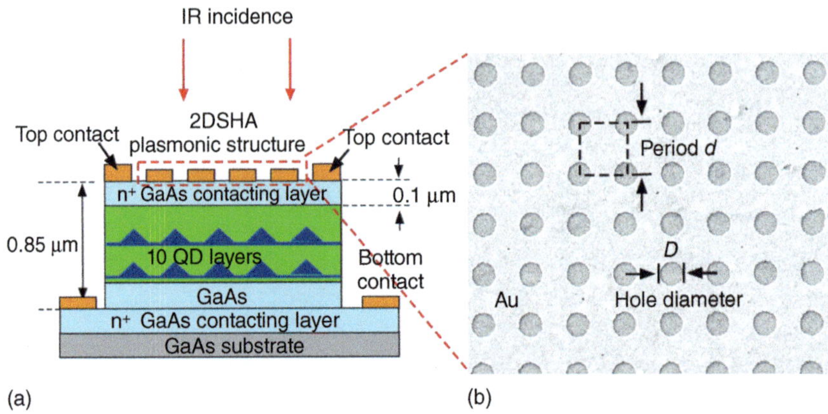

Figure 3.8 2DSHA structure-enhanced QDIP: (a) Schematic cross-section of the 2DSHA structure-enhanced QDIP. The 2DSHA is fabrication on top of the QD layers. The light illumination is from the top, i.e. top illumination; (b) SEM picture of the top 2DSHA structure.

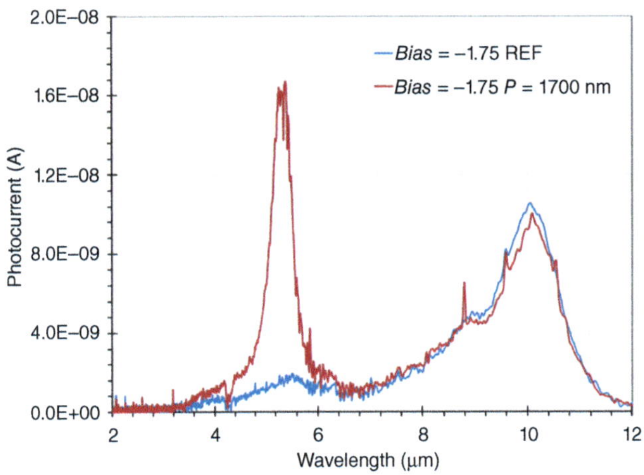

Figure 3.9 Comparison of the photocurrents of the 2DSHA structure-enhanced QDIP and the reference QDIP (REF). Red trace: photocurrent of the 2DSHA structure-enhanced QDIP; blue trace: REF QDIP. 2DSHA structure-enhanced QDIP shows a large photocurrent enhancement at the wavelength of 5.4 μm, which is consistent with the calculation using Eq. (3.52) with $i = 1$ and $j = 1$.

Figure 3.10 shows the measured photocurrent of the different 2DSHA-enhanced QDIP (pink curve) compared with the REF photodetector. The period of the 2DSHA plasmonic structure is reduced to 1450 nm. The measured enhancement peak wavelength is shifted to at 4.6 μm. From the experiments, it is confirmed that the plasmonic enhancement peak wavelengths can be engineered by designing the periods of the 2DSHA plasmonic structures.

Such selective spectral enhancement capability allows one to achieve a multi-spectral imaging array by fabricating the 2DSHA structures with different periods

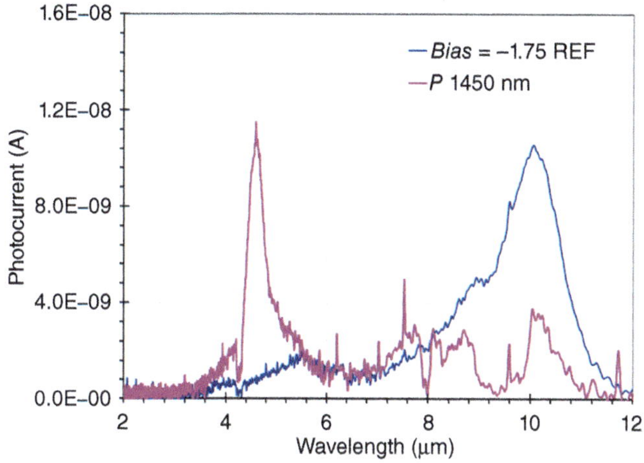

Figure 3.10 Photocurrents of a different 2DSHA structure-enhanced QDIP (pink trace) compared with the REF QDIP. The period of the 2DSHA structure is reduced to 1450 nm, whereas the photocurrent enhancement peak wavelength is shifted to 4.6 μm.

on top of a QDIP photodetector array. Since one can design a photomask containing 2DSHA structures with various periods, multispectral imaging can be fabricated in one-step photolithography, avoiding the yield reduction problems in typical multistep fabrications.

The spectral bands can be configured in a snapshot "mosaic" mode or in a snapshot "tile" mode. Figure 3.11a shows the layout for the snapshot mosaic mode. In this mode, $N \times N$ multispectral pixels with different spectral bands form a superpixel unit that can sense N^2 spectral bands. Figure 3.11b shows an example of a 4×4 superpixel unit that can obtain 16 bands. The 2048×2048 FPA will give 16-band spectral images

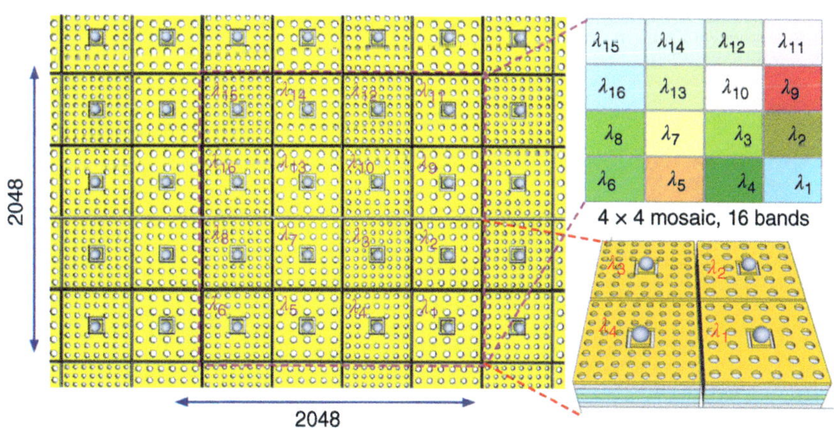

Figure 3.11 Example snapshot mosaic hyperspectral FPA. The 4×4 pixels form a unit that can sense 16 bands within the superpixel unit. The 2048×2048 FPA gives 16-band spectral images with a resolution of 512×512 for each band.

with a resolution of 512 × 512 for each band. Such a type of multispectral imaging FPA can be monolithically integrated devices without the need of optical filters. This can reduce the device's size and weight and enhance the device's reliability.

3.4 Localized Surface Plasmonic Wave-Enhanced QDIPs

3.4.1 Localized Surface Plasmonic Waves

When surface plasmonic waves propagate along small finite metallic surfaces such as nanodisks, nanowires, and nanoparticles, they are confined within the nanometal surfaces and form localized surface plasmonic waves (LSPR) [41], which propagate within the nanostructure and reflect at the boundary. Strong SPR can build up when the following resonant condition can be met.

$$k_{spp,m}d = m\pi \tag{3.53}$$

Equation (3.53) can also be expressed in terms of the resonant wavelength $\lambda_{spp,\,m}$:

$$\lambda_{spp,m} = \frac{2d}{m}\mathrm{Re}\left(\sqrt{\frac{\varepsilon_{rm}\varepsilon_{rd}}{\varepsilon_{rm}+\varepsilon_{rd}}}\right) \tag{3.54}$$

Figure 3.12a shows a schematic structure of the surface current induced by the surface plasmonic waves excited by the incident light in a metallic nanowire. Figure 3.12b shows the simulated current density along the nanowire with a width of 0.1 μm and a length of 2.0 μm. The resonant wavelengths of $\lambda_{spp,\,1} = 13.8$ μm and $\lambda_{spp,\,3} = 4.6$ μm are consistent with the calculation from Eq. (3.54).

3.4.2 Near-Field Distributions

Due to the sinusoidal surface electrical current distribution, the E-field profiles of the nanowire plasmonic structure follow the E-field distribution of a standard dipole antenna [42]. Figure 3.13a,b shows the simulated E-field distribution

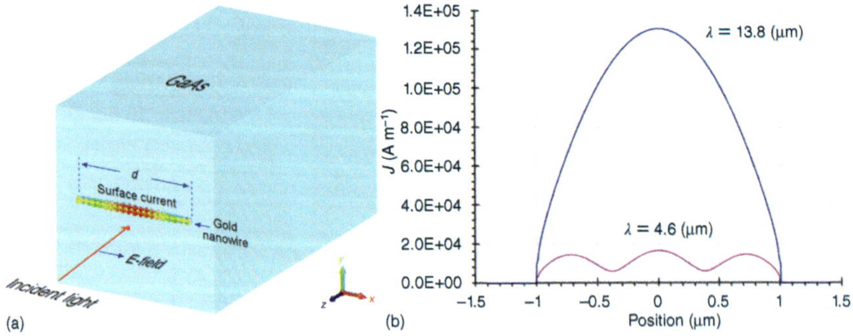

Figure 3.12 (a) Schematic structure of the surface current induced in a metallic nanowire; (b) simulated surface current density versus the positions of the nanowire for different incident wavelengths.

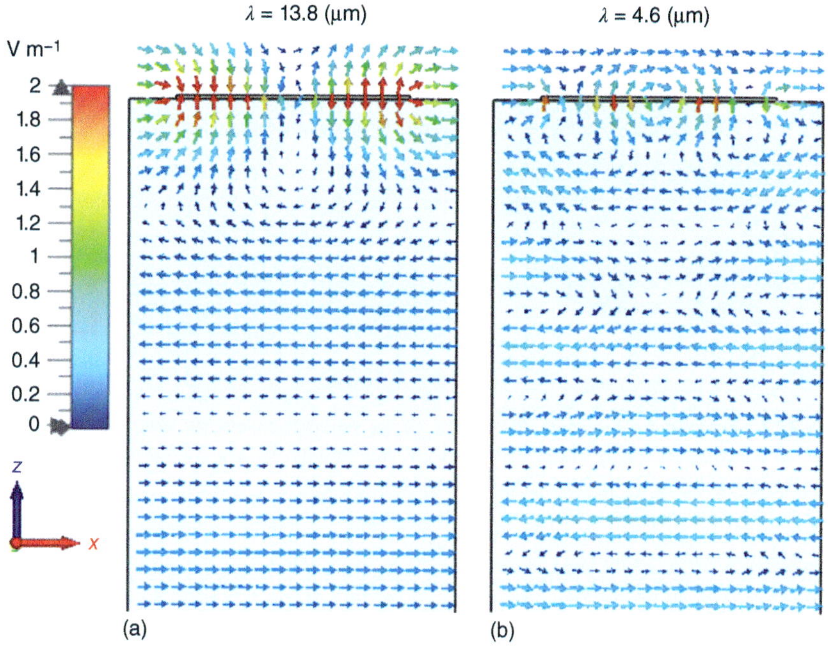

Figure 3.13 *E*-field distributions of the nanowire plasmonic structure at the $y = 0$ cross-section: (a) incident light wavelength $\lambda_{spp,1} = 13.8$ μm; (b) incident light wavelength $\lambda_{spp,3} = 4.6$ μm. The *E*-field is strong close to the plasmonic structure.

of the nanowire plasmonic structure at the $y = 0$ cross-section at the incident light wavelengths of $\lambda_{spp,1} = 13.8$ μm and $\lambda_{spp,3} = 4.6$ μm, respectively. The plasmonic structure changes the near-field *E*-field distribution and enhances the near-field close to the plasmonic structure. Similar to the 2DSHA plasmonic structure, such *E*-field enhancement provided by the localized plasmonic structure allows one to achieve strong optical absorption using a thin active absorption layer [9, 10].

3.4.3 Nanowire Pair

By placing two nanowires close to each other, one can form a nanowire pair array with localized plasmonic waves in each of the array elements. Figure 3.14a illustrates the schematic structure of a nanowire array aligned in the *x*-direction with a gap g between the array elements. Figure 3.14b shows the simulated *E*-field distribution at the $y = 0$ cross-section with the gap of $g = 0.4$ μm. The *E*-field is strong in the gap region.

Figure 3.15a,b shows the simulated amplitude of the *x*-component of the *E*-field (i.e. $|E_x|$) along the center of the nanowire (i.e. $y = 0$) for different gap values at the incident light wavelengths of $\lambda_{spp,1} = 13.8$ μm and $\lambda_{spp,3} = 4.6$ μm, respectively. The smaller gaps give a much higher *E*-field in the gap region. Such a strong *E*-field can be used to enhance the photodetector performance.

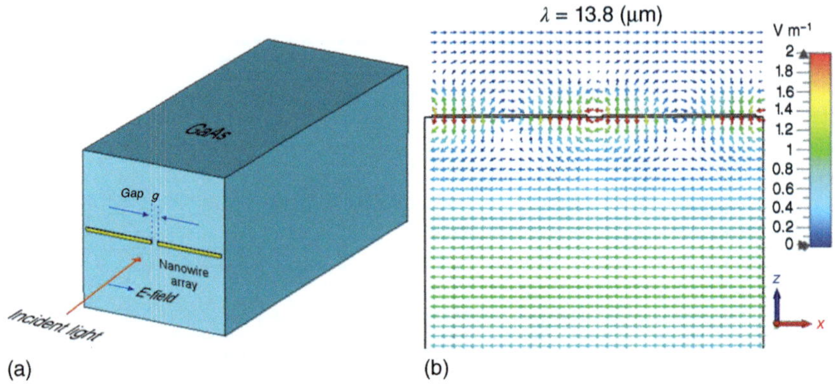

Figure 3.14 Nanowire plasmonic structure array: (a) schematic view; (b) simulated E-field at the $y = 0$ cross-sections with the incident light wavelength $\lambda_{spp,1} = 13.8$ μm and the gap dimension $g = 0.4$ μm. The E-field is strong in the gap region.

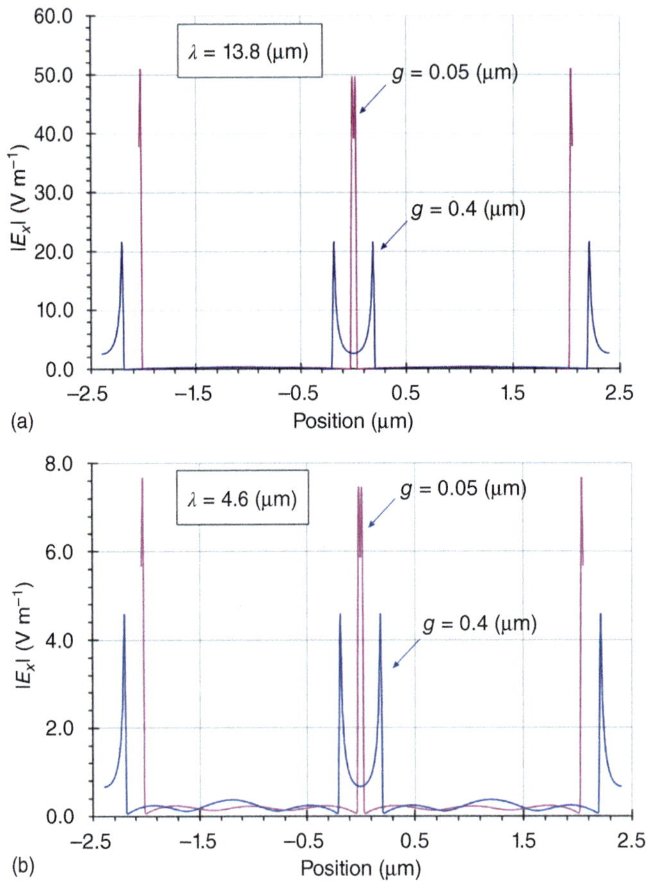

Figure 3.15 Simulated for different gap values (pink traces: $g = 0.05$ μm, blue traces: $g = 0.4$ μm). (a) The incident light wavelength $\lambda_{spp,1} = 13.8$ μm; (b) the incident light wavelength $\lambda_{spp,1} = 4.6$ μm. The smaller gap gives a much higher E-field.

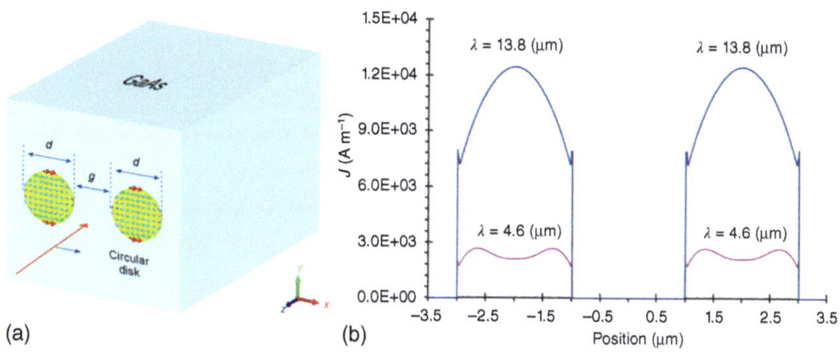

Figure 3.16 Localized surface plasmonic waves in circular disk array: (a) schematic structure of the metallic circular disk array; (b) simulated surface current density along the x-direction at $y = 0$, i.e. the center of the circular disk at different incident wavelengths (pink traces: incident light wavelength $\lambda = 13.8$ μm, blue trace: $\lambda = 4.6$ μm).

3.4.4 Circular Disk Array for Broadband IR Photodetector Enhancement

LSPRs can also be excited by periodic circular disk array plasmonic structures [9, 10]. Figure 3.16a shows the schematic structure of the circular disk array plasmonic structures. The incident light can excite surface plasmonic waves and surface currents on the circular disks. Figure 3.16b shows the simulated surface current along the x-direction at the position $y = 0$, i.e. the center of the circular disk.

For the incident light with the polarization in the x-direction, the plasmonic resonant condition of the circular disk is determined by the diameter of the circular:

$$k_{spp,m}d + \phi_0 = m\pi \tag{3.55}$$

where d is the diameter of the circular disk and ϕ_0 represents the phase shift at the edge of the circular disk. The first-order resonant wavelength $\lambda_{spp,\,1}$ can be written as:

$$\lambda_{spp,1} = \frac{2\pi d}{\pi - \phi_0}\mathrm{Re}\left[\sqrt{\frac{\varepsilon_{mr}\varepsilon_{dr}}{(\varepsilon_{dr} + \varepsilon_{mr})}}\right] \tag{3.56}$$

Figure 3.17 shows the simulated transmission profile of the circular disk array with the period $p = 5.0$ μm for various disk diameters. The resonant wavelengths (i.e. the dips on the transmission profile) shift to the short wavelengths as the diameters of the circular disk decrease.

Figure 3.18 shows the resonant wavelengths as a function of the disk diameters. A linear relation is obtained, which is consistent with Eq. (3.56).

The E-field radiation from the edge point of one circular disk can interact with the E-field of the adjacent disk. When the phase delay of the E-field radiation is $\phi = 2\pi$, the E-field is enhanced, whereas if the phase delay is $\phi = \pi$, the E-field enhancement is reduced. The radiation phase delay ϕ can be written as:

$$\phi = \frac{2\pi}{\lambda_0}g\sqrt{\varepsilon_{rd}} \tag{3.57}$$

where g is the gap between the two adjacent circular disks. Figure 3.19 shows the simulated x-component of the E-field along the x-direction at $y = 0$. For large gaps $g > 1$ μm, $g = 4$ μm gives a higher enhancement, corresponding to the radiation phase

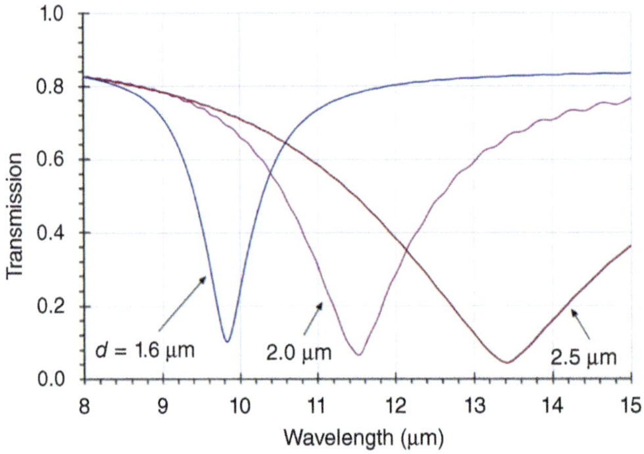

Figure 3.17 Transmission profiles of the circular disk arrays with various disk diameters. The resonant wavelengths (i.e. dips in the profiles) shift to short wavelengths with smaller disk diameters. The resonant wavelengths have a linear relationship with the disk diameter *d*.

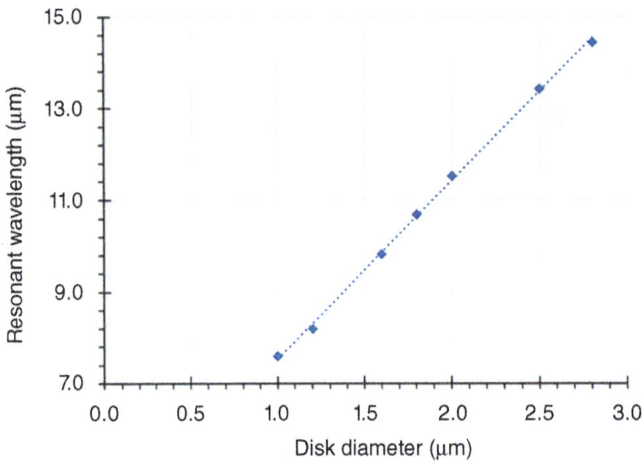

Figure 3.18 Resonant wavelengths as a function of the disk diameters. The resonant wavelengths vary linearly with the disk diameters *d*.

delay $\phi = 2\pi$, whereas $g = 2$ μm shows the lowest enhancement, corresponding to the radiation phase delay $\phi = \pi$ for destructive interference.

3.5 Plasmonic Perfect Absorber (PPA)

3.5.1 Introduction to Plasmonic Perfect Absorber

As discussed in the Sections 3.2.3 and 3.2.4, when the surface plasmonic wave is excited, the reflections from the plasmonic structures are significantly reduced

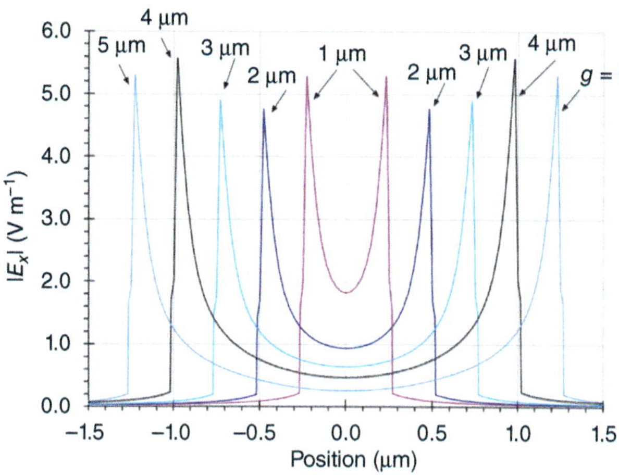

Figure 3.19 Simulated E_x field along the x-direction at $y = 0$. The gap $g = 4$ µm gives the radiation phase delay $\phi = 2\pi$ and thus shows the highest enhancement, whereas gap $g = 2$ µm provides the radiation phase delay $\phi = \pi$ and thus creates destructive interference between the adjacent disk and thus shows the lowest enhancement.

[7, 21], i.e. $R(\omega) \cong 0$, where ω is the angular frequency of the plasmonic resonant frequency. When a bottom reflector is added, a plasmonic perfector absorber (PPA) cavity [19, 20] is formed between the top plasmonic structure and the bottom reflector.

Figure 3.20a shows the schematic structure of the circular disk-based perfect absorber with the disk diameter d. The absorption spectrum depends on the disk diameter d. By varying the diameter d, one can tune the absorption spectrum. Figure 3.20b shows the simulated reflection light-intensity spectra of perfect absorbers with different diameter d. The bottom reflector makes the

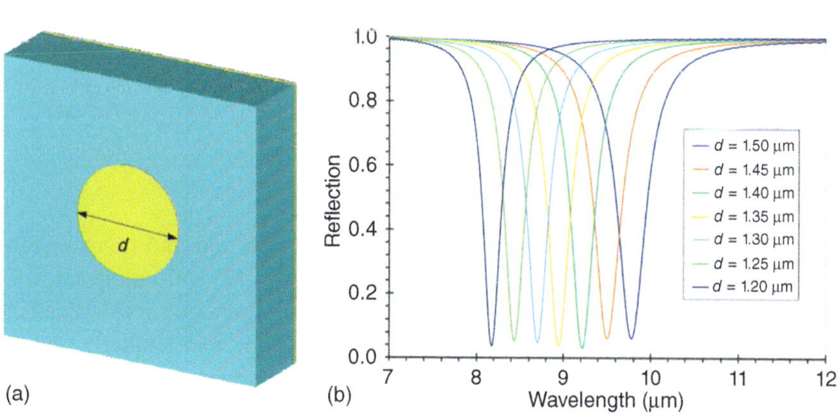

Figure 3.20 (a) Schematic structure of the perfect absorber; (b) simulated reflection spectra of perfect absorbers with different diameter d. The reflection spectra can be tuned by varying the disk diameter d.

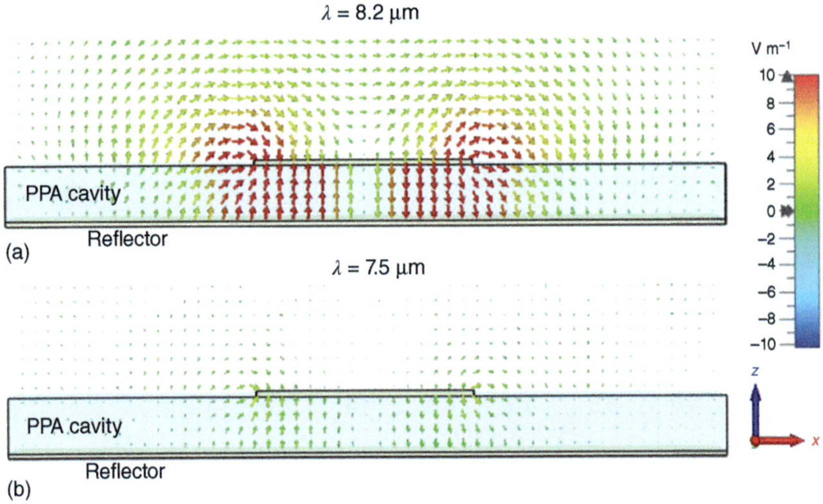

Figure 3.21 Simulated E-field in the PPA cavity with the period $p = 4$ μm and the disk diameter $d = 1.2$ μm: (a) at the resonant wavelength of $\lambda = 8.2$ μm; (b) at the off-resonant wavelength of $\lambda = 7.5$ μm. The E-field is significantly enhanced at the resonant wavelength. The primary E-field direction is in the z-direction inside the PPA cavity.

transmission through the cavity close to zero, i.e. $T(\omega) \cong 0$. Thus, the total absorption $A(\omega) = 1 - R(\omega) - T(\omega)$ is close to 100%, i.e. a perfect absorber [19].

Figure 3.21a,b shows the simulated E-fields of a PPA cavity with $d = 1.2$ μm and period $p = 4$ μm at the resonant wavelength (i.e. $\lambda = 8.2$ μm) and off-resonant wavelength (i.e. $\lambda = 7.5$ μm), respectively. The incident light is x-polarized with the magnitude of $|E| = 1\,\mathrm{V\,m^{-1}}$. At the resonant wavelength, the E-field is significantly enhanced. The E-field direction is changed from the x-direction to the z-direction inside the PPA cavity.

Figure 3.22 shows the simulated reflection light-intensity spectra of the perfect absorber with different period p. For $d = 1.2$ μm, period $p = 4.8$ μm gives the lowest reflection. This corresponds to $p = 4d$, i.e. the gap between the disks $g = 2d$. This is consistent with the analysis in Section 3.4.4, Figure 3.19, where $g = 2d$ gives the strongest plasmonic enhancement.

3.5.2 Plasmonic Perfect Absorber-Enhanced QDIP

Perfect absorbers [19, 43–46] with near-zero in both reflection and transmission have been developed in the microwave, tera-Hertz (THz), IR, and visible spectral regimes [19, 43–46]. Perfect absorbers have been used in numerous applications in thermal imaging, bolometers, photovoltaic, IR spectroscopy, and terahertz stealth [19, 47, 48]. PPA cavity-enhanced quantum dot IR photodetectors [20] can be obtained by embedding an MWIR/LWIR quantum dot IR photodetector inside a PPA cavity. The detailed fabrication steps have been reported [20]. Figure 3.23 shows the measured photocurrent spectra of the PPA cavity-enhanced LWIR QDIP compared with a reference QDIP without the PPA structure. Other than the PPA

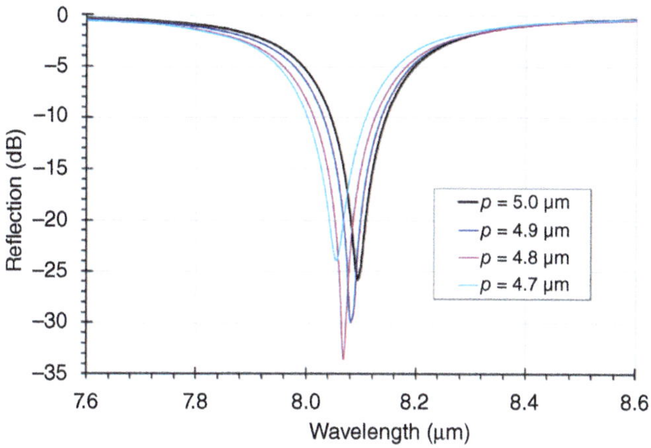

Figure 3.22 Simulated reflection spectra of perfect absorbers with different period p for the disk diameter $d = 1.2$ μm. Period $p = 4.8$ μm, i.e. $p = 4d$, gives the lowest reflection.

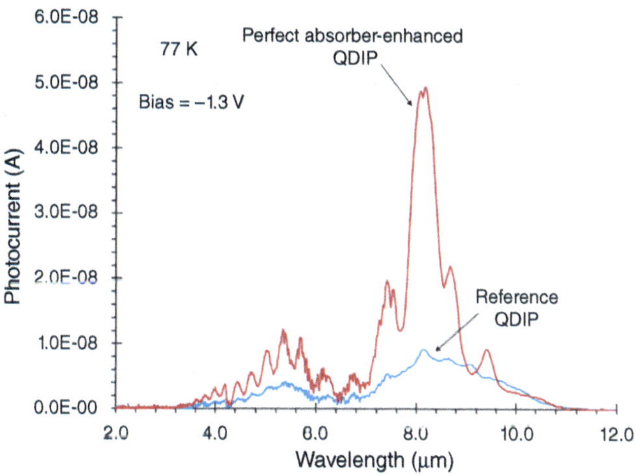

Figure 3.23 Measured photocurrent spectrum of the PPA QDIP compared to that of a reference QDIP with the PPA structure. The reference QDIP was made from the same material through the same fabrication processes. The PPA QDIP shows a much higher photocurrent than that of the reference QDIP.

structure, the material and the fabrication processes of the reference QDIP and the PPA QDIP were the same. The PPA QDIP shows a much higher photocurrent than that of the reference QDIP, indicating the effectiveness of the PPA cavity in enhancing the photodetection.

Due to the PPA enhancement, the PPA QDIP can achieve a high quantum efficiency (QE) of 50% [20]. The experiment measurement indicates the high absorption of the PPA cavity provides an effective approach to achieve high QE using a thin light absorption layer.

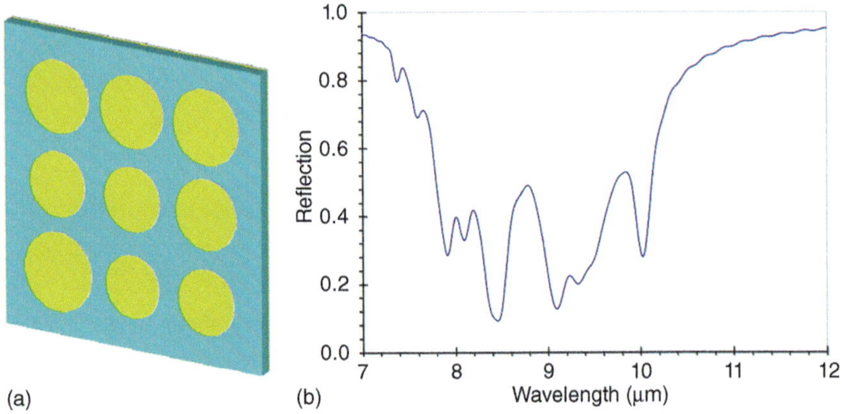

Figure 3.24 (a) Top view of the broadband absorber. Each unit contains circular disks with three different diameters; (b) reflection spectrum. It combines the low individual reflection profiles into a broadband low reflection.

3.5.3 Broadband Plasmonic Perfect Absorber

In addition to tuning the enhancement band by varying the circular disk diameter *d*, one can also achieve a broader band perfect absorber by incorporating multiple circular disks with different diameters d in one unit cell. Figure 3.24a shows the top view of the broadband perfect absorber. Each unit cell contains circular disks with different diameters. Figure 3.24b shows the simulated reflections of this perfect absorber. The perfect absorber combines the low reflection profiles into a broadband low reflection, i.e. high absorption.

3.5.4 2DSHA Plasmonic Perfect Absorber

In addition to the circular disk-based PPA, the 2DSHA plasmonic structure can also be used to form a PPA cavity by adding a reflector to the back of the 2DSHA structure. Figure 3.25a shows the schematic structure of the 2DSHA PPA cavity. Figure 3.25b shows the simulated reflection spectra of 2DSHA PPAs with various periods. Reflection minimums occur at the plasmonic resonances.

Figure 3.26 shows the plasmonic resonant wavelengths as a function of the periods *d* of the 2DSHA plasmonic structure. A linear curve is obtained.

Figure 3.27a,b shows the simulated $|E_z|$ and $|E_x|$ at the $y = 0$ cross-section of the 2DSHA PPA cavity at the resonant wavelength. The *E*-field inside the PPA cavity is primarily in the *z*-direction.

3.6 Chapter Summary

This chapter reviews basic concepts of SPRs and their excitation techniques in the IR spectral region, including prism and grating-coupling and the 2DSHA structure excitation. The SPRs in the 2DSHA structure show a Fano-type transmission

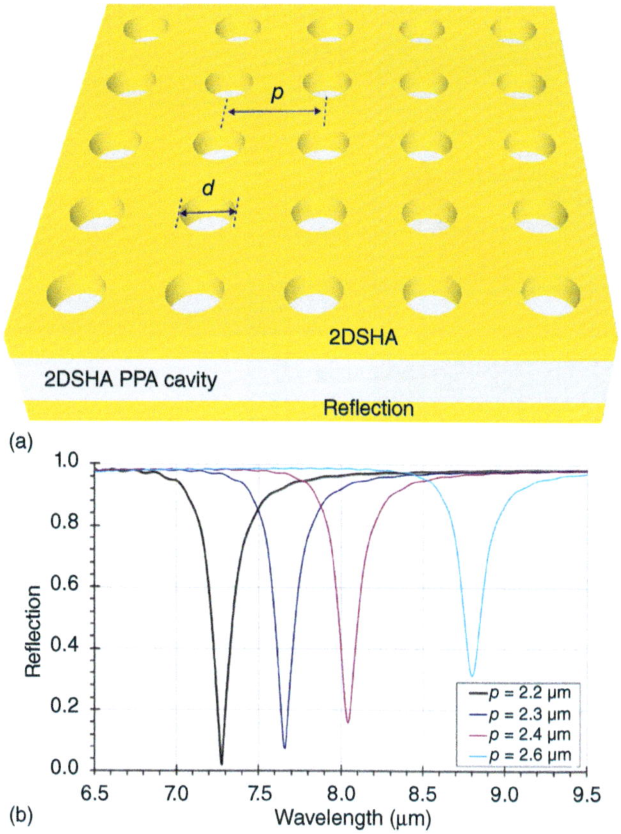

Figure 3.25 (a) Schematic structure of the 2DSHA perfect absorber; (b) simulated reflection spectra for 2DSHA PPA with various periods.

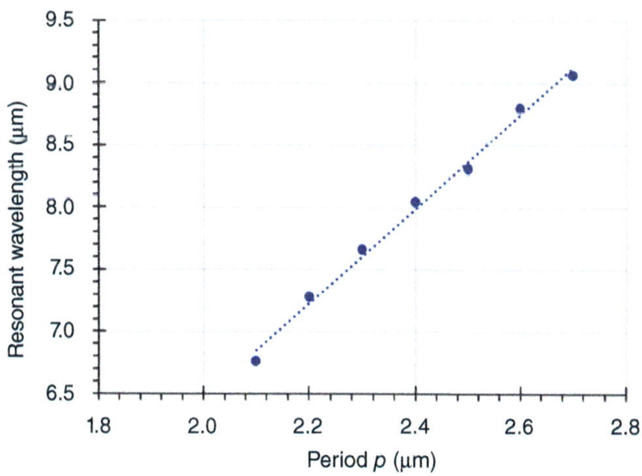

Figure 3.26 Resonant wavelength versus the period of the 2DSHA PPA plot. The Resonant wavelengths vary linearly with the periods of the 2DSHA PPA.

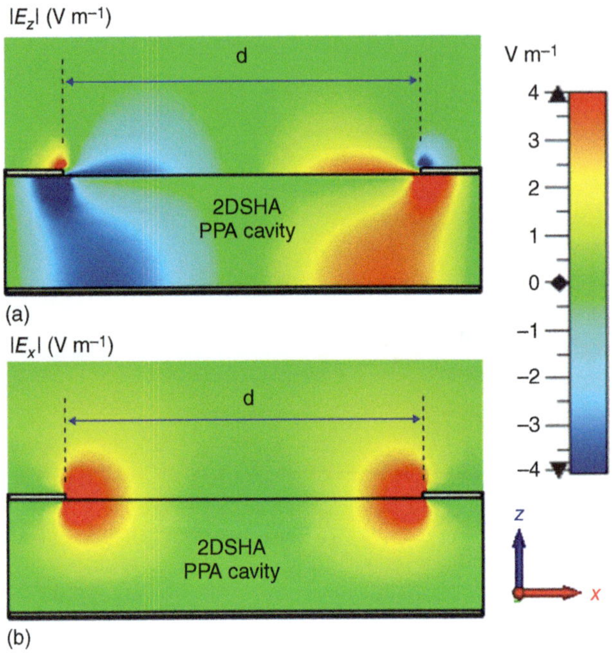

Figure 3.27 Simulated E fields at the $y = 0$ cross section of the 2DSHA PPA cavity at the resonant wavelength: (a) $|E_z|$; (b) $|E_x|$.

profile, which is explained using the standard EM theory. LSPRs in nanowires, nanodisks, and the array are discussed together with the resonant conditions and the near-field profiles. The SPR- and LSPR-enhanced QDIP photodetectors are presented with the experiment results, which verify the enhancement by the SPRs and LSPRs. The chapter also discusses two types of PPA cavities, namely the nanodisk PPA cavity and the 2DSHA PPA cavity together with their resonant conditions and the near-field profiles. The nanodisk PPA-enhanced QDIP is also presented with experimental data, which show the effectiveness of the PPA cavity in the photodetector performance enhancement.

References

1 Arnold, M., Zimin, D., and Zogg, H. (2005). Resonant-cavity-enhanced photodetectors for the mid-infrared. *Appl. Phys. Lett.* 87 (14): 141103. https://doi.org/10.1063/1.2061855.

2 Letka, V., Bainbridge, A., Craig, A.P. et al. (2019). Resonant cavity-enhanced photodetector incorporating a type-II superlattice to extend MWIR sensitivity. *Opt. Express* 27 (17): 23970–23980. https://doi.org/10.1364/OE.27.023970.

3 Murtaza, S.S., Anselm, K.A., Hu, C. et al. (1995). Resonant-cavity enhanced (RCE) separate absorption and multiplication (SAM) avalanche photodetector (APD). *IEEE Photonics Technol. Lett.* 7 (12): 1486. https://doi.org/10.1109/68.477291.

4 Vasinajindakaw, P., Gu, G., and Lu, X. (2013). Broadband photocurrent enhancement in longwave infrared quantum dot photodetectors by sub-wavelength surface gratings. *Opt. Photonics Lett.* 6 (1): 1350002. https://doi.org/10.1142/s1793528813500020.

5 Gu, G., Vaillancourt, J., Vasinajindakaw, P., and Lu, X. (2013). Backside-configured surface plasmonic structure with over 40 times photocurrent enhancement. *Semicond. Sci. Technol.* 28 (10): 105005.

6 Vaillancourt, J., Mojaverian, N., and Lu, X. (2014). A longwave infrared focal plane array enhanced by backside-configured plasmonic structures. *IEEE Photonics Technol. Lett.* 26 (8): 745–748. https://doi.org/10.1109/LPT.2014.2303118.

7 Ebbesen, T.W., Lezec, H.J., Ghaemi, H.F. et al. (1998). Extraordinary optical transmission through sub-wavelength hole arrays. *Nature* 391 (6668): 667–669. https://doi.org/10.1038/35570.

8 Vasinajindakaw, P., Vaillancourt, J., Gu, G. et al. (2011). A Fano-type interference enhanced quantum dot infrared photodetector. *Appl. Phys. Lett.* 98 (21): 211111. https://doi.org/10.1063/1.3593128.

9 Gu, G., Vaillancourt, J., and Lu, X. (2014). Analysis of near-field components of a plasmonic optical antenna and their contribution to quantum dot infrared photodetector enhancement. *Opt. Express* 22 (21): 24970–24976. https://doi.org/10.1364/OE.22.024970.

10 Kemsri, T., Gu, G., Zhang, Y. et al. (2017). Angular-dependent photodetection enhancement by a metallic circular disk optical antenna. *AIP Adv.* 7 (2): 025013. https://doi.org/10.1063/1.4975705.

11 Lee, S.J., Ku, Z., Barve, A. et al. (2011). A monolithically integrated plasmonic infrared quantum dot camera. *Nat. Commun.* 2 (1): 286. https://doi.org/10.1038/ncomms1283.

12 Liu, R., Vasinajindakaw, P., Gu, G. et al. (2012). Optimizing light absorption in quantum dot infrared photodetectors by tuning surface confinement of surface plasmonic waves. *J. Phys. D: Appl. Phys.* 46 (1): 015102. https://doi.org/10.1088/0022-3727/46/1/015102.

13 Krasnok, A.E., Maksymov, I.S., Denisyuk, A.I. et al. (2013). Optical nanoantennas. *Phys. Usp.* 56 (6): 539. http://stacks.iop.org/1063-7869/56/i=6/a=539.

14 Neda, M., Guiru, G., and Xuejun, L. (2015). A plasmonic dipole optical antenna coupled quantum dot infrared photodetector. *J. Phys. D: Appl. Phys.* 48 (47): 475102. http://stacks.iop.org/0022-3727/48/i=47/a=475102.

15 Xuejun, L., Neda, M., Lin, L. et al. (2017). A back side configured pointed dipole plasmonic optical antenna array enhanced quantum dot infrared photodetector. *Semicond. Sci. Technol.* 32 (12): 125017. http://stacks.iop.org/0268-1242/32/i=12/a=125017.

16 GhafaryAghdam, A., Xiang, B., Li, L. et al. (2018). Double-pointed optical antenna in the longwave infrared (LWIR) spectral regime. In: *NAECON 2018 – IEEE National Aerospace and Electronics Conference*, Dayton, OH, USA (23–26 July 2018), 465–469. https://doi.org/10.1109/NAECON.2018.8556820.

17 Novotny, L. and van Hulst, N. (2011). Antennas for light. *Nat. Photonics* 5 (2): 83–90. https://doi.org/10.1038/nphoton.2010.237.

18 Giannini, V., Fernández-Domínguez, A.I., Heck, S.C., and Maier, S.A. (2011). Plasmonic nanoantennas: fundamentals and their use in controlling the radiative properties of nanoemitters. *Chem. Rev.* 111 (6): 3888–3912. https://doi.org/10.1021/cr1002672.

19 Liu, N., Mesch, M., Weiss, T. et al. (2010). Infrared perfect absorber and its application as plasmonic sensor. *Nano Lett.* 10 (7): 2342–2348. https://doi.org/10.1021/nl9041033.

20 Lu, X., Vaillancourt, J., and Gu, G. (2017). A plasmonic perfect absorber enhanced longwave infrared quantum dot infrared photodetector with high quantum efficiency. *J. Phys. D: Appl. Phys.* 50 (13): 135101. https://doi.org/10.1088/1361-6463/aa5126.

21 Ritchie, R.H. (1957). Plasma losses by fast electrons in thin films. *Phys. Rev.* 106 (5): 874–881. https://doi.org/10.1103/PhysRev.106.874.

22 Lu, L., Kobayashi, A., Tawa, K., and Ozaki, Y. (2006). Silver nanoplates with special shapes: controlled synthesis and their surface plasmon resonance and surface-enhanced Raman scattering properties. *Chem. Mater.* 18 (20): 4894–4901. https://doi.org/10.1021/cm0615875.

23 Nie, S. and Emory, S.R. (1997). Probing single molecules and single nanoparticles by surface-enhanced Raman scattering. *Science* 275 (5303): 1102–1106. https://doi.org/10.1126/science.275.5303.1102.

24 Campion, A. and Kambhampati, P.Y. (1998). Surface-enhanced Raman scattering. *Chem. Soc. Rev.* 27 (4): 241–250. https://doi.org/10.1039/a827241z.

25 Chang, C.-C., Sharma, Y.D., Kim, Y.-S. et al. (2010). A surface plasmon enhanced infrared photodetector based on InAs quantum dots. *Nano Lett.* 10 (5): 1704–1709.

26 Vasinajindakaw, P., Vaillancourt, J., Gu, G., and Lu, X. (2012). Surface plasmonic enhanced polarimetric longwave infrared photodetection with band pass spectral filtering. *Semicond. Sci. Technol.* 27 (6): 65005–65009.

27 Gu, G., Mojaverian, N., Vaillancourt, J., and Lu, X. (2014). Surface plasmonic resonance induced near-field vectors and their contribution to quantum dot infrared photodetector enhancement. *J. Phys. D: Appl. Phys.* 47 (43): 435106. http://stacks.iop.org/0022-3727/47/i=43/a=435106.

28 Guiru Gu, X.L., Kemsri, T., and Zhang, Y. (2017). Surface plasmonics and its applications in infrared sensing. In: *Nanoplasmonics Fundamentals and Applications* (Chapter 9) (ed. G. Barbillon), 195–214. London: IntechOpen Limited.

29 Drude, P. (1900). Zur Elektronentheorie der Metalle. *Ann. Phys.* 306 (3): 566–613. https://doi.org/10.1002/andp.19003060312.

30 Reshef, O., De Leon, I., Alam, M.Z., and Boyd, R.W. (2019). Nonlinear optical effects in epsilon-near-zero media. *Nat. Rev. Mater.* 4: 535–551.

31 Niu, X., Hu, X., Chu, S., and Gong, Q. (2018). Epsilon-near-zero photonics: a new platform for integrated devices. *Adv. Opt. Mater.* 6 (10): 1701292. https://doi.org/10.1002/adom.201701292.

32 Alú, A., Silveirinha, M.G., Salandrino, A., and Engheta, N. (2007). Epsilon-near-zero metamaterials and electromagnetic sources: tailoring the radiation phase pattern. *Phys. Rev. B: Condens. Matter* 75: 155410.

33 Raether, H. (1988). *Surface Plasmons on Smooth Surfaces*. Berlin: Springer-Verlag Berlin Heidelberg.

34 Ordal, M.A., Long, L.L., Bell, R.J. et al. (1983). Optical properties of the metals Al, Co, Cu, Au, Fe, Pb, Ni, Pd, Pt, Ag, Ti, and W in the infrared and far infrared. *Appl. Opt.* 22 (7): 1099–1119. https://doi.org/10.1364/AO.22.001099.

35 Otto, A. (1969). Streustrahlung von Silber durch Anregung von Oberflächenplasmaschwingungen. *Z. Angew. Phys.* 224 (1): 65–73. https://doi.org/10.1007/BF01392234.

36 Shalabney, A. and Abdulhalim, I. (2011). Sensitivity-enhancement methods for surface plasmon sensors. *Laser Photonics Rev.* 5 (4): 571–606. https://doi.org/10.1002/lpor.201000009.

37 Fano, U. (1961). Effects of configuration interaction on intensities and phase shifts. *Phys. Rev.* 124 (6): 1866–1878. https://doi.org/10.1103/PhysRev.124.1866.

38 Genet, C., van Exter, M.P., and Woerdman, J.P. (2003). Fano-type interpretation of red shifts and red tails in hole array transmission spectra. *Opt. Commun.* 225 (4): 331–336. https://doi.org/10.1016/j.optcom.2003.07.037.

39 Limonov, M.F., Rybin, M.V., Poddubny, A.N., and Kivshar, Y.S. (2017). Fano resonances in photonics. *Nat. Photonics* 11 (9): 543–554. https://doi.org/10.1038/nphoton.2017.142.

40 Vasinajindakaw, P., Lu, X., Qian, X., et al. (2011). High-operating temperature (HOT) broadband quantum-dot infrared photodetector. *SPIE Defense, Security, and Sensing*, Orlando, FL. SPIE.

41 Mayer, K.M. and Hafner, J.H. (2011). Localized surface plasmon resonance sensors. *Chem. Rev.* 111 (6): 3828–3857. https://doi.org/10.1021/cr100313v.

42 Jackson, J.D. (1998). *Classical Electrodynamics*. Wiley.

43 Fante, R.L. and McCormack, M.T. (1988). Reflection properties of the Salisbury screen. *IEEE Trans. Antennas Propag.* 36 (10): 1443–1454. https://doi.org/10.1109/8.8632.

44 Hu, C., Liu, L., Zhao, Z. et al. (2009). Mixed plasmons coupling for expanding the bandwidth of near-perfect absorption at visible frequencies. *Opt. Express* 17 (19): 16745–16749. https://doi.org/10.1364/OE.17.016745.

45 Landy, N.I., Sajuyigbe, S., Mock, J.J. et al. (2008). Perfect metamaterial absorber. *Phys. Rev. Lett.* 100 (20): 207402. http://link.aps.org/doi/10.1103/PhysRevLett.100.207402.

46 Kats, M.A., Sharma, D., Lin, J. et al. (2012). Ultra-thin perfect absorber employing a tunable phase change material. *Appl. Phys. Lett.* 101 (22): 221101. https://doi.org/10.1063/1.4767646.

47 Chen, K., Adato, R., and Altug, H. (2012). Dual-band perfect absorber for multispectral plasmon-enhanced infrared spectroscopy. *ACS Nano* 6 (9): 7998–8006. https://doi.org/10.1021/nn3026468.

48 Liu, K., Hu, H., Song, H. et al. (2013). Wide-angle and polarization-insensitive perfect absorber for organic photovoltaic layers. *IEEE Photonics Technol. Lett.* 25 (13): 1266–1269. https://doi.org/10.1109/LPT.2013.2264047.

4

Optical Resistance Switch for Optical Sensing

Shiva Khani[1], Ali Farmani[2], and Pejman Rezaei[1]

[1] *Semnan University, Faculty of Electrical and Computer Engineering, HCRJ+4F9, Semnan, 35131-19111, Iran*
[2] *Lorestan University, School of Electrical and Computer Engineering, Falakol' Aflak Street, Khorramabad, Lorestan, 44316-68151, Iran*

4.1 Introduction

Optical components [1–3] are superior to semiconductor electronic ones [4, 5] in terms of higher computational speeds and lower noise values. Consequently, the development of optical devices is growing in prosperity. Photonic crystal (PC) structures [6–8] can be a suitable candidate for the realization of different passive and active optical devices such as PC filters [9, 10], sensors [11–13], splitters [14, 15], slow lights [16], demultiplexers [17, 18], switches [19–21], modulators [22, 23], and logic gates [24, 25]. PCs are periodic arrangements of dielectric layers with alternating refractive indexes. The distribution of the dielectric layers' refractive index is periodic in one, two, or three dimensions [26]. Due to their remarkable capability to appear as photonic band gaps (PBGs) with sharp transient edges, PCs are considered an efficient basis for the realization of optical switches and sensors [27]. In the frequency range of the PBG, no optical wave is allowed to propagate through the PC structure. The refractive index and structural parameters of the PC structure determine the PBG frequency range [28]. The relatively large size (micrometer size) of the PC structures is their main drawback compared to other optical structures, such as plasmonic [29–31] and graphene [32, 33] structures.

Surface plasmon polaritons (SPPs), due to their remarkable capabilities to overcome the diffraction limit and manipulate light in a nanoscale domain, are considered as an efficient basis for the realization of highly integrated optical circuits [34–36]. SPPs are created at the interface between an insulator and a metal layer with different signs of dielectric constants through the interaction of surface electrons and incident photons [37]. Accordingly, all of the conventional PC components can be redesigned based on SPPs with more miniaturized footprints. Over the last few years, a variety of devices based on SPPs have been proposed, i.e.

Optical Imaging and Sensing: Materials, Devices, and Applications, First Edition.
Edited by Jiang Wu and Hao Xu.

plasmonic filters [38, 39], Bragg reflectors [40, 41], sensors [42–44], demultiplexers [45, 46], Mach–Zehnder interferometers (MZIs) [47], and switches [48–51]. Another advantage of metal–insulator–metal (MIM) plasmonic structures is the possibility of integrating such configurations with microwave circuits [52, 53]. It is because MIM structures have also been used for the realization of previous microwave microstrip components [54, 55].

Due to the large negative real part and small imaginary part of permittivity of noble metals such as silver and gold in optical frequencies, such metal materials have been widely used in plasmonic devices [56]. However, plasmonic components designed using these metals have some drawbacks. First of all, such structures are inherently lossy [57]. Secondly, these structures cannot be tuned freely when their geometry parameters are fixed [58–60]. This motivates researchers to use other materials with low ohmic losses and tunability property such as graphene structures [33, 61, 62].

Graphene is a two-dimensional (2D) monolayer sheet of carbon atoms packed into a honeycomb lattice [63, 64]. Graphene can behave like a thin metal film with negative permittivity by means of chemical doping, electric and magnetic fields [65, 66]. Meanwhile, graphene structures have remarkable advantages compared to noble metals. Such features include ultralow ohmic losses of graphene, much thinner thickness of graphene than metal layers, tunability of graphene's complex conductivity, and mechanical strength of graphene structures. The mentioned properties of graphene make it an appropriate choice to design highly integrated graphene-plasmonic (GP) devices [67–70].

Among the various optical devices, optical switches [71, 72] and sensors [73–75] are popular topics and have a wide range of applications. Up to now, various approaches based on different structures, such as PC [76, 77], plasmonic [78, 79], and graphene [80, 81], have been adopted to design optical switches and sensors. An all-optical switch based on a directional coupler structure in a 2D PC lattice with Kerr optical nonlinearity has been proposed in [21]. In the proposed structure reported in [21], the high switching speed and low reflection coefficient for the control signal make the presented switch suitable for all-optical integration purposes. In other approaches [2, 29, 48, 50], novel plasmonic topologies using two isolated MIM waveguides to pass the data and control signals have been proposed for all-optical switching applications. Using such a method (isolated data and control paths) prevents the intermodulation between the data and control signals and reduces harmonic distortion. Other methods, including optical switches using a deposited graphene nanoribbon on a silica (SiO_2)/silicon (Si) substrate [82], a terahertz switch based on a graphene monolayer, poly-methyl methacrylate (PMMA) [83], and an all-optical multi-channel switch using square ring resonators [84], have also been presented.

Over the last few years, there have also been various suggestions for realizing optical sensors using different structures. Due to the sharp edge of the Fano resonance shape, such a resonance mode can improve the sensitivity and figure of merit (FOM) of optical sensors. As a result, many resonators and optical configurations have been presented to obtain the Fano resonance shape, including an asymmetric MIM waveguide structure [85], splitting ring resonator and tooth-shaped resonator [86], and dual ring resonators [87]. Another method that has been used in [88, 89] is using optical structures with plasmon-induced transparency (PIT) resonance modes.

In addition to the mentioned published optical switch and sensor structures, there are also reports on structures to realize the function of both optical switches and sensors simultaneously. In such structures, multifunctional devices are used as optical switches and sensors using two different insulator materials [81, 90–93]. The extensive design of dual-functional devices (optical switch and sensor) includes plasmonic nanodisk resonators [90], dumbbell-shaped cavity slots [91], rectangular resonators [92], two pairs of graphene nanorings and a graphene nanoribbon [93], and a plasmonic graphene-based structure [81].

In this chapter, first as a case study, graphene and nanomaterial heterostructures (combination of plasmonic and PC structures) are used to design optical switches. Thereafter, to study the modulation characteristics, an optical modulator has been designed using plasmonic structures. Finally, the summary and conclusions are discussed in Section 4.5.

4.2 Graphene Optical Switch

Optical switches are used for light routing and switching. Accordingly, they have found wide applications in many optical systems. On the other hand, optical switches can also be used for optical sensing applications. It is mainly due to the high wavelength shifts of optical switches. Such devices have been presented in [94–96]. Among the most important optical structures that have found wide application for switching mechanisms are graphene structures.

The optical features of graphene can be tuned by varying its Fermi level. The Fermi level and therefore the chemical potential (μ_c) and the surface conductivity of graphene can be changed by applying an external voltage. When the chemical potential is higher than the threshold value, graphene acts as a metal layer, while graphene can be considered as a dielectric layer for the chemical potential lower than the threshold value. The surface conductivity of graphene (σ), which is composed of intraband and interband parts, can be calculated by the Kubo formula [97, 98].

$$\sigma(\omega) = \sigma_{intra}(\omega) + \sigma_{inter}(\omega) \tag{4.1}$$

where σ_{intra} and σ_{inter} are the intraband and interband transitions, respectively, and they can be expressed as:

$$\sigma_{intra} = \frac{2ie^2 k_B T}{\pi\hbar^2(\omega + i\tau^{-1})} \ln\left(2\cosh\left(\frac{E_f}{2k_B T}\right)\right) \tag{4.2}$$

$$\sigma_{inter} = \frac{e^2}{4\hbar}\left\{\frac{1}{2} + \frac{1}{\pi}\arctan\left(\frac{\hbar\omega - 2E_f}{2k_B T}\right) - \frac{i}{2\pi}\ln\left(\frac{(\hbar\omega + 2E_f)^2}{(\hbar\omega - 2E_f)^2 + (2k_B T)^2}\right)\right\} \tag{4.3}$$

where k_B is the Boltzmann constant, E_f demonstrates the Fermi level of graphene, $\hbar$ shows the reduced Planck constant, and τ is the carrier relaxation time. In [99],

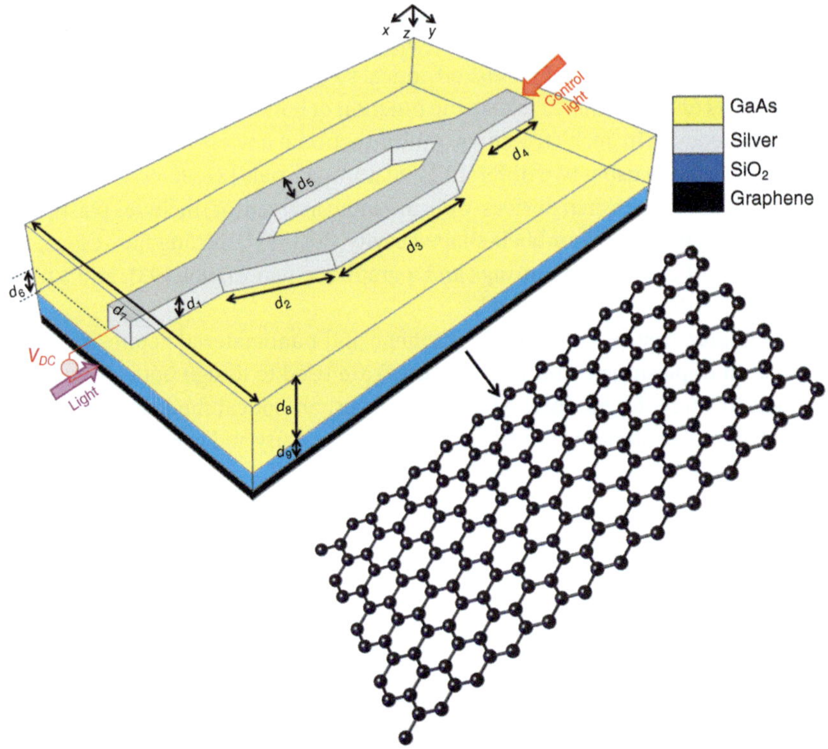

Figure 4.1 General structure of the proposed all-optical graphene plasmon switch.

all-optical switches based on graphene plasmon structures were proposed. The general structure of the designed graphene plasmon switch is shown in Figure 4.1. This structure is composed of a novel combination of insulator–metal–insulator (IMI) and graphene waveguides designed in the form of a MZI topology. In this structure, the metal material in the MZI structure is silver. The complex relative permittivity of silver is described by the accurate Drude–Lorentz model [100]:

$$\varepsilon_m(\omega) = 1 - \frac{\omega_p^2}{\omega(\omega + j\gamma)} + \sum_{n=1}^{5} \frac{f_n \omega_n^2}{\omega_n^2 - \omega^2 - j\omega\gamma_n} \tag{4.4}$$

where $\omega_p = 2002.6 \times 10^{12}$ Hz is the bulk plasma frequency of metal, $\gamma = 11.61 \times 10^{12}$ Hz is a damping constant, and ω is the angular frequency of the incident light. Also, values of resonant frequencies (ω_n), damping constants (γ_n), and weights (f_n) are given in Table 4.1. The real and imaginary parts of the silver permittivity are shown in Figure 4.2a.

The insulator material around the MZI structure is a Kerr nonlinear material, which is chosen to be GaAs. The bottom layer (black area) is a graphene layer that is placed on a silica substrate. The graphene permittivity is characterized by [66]:

$$\varepsilon_g(\omega) = 1 + i\left(\frac{\sigma(\omega)}{\varepsilon_0 \omega t_g}\right) \tag{4.5}$$

Table 4.1 Parameters of the Drude–Lorentz model for silver.

N	ω_n (THz)	γ_n(THz)	f_n
1	197.3	939.62	7.9247
2	1083.5	109.29	0.5013
3	1979.1	15.71	0.0133
4	4392.5	221.49	0.8266
5	9812.1	584.91	1.1133

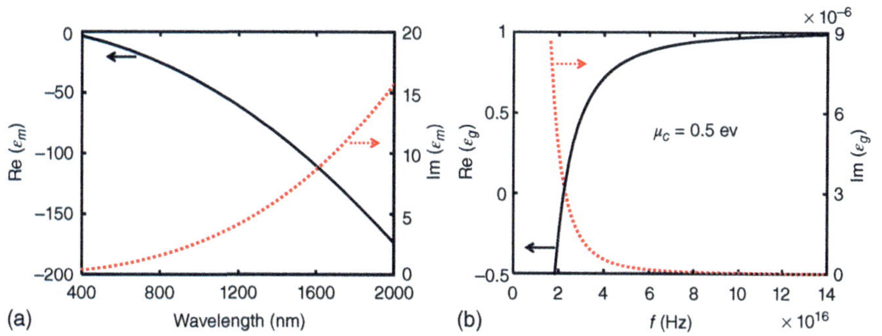

Figure 4.2 Real and imaginary parts of (a) silver permittivity and (b) graphene permittivity.

where t_g is the graphene thickness and ε_0 is the permittivity of vacuum. Figure 4.2b shows the real and imaginary parts of the graphene permittivity. The structural parameters of the proposed optical switch (Figure 4.1) are as follows: $d_1 = 300$, $d_2 = 3000$, $d_3 = 3000$, $d_4 = 350$, $d_5 = 50$, $d_6 = 500$, $d_7 = 600$, $d_8 = 1000$, and $d_9 = 100$ (all in nm).

Figure 4.1 shows that there is a metal layer in the shape of an MZI above the graphene sheet so that the graphene region under the metal plate turns into waveguides by applying the DC voltage. It means that when the DC voltage is applied between two layers (metal and graphene), a capacitive structure in the form of the MZI is generated. The created capacitor is studied in two modes of DC and AC in the following. After applying the DC voltage, the chemical potential of the created MZI structure on the graphene layer is modulated from 0 to 0.31 eV. Accordingly, surface plasmon waves can propagate in the graphene MZI structure, while these waves cannot excite in other sections of the graphene layer ($\mu_c = 0$ eV).

4.2.1 DC Mode of the Gate Capacitor

Figure 4.3a shows the order of the created capacitor layers in Figure 4.1. As seen in Figure 4.1, two insulator materials of SiO_2 and GaAs exist between the two capacitor plates (metal and graphene plates). It is because graphene is created by the plasma-enhanced chemical vapor deposition (PECVD) method on a silica

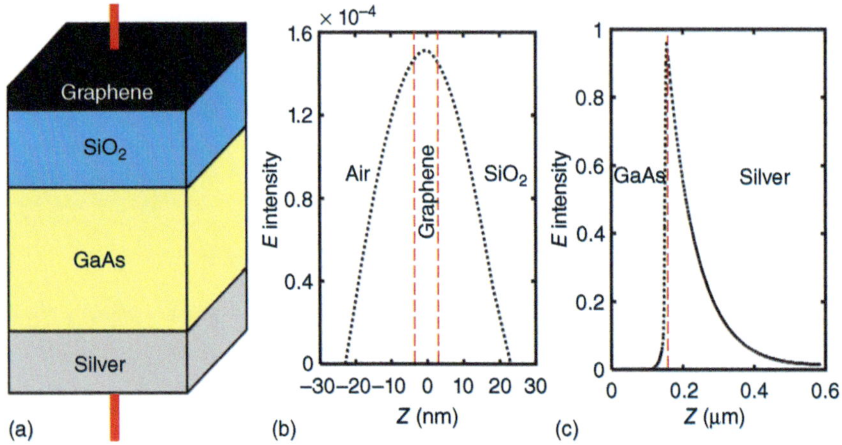

Figure 4.3 (a) Order of the created capacitor layers in Figure 4.1, (b) light penetration depth of the graphene (main) waveguide, and (c) light penetration depth of the metal (control) waveguide.

layer, and the nonlinear material of GaAs is used for light modulation. As a result, the equivalent capacitance includes two series capacitors. Such an equivalent capacitance can be obtained as:

$$C_{ox} = \frac{1}{\dfrac{1}{C_{SiO_2}} + \dfrac{1}{C_{GaAs}}} = \frac{\varepsilon_0}{\dfrac{d^{nm}_{SiO_2}}{\varepsilon_r^{SiO_2}} + \dfrac{d^{nm}_{GaAs}}{\varepsilon_r^{GaAs}}} \tag{4.6}$$

As seen in Eq. (4.6), the capacitance value depends on four parameters of the heights of the insulator layers ($d^{nm}_{SiO_2}$ and d^{nm}_{GaAs}) and the permittivity values of the insulator materials ($\varepsilon_r^{SiO_2}$ and ε_r^{GaAs}). To achieve the appropriate dimensions of the two insulator materials ($d^{nm}_{SiO_2}$ and d^{nm}_{GaAs}), the vertical propagation length of the generated surface plasmon waves in the main and control waveguides (graphene and metal waveguides) should be considered. The penetration depth in z-direction for two waveguides is shown in Figure 4.3b,c. According to Figure 4.3b,c, values more than 500 and 20 nm should be considered for the heights of GaAs and SiO_2 layers, respectively. It is because of two reasons. First, there exists the possibility of light modulation in the metal waveguide and secondly to avoid interferences between the two main and control waveguides.

As mentioned, two parameters of the permittivity values of SiO_2 and GaAs also affect the capacitance value. As a result, the $\varepsilon_r^{SiO_2}$ and ε_r^{GaAs} values should be determined. In all conditions, the permittivity of SiO_2 is constant and equal to 3.9. Therefore, the capacitance value can be expressed as:

$$C_{ox} = \frac{\varepsilon_r^{GaAs}\varepsilon_0}{\dfrac{20\,\text{nm} + d_1}{3.9} \times \varepsilon_r^{GaAs} + 500\,\text{nm} + d_2} \tag{4.7}$$

When light is not modulated inside the GaAs material, the permittivity of GaAs is equal to 12.9. By considering $d^{nm}_{SiO_2} = 100$ and $d^{nm}_{GaAs} = 500\text{nm}$, the DC gate capacitor can be calculated from Eq. (4.7). The obtained C_{ox}^{DC} is equal to 137.4 μF. The calculation procedure shows that the material type of the nongraphene gate is not important in this section. To achieve a chemical potential of 0.31 eV in graphene MZI structure, the applied voltage value can be calculated from Eq. (4.8), which is equal to 82 V.

$$|V_{DC}| = \left(\frac{e}{\pi C_{ox}^{DC}}\right)\left(\frac{\mu_c}{\hbar v_f}\right)^2 \tag{4.8}$$

In addition, by applying the obtained DC voltage, the gate capacitance should be equal to 128.8 μF for the chemical potential of graphene to be equal to 0.3 eV. This change in capacitance value is created by the light modulation in the AC mode of the gate capacitor.

4.2.2 AC Mode of the Gate Capacitor

The use of surface plasmons between the nongraphene capacitor plate and insulator material is one of the techniques to change the capacitance value. The important point is that the calculation of the capacitance which changes the chemical potential level in switching structure. In the micrometer frequency range, the refractive index of GaAs changes as:

$$n^2 - 1 = 4.372\,514 + \frac{5.466\,742\lambda^2}{\lambda^2 - (0.443\,130\,7)^2} + \frac{0.024\,299\,6\,\lambda^2}{\lambda^2 - (0.874\,645\,3)^2} + \frac{1.957\,522\,\lambda^2}{\lambda^2 - (36.9166)^2} \tag{4.9}$$

The permittivity variations of GaAs versus wavelength are shown in Figure 4.4a. Figure 4.4a is obtained from Eq. (4.9). The wavelength range of 1150–1550 nm is among the most important frequency ranges in telecommunication applications and optical integrated circuits. Consequently, this range is considered for GaAs permittivity variations. The capacitance value in the range of 1150–1550 nm is calculated using Eq. (4.8), and its variations are shown in Figure 4.4b. Also, Figure 4.4c shows the relationship between the wavelength of the control waves and the variation of

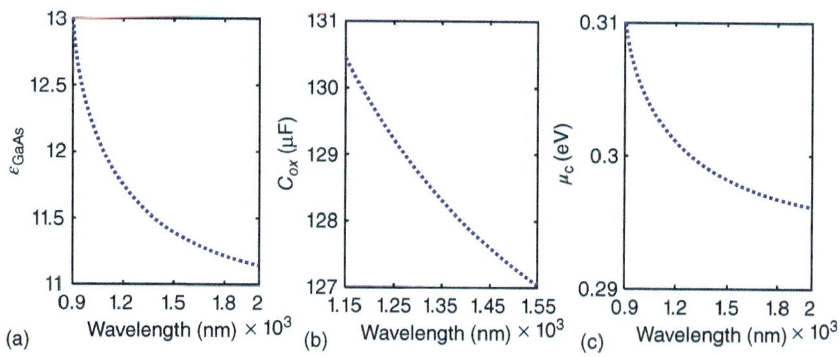

Figure 4.4 (a) GaAs permittivity, (b) variations of the waveguide capacitance, and (c) variations of the chemical potential for the proposed switch.

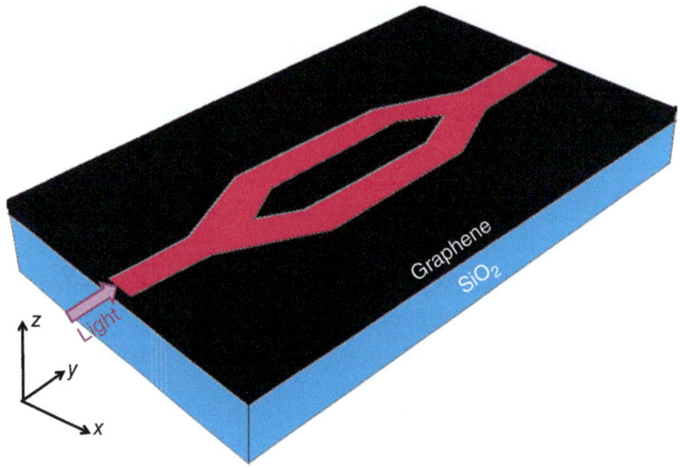

Figure 4.5 Main waveguide of the optical switch (Figure 4.1).

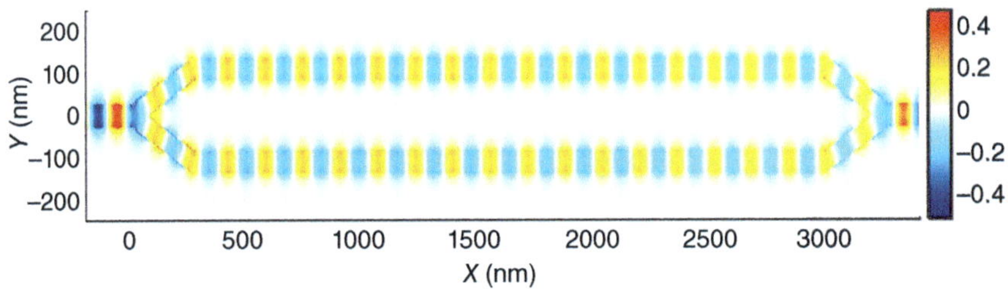

Figure 4.6 Field profile of E_x at 1.42 μm for the main waveguide (with the chemical potential of 0.31 eV for both arms).

the chemical potential. Based on the obtained results, the best wavelength for the control light is equal to 1300 nm.

After studying the created capacitor in two DC and AC modes, the operation of the proposed optical switch based on the main and control waveguides is investigated in the following. As mentioned above, the main waveguide includes air, graphene, and silica layers. Figure 4.5 shows this structure. In Figure 4.5, control light is not applied to the structure. Accordingly, the chemical potential of the total MZI structure is 0.31 eV. It is worth mentioning that the chemical potential of other places (black places) in the graphene layer is equal to 0 eV.

Figure 4.6 shows the field profile of the main waveguide by applying the main light source. As seen in Figure 4.6, when the chemical potential level of two arms of the MZI structure is the same, the phase difference of waves between these two arms is 0°. Consequently, the waves can transmit to the output port of the MZI structure.

The next is that the main and control light sources be applied simultaneously. In this case, the control signal induces the Kerr effect in the nonlinear medium (GaAs material), and one of the arms has a different chemical potential compared to the whole main waveguide. Figure 4.7 shows the mechanism of wave propagation for the

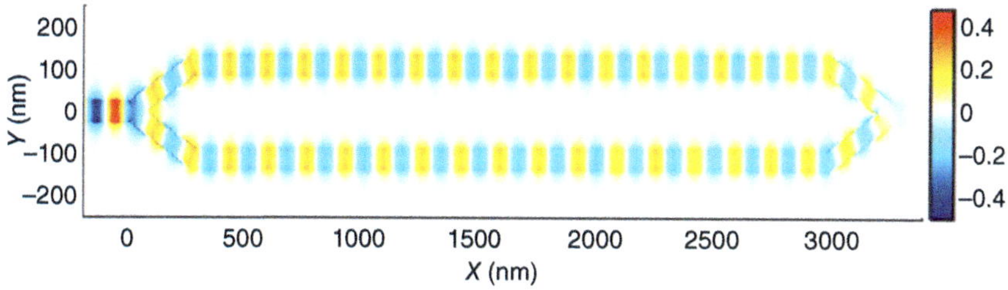

Figure 4.7 Field profile of E_x at 1.42 μm for the main waveguide (with the chemical potential of 0.31 eV for the upper arm and 0.3 eV for the lower arm).

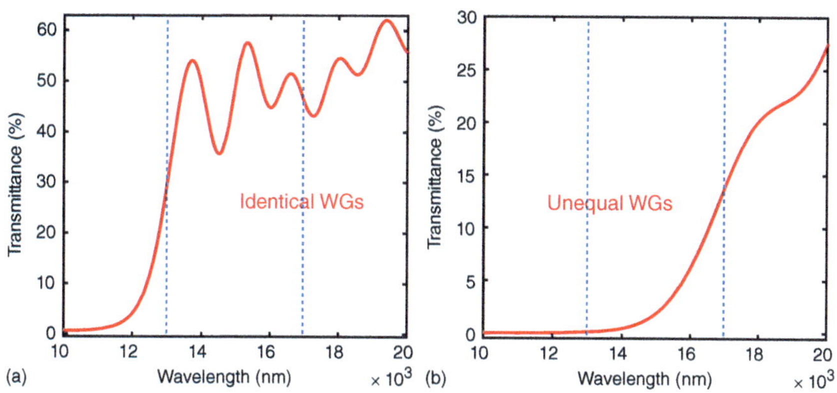

Figure 4.8 Transmission spectrum of main waveguide for (a) the same chemical potentials of two arms and (b) the different chemical potentials of two arms.

different chemical potential levels for the two arms of the main waveguide (0.31 eV for the upper arm and 0.3 eV for the lower arm). As seen, after spending almost 2 μm from the arms, the phase difference of waves between two arms is equal to 180°. Accordingly, the waves cannot be transmitted to the output port.

To provide a better view of the switching operation of the proposed optical switch, the transmission spectra of the main waveguide for the same and different chemical potentials of two arms have also been presented in Figure 4.8. By comparison of two cases (Figure 4.8a,b), the wavelength range of 1300–1700 nm is the best range for the main waves. It is because, in this wavelength range, the transmittance values are more than 40% and less than 10% for the identical and unequal waveguides, respectively.

Based on the obtained results, to achieve a phase difference of 180° between two arms of the graphene MZI structure, it is necessary that a suitable control waveguide be designed. In other words, the control light should affect only one of the arms. For this purpose, two different techniques are proposed. The first one is that by grating one of the arms, the propagation of the surface plasmons is prevented in this arm. Figure 4.9a shows this structure (switch I). The presented structural parameters in Figure 4.9a are as follows: $d_{10} = 500$, $d_{11} = 750$, $d_{12} = 20$, and $d_{13} = 500$ (all in nm). In the second technique, different plasmon frequencies are used for two

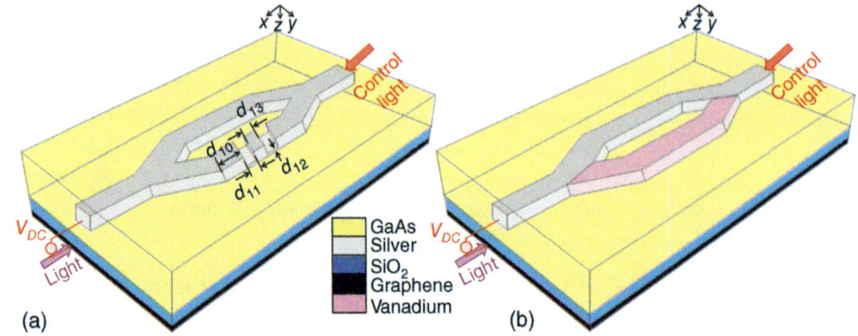

Figure 4.9 The proposed all-optical graphene plasmon switch using a control waveguide with (a) a grating structure in one of the arms (switch I) and (b) different metal material in one of the arms (switch II).

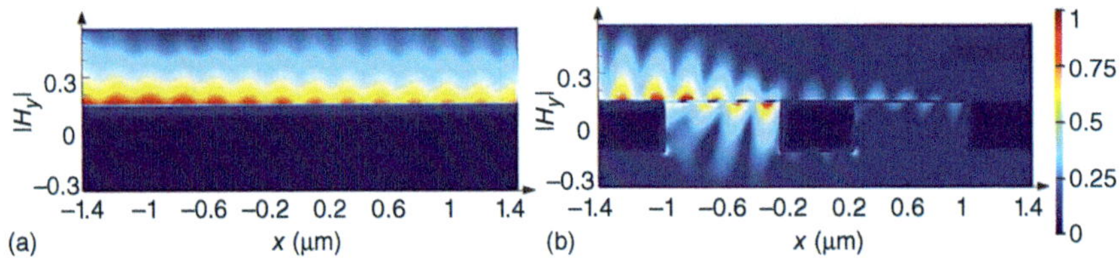

Figure 4.10 Field profile of $|H_y|$ for the (a) simple arm and (b) grating arm.

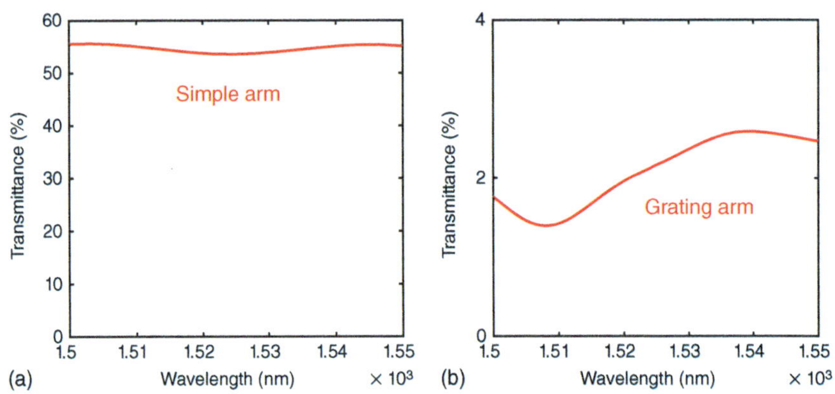

Figure 4.11 Transmission spectrum of the main waveguide for the (a) simple arm and (b) grating arm.

arms of the metal MZI structure. Accordingly, one of the MZI arms is filled with vanadium material. The proposed structure (switch II) is shown in Figure 4.9b. In the following, each of the mentioned methods has been studied separately.

Figure 4.10 shows the field profiles of the proposed switch I for both arms (simple and grating arms) with two main and control light sources. As seen in Figure 4.10a, the waves can propagate through the simple arm. For the grating arm, the incident light cannot reach the end of the arm. Also, Figure 4.11 shows

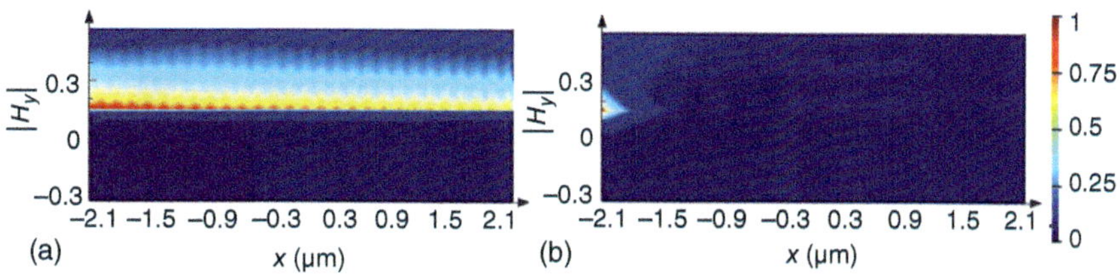

Figure 4.12 Field profile of $|H_y|$ for the (a) silver arm and (b) vanadium arm.

the transmission spectra of the main waveguide for the simple and grating arms. As seen, the transmission values for the simple and grating arms are almost equal to 55% and 2%, respectively. Although the first proposed structure (switch I) has a suitable switching operation, it should be noted that the designed structure has some drawbacks. First of all, light propagation cannot be stopped at the beginning of the grating arm, and it passes through about 1 μm of the arm. Secondly, due to the small size of d_{12}, the fabrication process for this structure is difficult. Therefore, the second method has also been proposed (Figure 4.9b).

As mentioned above, in the second method (switch II), two different metal materials of silver and vanadium, which have different plasmon frequencies, are used for two arms of the control waveguide. Due to the good electrical conductivity of vanadium, this metal is replaced in one of the silver arms. This causes the incident light to not propagate at the boundary of GaAs–vanadium. Figure 4.12 shows the field profile of this case. As seen, light propagation can be stopped at the beginning of the vanadium MZI arm.

4.3 Nanomaterial Heterostructures-Based Switch

To benefit from the advantages of different optical structures, the combination of such structures can be used to design optical switches. The idea of using two different optical structures (plasmonic and PC configurations) has been used to design an all-optical switch in [101]. The design process of the proposed switch in [101] is discussed in the following.

The proposed heterostructure all-optical switch is shown in Figure 4.13. As seen in Figure 4.13, the 1D PC structure used in this configuration is located at the intersection of MIM plasmonic and insulator waveguides. The insulator and metal materials in the horizontal MIM plasmonic waveguides are air and silver, respectively. The complex relative permittivity of silver is described by the Drude–Lorentz model (Eq. (4.4)) [102]. In addition, the material used in the insulator waveguides is assumed to be SiO_2.

The zoomed view of the central section of the proposed switch (PC section) is shown in Figure 4.14a. This 1D PC structure is composed of the alternating nonlinear insulator layers with the high refractive index of $n_H = 3.4$ (layers with the thickness of l_1) and the low refractive index of $n_L = 1.77$ (layers with the thickness of l_2). The structural parameters of the proposed switch are as follows: $l_1 = 120$, $l_2 = 290$,

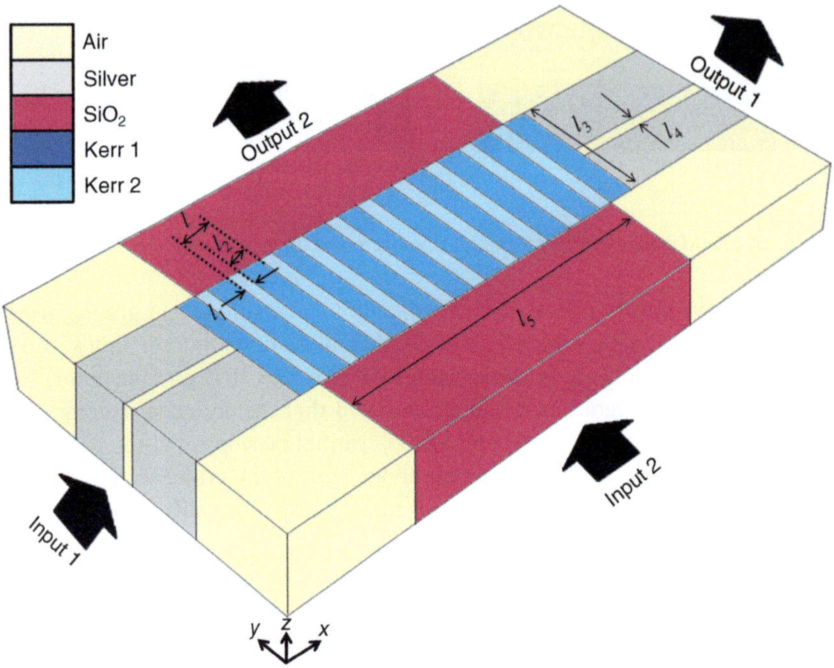

Figure 4.13 Proposed heterostructure all-optical switch.

$l_3 = 700$, $l_4 = 100$, and $l_5 = 3836$ (all in nm). Also, the lattice constant of the 1D PC structure is l, which is equal to $l = l_1 + l_2$.

The obtained band diagram of the used 1D PC structure using the plane wave expansion (PWE) method is shown in Figure 4.14b. As seen, there is a PBG at the wavelength range of extended from 1534 to 2320 nm. To verify the performance of the used 1D PC structure, its transmission spectrum is also obtained using the finite-difference time-domain (FDTD) method (Figure 4.14c). It can be seen that the transmission value is equal to zero at the mentioned PBG wavelength range. Accordingly, the sharp edges of the created PBG can be used for the switching mechanism. Based on the obtained results, the horizontal and vertical waveguides (MIM plasmonic and insulator waveguides) are used for data and control signals, respectively. Suitable isolation between two signals can be achieved using this method.

After investigating the performance of the used 1D PC structure located at the center of the proposed topology (Figure 4.13), the switching mechanism of the designed all-optical switch is studied. For this purpose, the transmission spectra of the designed switch for data (from input 1 to output 1) and control signals (from input 2 to output 2) are obtained. Figure 4.15a shows the transmission spectrum of the data signal when only the data signal source (located at input 1) is "on." As seen, the created sharp PBG edge (low edge) can be selected for the data wavelength. Also, the transmission spectra of the control signals for transverse magnetic (TM) and transverse electric (TE) modes are shown in Figure 4.15b and c, respectively. In this case, only the control signal source (located at input 2) is applied. As seen in Figure 4.15b,c, the wavelengths of 1606 and 1427 nm for fundamental TM and TE

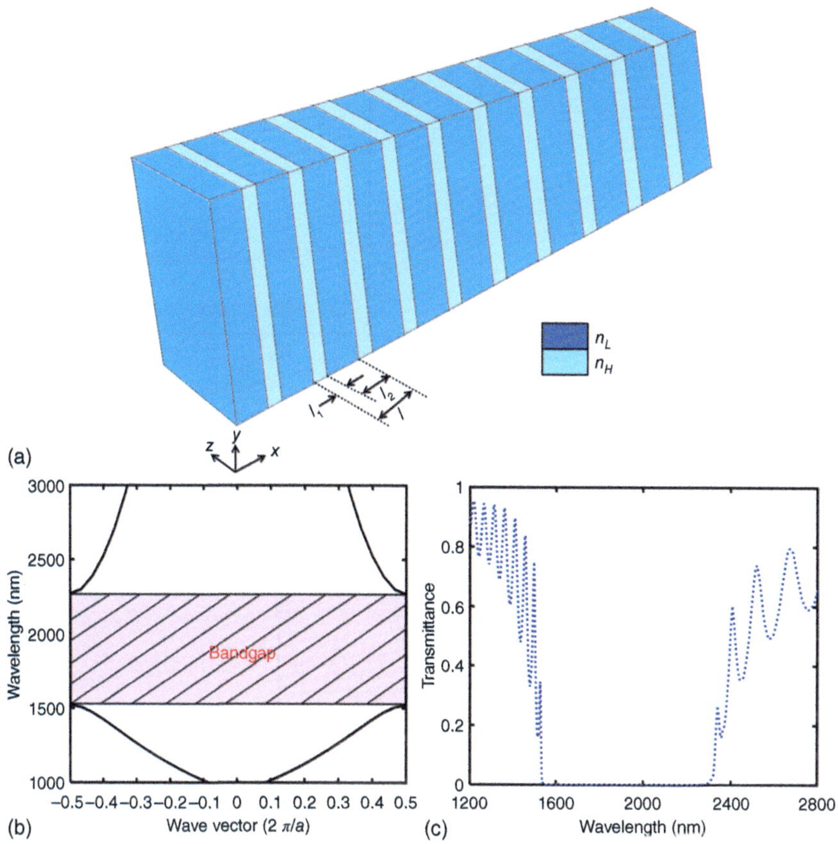

Figure 4.14 (a) Proposed structure of 1D PC, (b) its band diagram, and (c) its transmission spectrum.

modes can be suitable choices for the control wavelengths. It is because these two wavelengths can propagate from input 2 to output 2 almost completely, and they are located at the PBG of the data signal.

At the next step, both data and control signal sources are simultaneously applied to inputs 1 and 2, respectively. In this case, the refractive indexes of Kerr materials (Kerr 1 and 2) change, and the transmission spectrum of the data signal shifts to higher wavelengths. According to the used Kerr material type, three situations can occur. These three situations are as follows.

4.3.1 Situation 1: $n_2L \gg n_2H$

When the nonlinear refractive index coefficient (n_2) of the wide layers (layers with the thickness of l_2 and the refractive index of n_L) is much more than that of the narrow layers (layers with the thickness of l_1 and the refractive index of n_H), only the refractive indexes of the wide layers change. As a result, in this situation, the data transmittance shift is only due to the refractive index changes of n_L. This case is shown in Figure 4.16a. As seen by changing the refractive index of n_L ($\Delta n_L = 0.01$

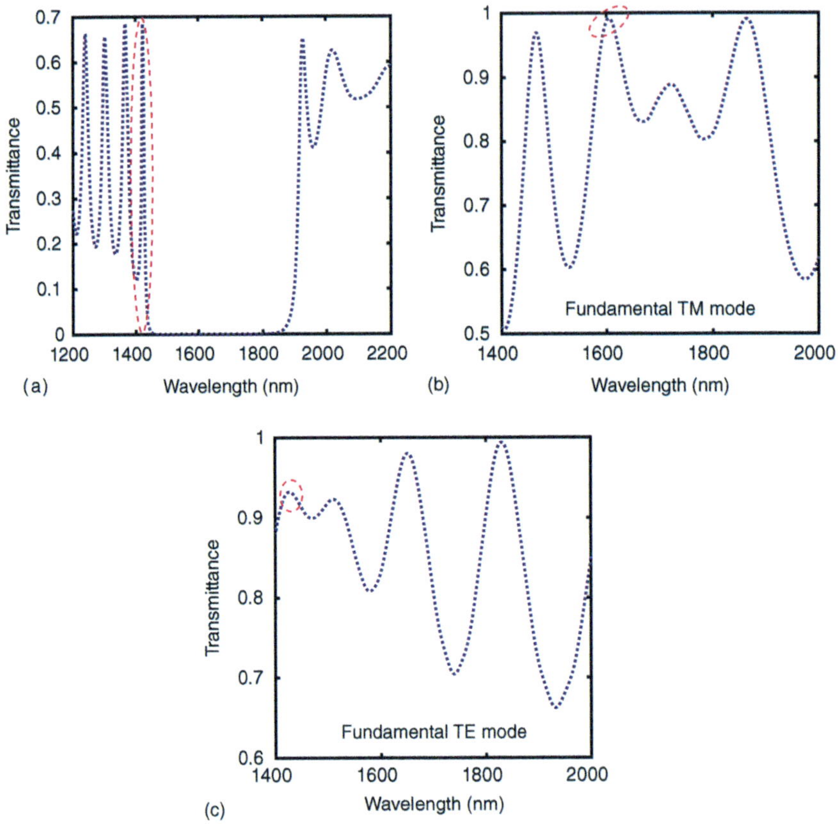

Figure 4.15 Transmission spectrum of the proposed switch (Figure 4.13) (a) from input 1 to output 1, (b) from input 2 to output 2 for TM mode, and (c) from input 2 to output 2 for TE mode.

and $\Delta n_H = 0$), the data signal can propagate to the output port. In other words, the transmittance value at the data wavelength (1429 nm) increases from 0.15 (blue curve) to 0.698 (red curve). As a result, the switching operation can occur in this case.

4.3.2 Situation 2: $n_2H \gg n_2L$

Based on the mentioned explanations in the previous case, when $n_{2H} \gg n_{2L}$, the data transmittance shift is only caused by the refractive index changes of n_H. Figure 4.16b shows this case ($\Delta n_H = 0.01$ and $\Delta n_L = 0$). As seen in Figure 4.16b, by increasing the refractive index of n_H, the data signal shift is very imperceptible. Accordingly, the switching operation cannot occur in this case.

4.3.3 Situation 3: $n_2H \simeq n_2L$

In the last situation, the nonlinear effect in the narrow layers is equal to the nonlinear effect in the wide layers of the PC structure. Consequently, by applying the control

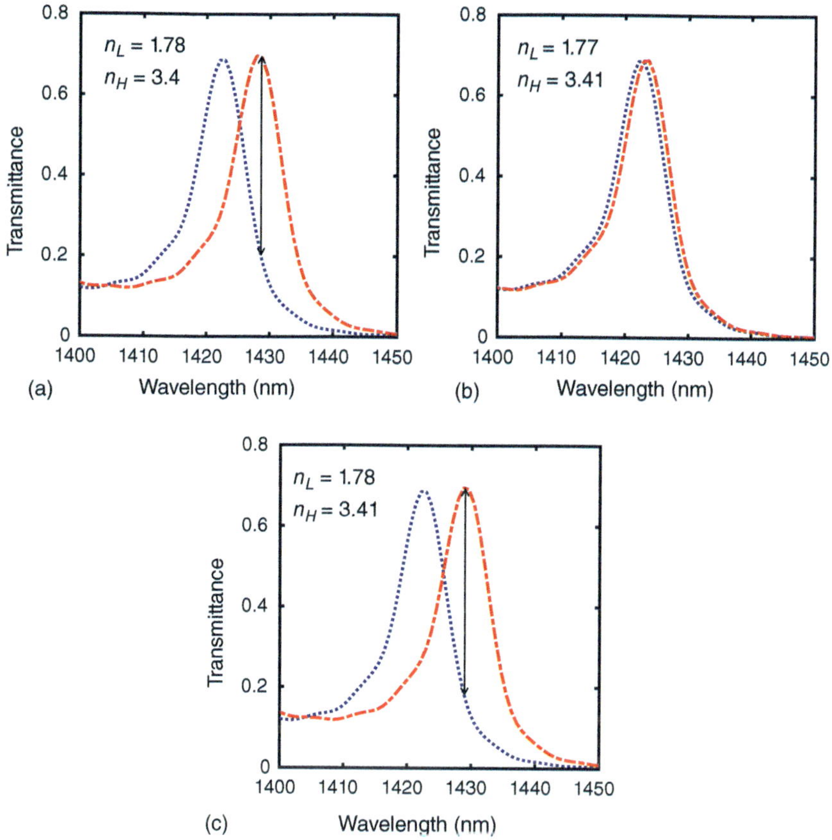

Figure 4.16 Transmission spectra of the proposed switch (Figure 4.13) without and with control signal source for situations (a) 1, (b) 2, and (c) 3.

signal source, both refractive indexes of n_H and n_L increase simultaneously. This case is shown in Figure 4.16c ($\Delta n_H = 0.01$ and $\Delta n_L = 0.01$). Figure 4.16c shows that the transmittance value of the data signal increases from 0.15 to 0.7, and the switching operation occurs.

To give a better view of the refractive index changes on the switching mechanism of the proposed structure, the field profiles of E_y magnitude for two moods of "off" and "on" states of the designed switch are presented in Figure 4.17. As seen in Figure 4.17a, when only the data signal source is applied, the data wavelength cannot propagate from input 1 to output 1. In other words, the designed switch is at the "off" state. As reviewed, by applying two data and control signal sources for two situations 1 and 3, the switching operation can occur. The field profiles of the two situations are shown in Figure 4.17b and c, respectively. As seen in Figure 4.17b,c, the incident light can pass through the structure in these cases.

After investigating the general switching operation of the proposed structure, its switching mechanism is studied using practical Kerr materials with the mentioned properties. Transmission spectra of the designed switch using different practical

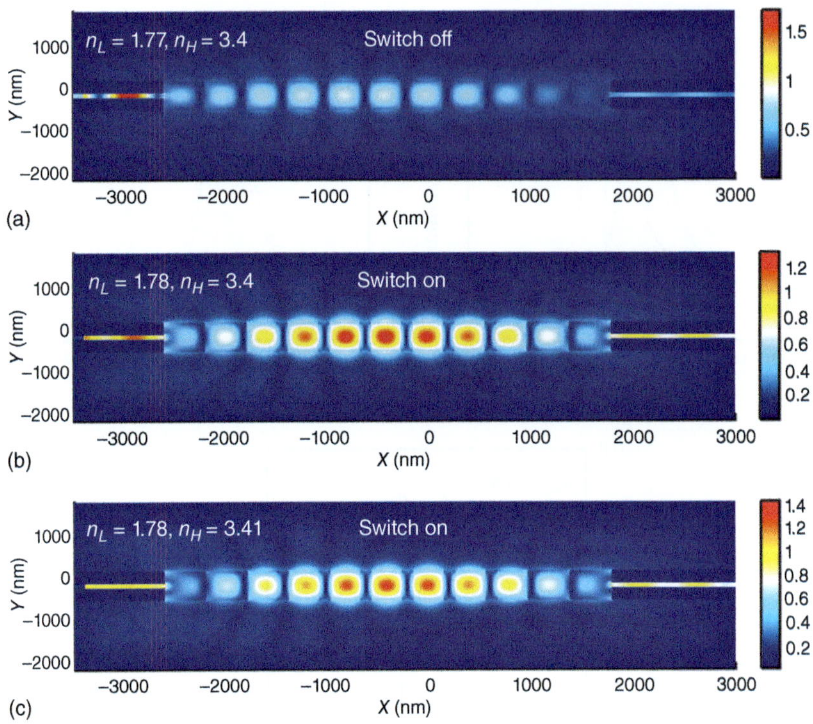

Figure 4.17 Field profile of $|E_y|$ at the data wavelength (1429 nm) (a) with only data signal source ("off state"), (b) with both data and control signal sources for situation 1 ("on state"), and (c) with both data and control signal sources for situation 3 ("on state").

Kerr materials are shown in Figure 4.18. In the first, second, and third switches (S_1, S_2, and S_3), the Kerr materials with the nonlinear properties of $n_{2(CS_2)} \simeq n_{2(Si)}$, $n_{2(AuSiO_2)} \gg n_{2(GaAs)}$, and $n_{2(GGS\text{-}Au_{10})} \simeq n_{2(Si)}$ are used, respectively. Also, Table 4.2 shows the features of such switches. As seen in Table 4.2, based on the transmittance values of the "on" and "off" states of such switches (T_{on} and T_{off}), the switching operation occurs in all three cases.

To provide a better view of the proposed switch operation, its time-domain behavior is also presented. In this section, two nonlinear Kerr materials with the refractive indexes of 1.77 and 3.4 and the $\chi^{(3)}$ values of 2×10^{-18} and 4.8×10^{-18} m^2 V^{-2} are used to fill the PC structure to achieve an all-optical switch. Figure 4.19a shows the time-domain data and control signal sources inserted into the input ports (inputs 1 and 2) of the proposed switch. Here, the wavelengths of the data signal and the control signal (fundamental TE mode) are chosen to be 1926 and 1427 nm, respectively.

As seen in Figure 4.19a, the data signal source is a continuous-wave signal. Also, the time-domain control signal source is a continuous-wave signal with an input power of 29.3 MW cm^{-1}. This signal is at the "on" state from 2.75 to 4.25 ps. The transition time of 500 fs is considered for the rising and falling edges. Figure 4.19b shows the time-domain output signal derived from data and control input sources. As seen, when only the data signal source is on, the output power is more than 0.6.

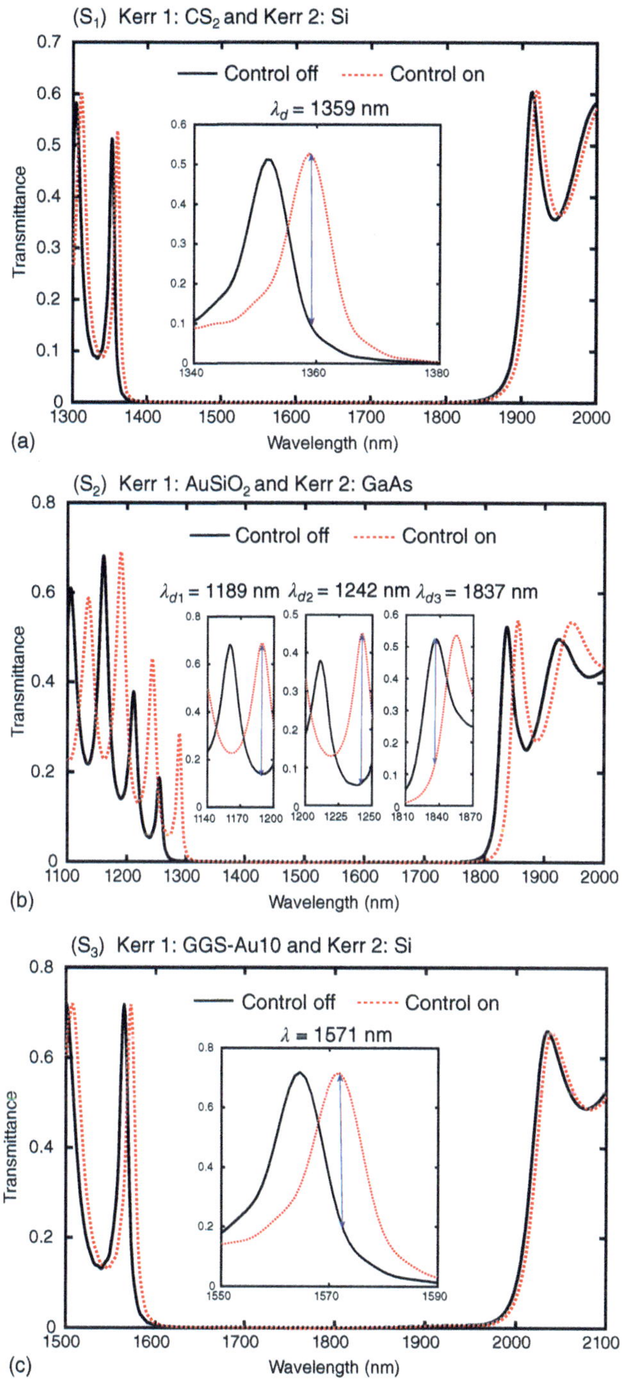

Figure 4.18 Transmission spectra of the proposed switch (Figure 4.13) without and with control signal source for two Kerr materials of (a) CS_2, Si, (b) $AuSiO_2$, GaAs, and (c) GGS-Au10, Si.

Table 4.2 Mechanism of the heterostructure all-optical switch (Figure 4.13) with different Kerr materials.

Switch	Kerr material	n_0	n_2 (cm^2 W^{-1})	$\chi^{(3)}$ (m^2 V^{-2})	λ_d (nm)	T_{on}	T_{off}	I (MW cm^{-2})
S_1	CS_2	1.63	0.3×10^{-13}	4×10^{-20}	1359	0.525	0.09	232 000
	Si	3.5	0.43×10^{-13}	2.8×10^{-19}				
S_2	$AuSiO_2$	1.47	2.7×10^{-9}	2.37×10^{-15}	1189			22.2
	GaAs	3.4	1.59×10^{-13}	9.7×10^{-19}	1242			
					1837	0.692	0.14	
						0.454 0.53	0.06	
							0.16	
S_3	GGS-Au10	2	1.8×10^{-13}	3.8×10^{-19}	1571	0.72	0.19	66 600
	Si	3.5	0.43×10^{-13}	2.8×10^{-19}				

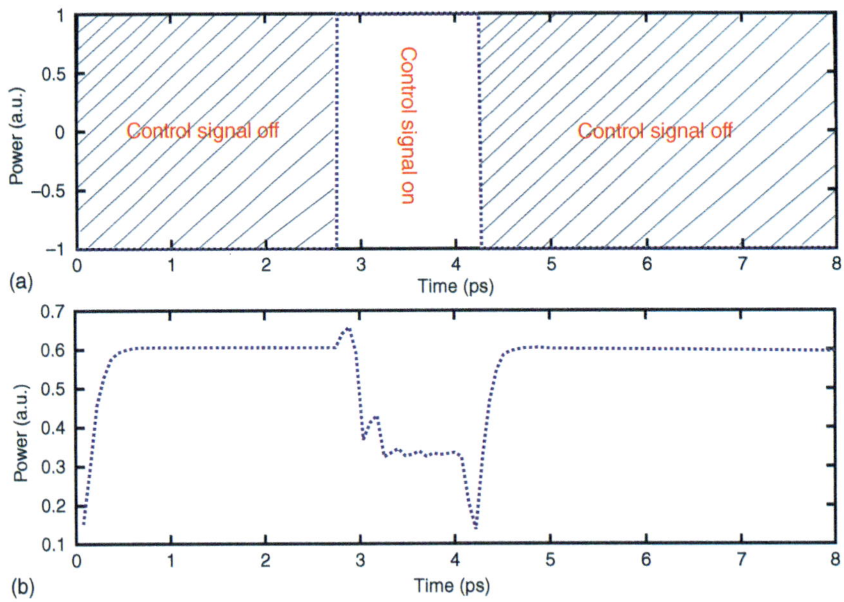

Figure 4.19 Time-domain representations of (a) data and control signals and (b) output signal for the proposed heterostructure all-optical switch.

In 2.75–4.25 ps (when both data and control signal sources are on), the transmitted power to the output port decreases. The main drawback of the time-domain output signal is its low contrast ratio value. To improve the contrast ratio value, a suppression filter should be added to the output port, which is discussed in the following.

Figure 4.20 shows the proposed heterostructure all-optical switch with its suppression filter. The designed filter is composed of seven unequal stubs coupled to a MIM plasmonic waveguide. The structural parameters of the designed suppression

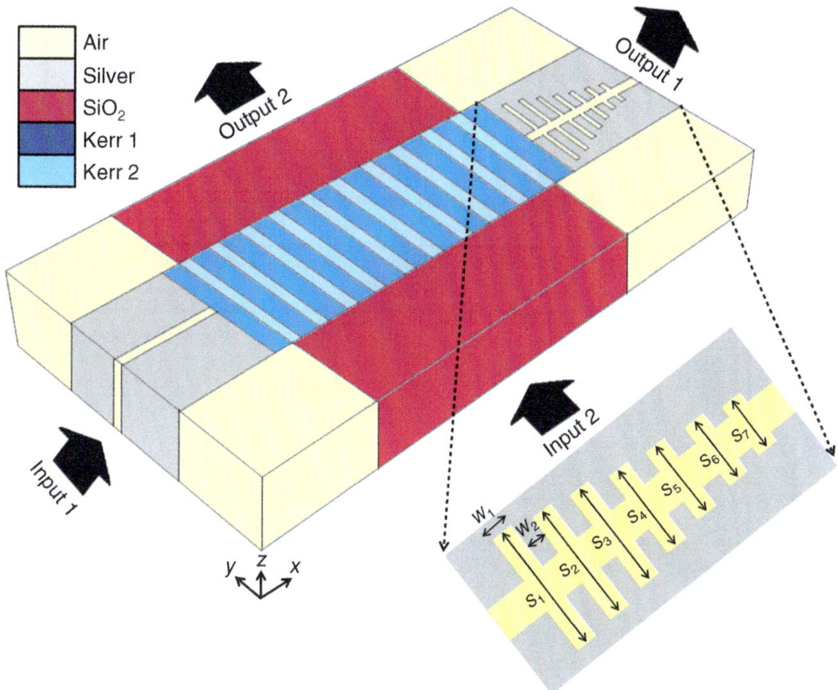

Figure 4.20 Proposed heterostructure all-optical switch with the suppression filter at the output port.

filter are as follows: $s_1 = 400$, $s_2 = 360$, $s_3 = 320$, $s_4 = 280$, $s_5 = 240$, $s_6 = 200$, $s_7 = 160$, $w_1 = w_2 = 50$ (all in nm).

Figure 4.21 shows the time-domain input and output signals of the proposed heterostructure all-optical switch with the suppression filter at the output port (Figure 4.20). It is worth mentioning that the data and control input signals are similar to Figure 4.19a. As seen, here the switching operation also occurs, but the output power for the "off" state reduces due to the existence of the suppression filter. As a result, the contrast value increases. Figure 4.21b shows that the values of the rise and fall times are equal to 250 fs.

To achieve higher transmittance values in edges and out of the PBG, tapered resonators are added to the first proposed all-optical switch. The schematic of the improved proposed all-optical switch is shown in Figure 4.22. The values of the structural parameters of the tapered resonators are equal to $d_6 = 117$ and $d_7 = 250$ nm. Other parameters have already been explained.

The transmittance spectra of the data and control signals are shown in Figure 4.23. According to Figure 4.23a, the transmittance values in all of the wavelengths (obviously, except for the PBG) increase. Increasing the transmittance values in the PBG edges provides all-optical switches with more transmittance values for the "on" state, which is because there is more coupling strength between MIM plasmonic waveguides and the PC structure in this all-optical switch. As known, there is a trade-off

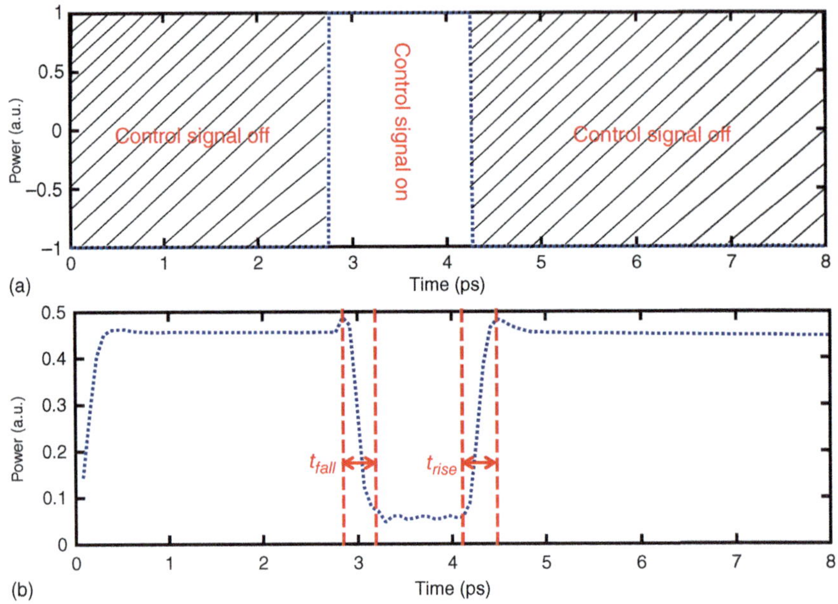

Figure 4.21 Time-domain representations of (a) data and control signals and (b) output signal for the proposed heterostructure all-optical switch with the suppression filter at the output port.

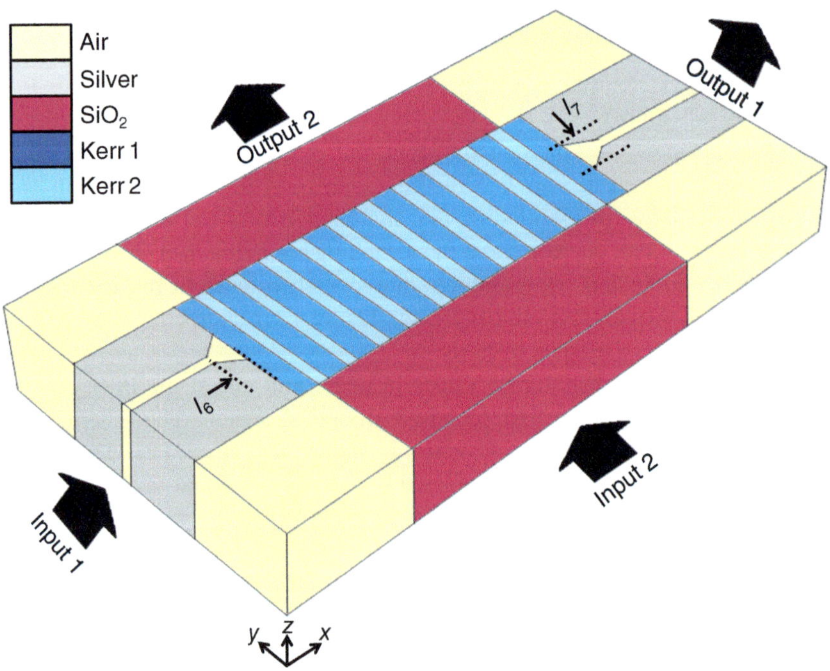

Figure 4.22 Improved proposed heterostructure all-optical switch.

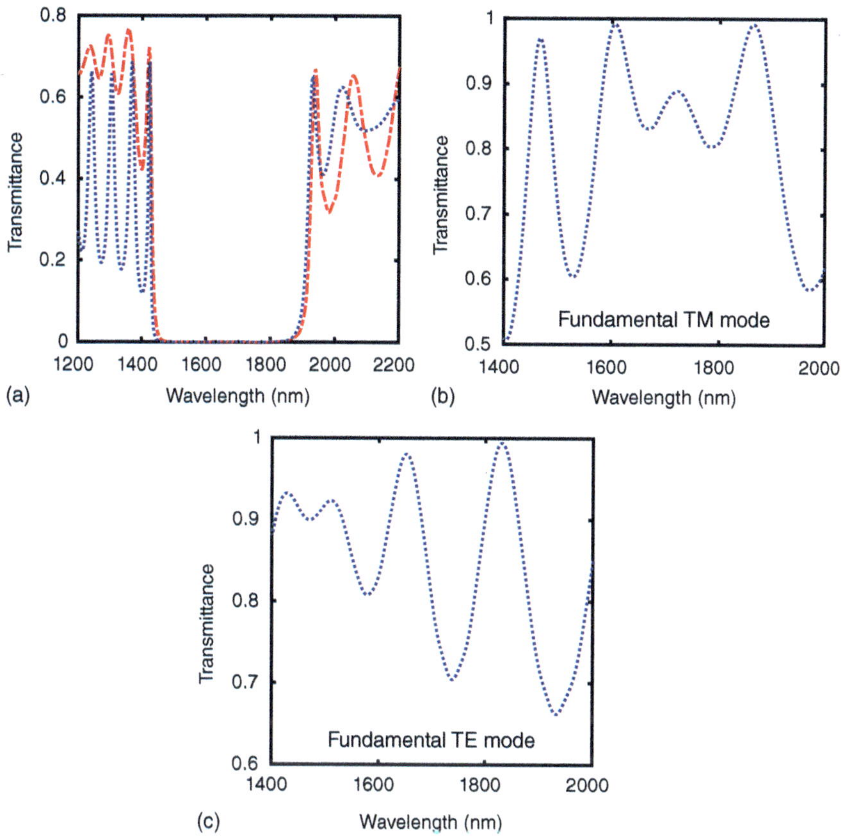

Figure 4.23 Transmission spectrum of the improved proposed switch (Figure 4.22) (a) from input 1 to output 1, (b) from input 2 to output 2 for TM mode, and (c) from input 2 to output 2 for TE mode.

between designing parameters of all-optical switches. Increasing the coupling effect between MIM plasmonic waveguides and PC leads to slower sharpness in transition from the maximum transmittance to the minimum transmittance.

Since the existing coupling between the insulator waveguides and PC structure does not change, the transmittance spectrum of the pump signal (from input 2 to output 2) is similar to the previous all-optical switch (Figure 4.15b,c). Here the switching mechanism of the improved proposed all-optical switch for different three situations is also investigated.

Figure 4.24 shows the transmittance spectra for three different situations, including $n_{2L} \gg n_{2H}$, $n_{2H} \gg n_{2L}$, and $n_{2L} \approx n_{2H}$ for the improved proposed all-optical switch. As mentioned, for two modes of $n_{2L} \gg n_{2H}$ (Figure 4.24a) and $n_{2L} \approx n_{2H}$ (Figure 4.24c), the switching operation occurs. On the contrary, the switching operation for $n_{2H} \gg n_{2L}$ cannot be obtained (Figure 4.24b) because the higher refractive index layers have a lower effect on the data signal transferring from input 1 to output 1. It should be noted that due to the lower sharpness of the PBG edges

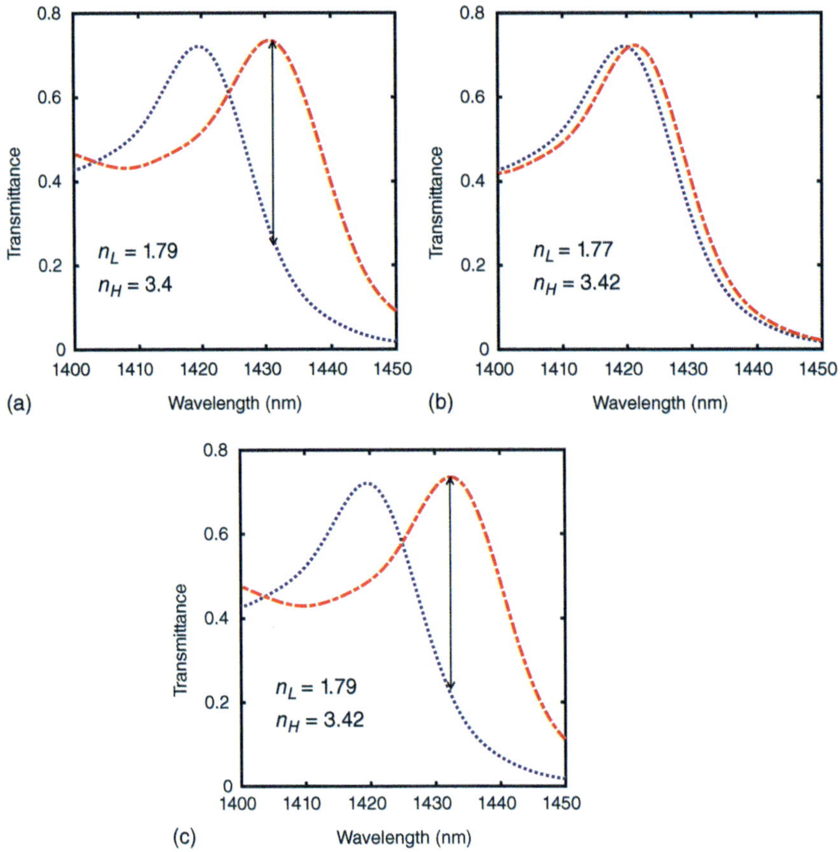

Figure 4.24 Transmission spectra of the improved proposed switch (Figure 4.13) without and with control signal source for situations (a) 1, (b) 2, and (c) 3.

for the improved proposed all-optical switch, more control light intensity is needed to increase the refractive index of Kerr layers by a value of 0.02.

The switching wavelength for the improved proposed all-optical switch in Figure 4.24a,c is equal to 1432 nm. The maximum transmittance values at the "on" state of the improved all-optical switch are equal to 0.74 and 0.735 for situations 1 (Figure 4.24a) and 2 (Figure 4.24c), respectively. Also, the transmittance value at the "off" state of the switch is equal to 0.185, which is close to zero. Finally, providing suitable isolation between data and control signals, and requiring lower control intensities cause that the proposed structure can be a good candidate for use in other integrated optical circuits.

4.4 Modulation Characteristics

To bring electrical signals to the optical domain, optical systems usually require an optical modulator. As a result, optical modulators have found wide application in many optical systems. By applying the control signal, the phase, frequency, or amplitude of the data signal can be modulated in optical modulators. This mechanism can produce phase modulators, frequency modulators, and amplitude modulators, respectively. In addition, the classification of optical modulators is performed based on the different methods that are used to design them. These methods are the electro-optic method [103, 104], thermo-optic method [105], and exploration of nonlinear effects [106]. Since the optical nonlinearity is used for the active control of optical signals in all-optical modulators, such modulators have high modulation speeds. Consequently, numerous optical topologies have been used to design all-optical modulators so far [107–109]. The main drawbacks of the most designed optical modulators in the literature can be summarized as follows. First, in most of them, the time-domain behaviors of modulators have not been investigated. However, to prove the correct operation of modulators, it is necessary to present time-domain simulations when the nonlinear Kerr effect is used. Also, most of them use a common path for both data and control signals, whereas an ideal modulator should be able to provide suitable isolation between the data and control signals.

One of the most applied optical structures to design optical modulators is plasmonic structures. In [110], a novel and simple topology has been proposed to design a plasmonic all-optical amplitude modulator. This structure tried to reform the mentioned disadvantages. The step-by-step designing procedure of the designed all-optical amplitude modulator is investigated in the following.

Figure 4.25 shows the proposed structure of the designed basic filter, which is used in the realization of the proposed all-optical modulator. As seen in Figure 4.25, the filter structure consists of a nonlinear nanodisk resonator, surrounded by a silicon ring-shaped resonator located inside a circular air resonator, and a horizontal MIM plasmonic waveguide. The metal and Kerr materials used in this structure are silver and InGaAsP, respectively. Silver is characterized by a comprehensive and accurate Drude–Lorentz model (Eq. (4.4)). Also, the third-order nonlinear susceptibility

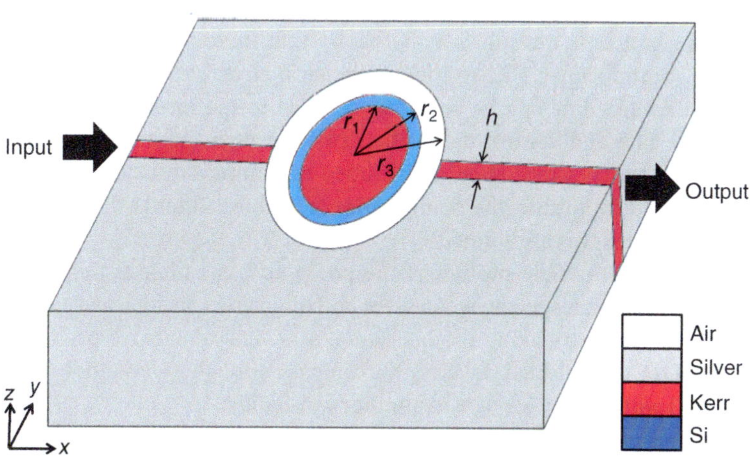

Figure 4.25 Proposed structure of the designed basic filter.

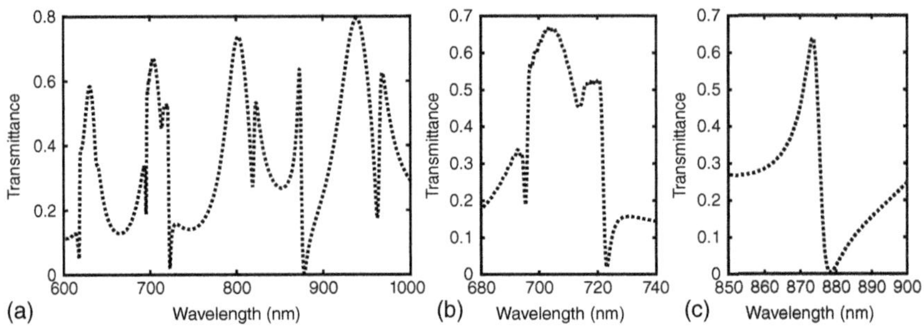

Figure 4.26 (a) Transmission spectrum of the designed basic filter, (b) zoomed view of the selected mode for the control wavelength, and (c) zoomed view of the selected mode for the data wavelength.

($\chi^{(3)}$) and the linear dielectric constant (ε_0) of the Kerr material (InGaAsP) are equal to 1×10^{-18} m^2 V^{-2} and 2.25, respectively. The structural parameters values of the proposed filter are as follows: $r_1 = 517$, $r_2 = 570$, $r_3 = 650$, and $w = 50$ (all in nm).

Figure 4.26a shows the transmission spectrum of the designed basic filter using FDTD simulations in the wavelength range of 600–1000 nm. As seen in Figure 4.26a, the transmission spectrum has several resonance modes due to the presence of different resonance wavelengths at the designed resonator. Based on the obtained results, two of the generated resonance modes should be chosen for the data and control signals. Since it is necessary that the control signal should pass almost completely through the structure and have gradual transitions, the resonance mode located at the wavelength of 702 nm can be a suitable choice for the control signal. Figure 4.26b shows the zoomed view of the selected control mode. Furthermore, the data signal should have a sharp edge and a high difference between its notch and peak transmittance values. Therefore, the Fano-resonance mode located at the wavelength of 702 nm is selected for the data signal. The zoomed view of this signal is also shown in Figure 4.26c.

As mentioned above, the goal of this design is that two isolated paths be used for data and control signals. As a result, it is necessary that there be another waveguide to pass the control signal. Due to the symmetrical structure of the proposed resonator, another waveguide can be vertically coupled to the resonator. Accordingly, the selected data and control modes that pass through the horizontal and vertical waveguides, respectively, should guarantee minimum cross-talk between two waveguides. For this purpose, the field profiles of the basic filter at the selected data and control wavelengths are studied.

Figure 4.27a,b shows the field profiles of the proposed basic filter at the wavelengths of 702 and 878 nm, respectively. As seen, both data and control wavelengths have even and odd symmetry along the horizontal and vertical axes, respectively. Therefore, each of the two selected wavelengths can pass through a waveguide (horizontal or vertical) without any leakage to another waveguide.

By adding the vertical waveguide to the basic filter, an all-optical amplitude modulator can be designed. Figure 4.28 shows the general structure of the proposed all-optical amplitude modulator. As seen in Figure 4.28, the horizontal and vertical

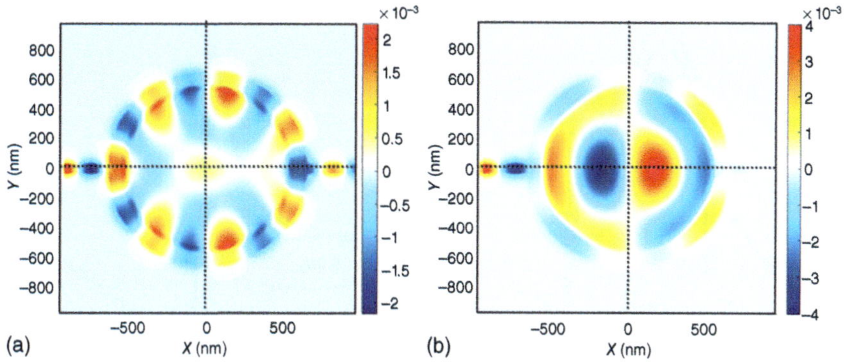

Figure 4.27 Field profile of Re(H_z) for the basic filter at the resonance modes of (a) 702 nm and (b) 878 nm.

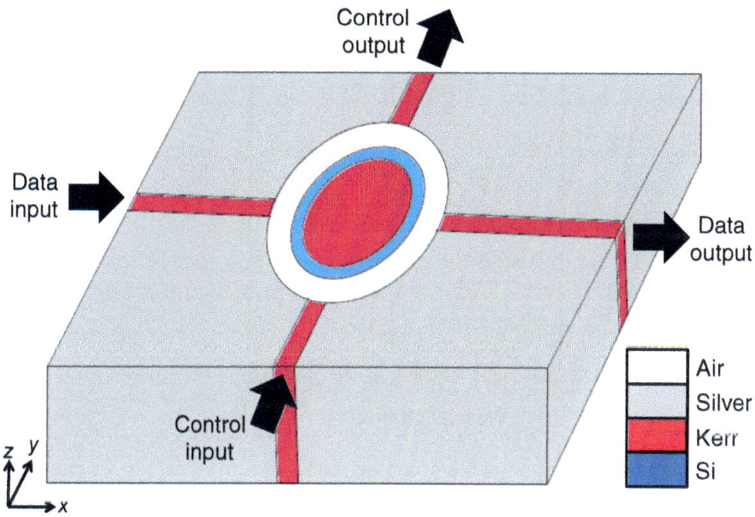

Figure 4.28 Proposed structure of the all-optical amplitude modulator.

waveguides are considered to pass the data and control signals, respectively. Using this method, an all-optical amplitude modulator with suitable isolation between the data and control signals can be achieved.

The transmission spectra of the data signal (from data input to data output port) with only the data signal source and control signal (from control input to control output port) with only the control signal source are shown in Figure 4.29a and b, respectively. It is worth mentioning that the transmittances of the data and control signals be compared with the selected data and control signals of the basic filter in Figure 4.29a,b. As seen, the transmission spectra of the designed modulator almost match the transmittances of the basic filter. In other words, the data and control wavelengths pass through the horizontal and vertical waveguides, respectively, without any leakage. The next step to investigate the modulation operation of

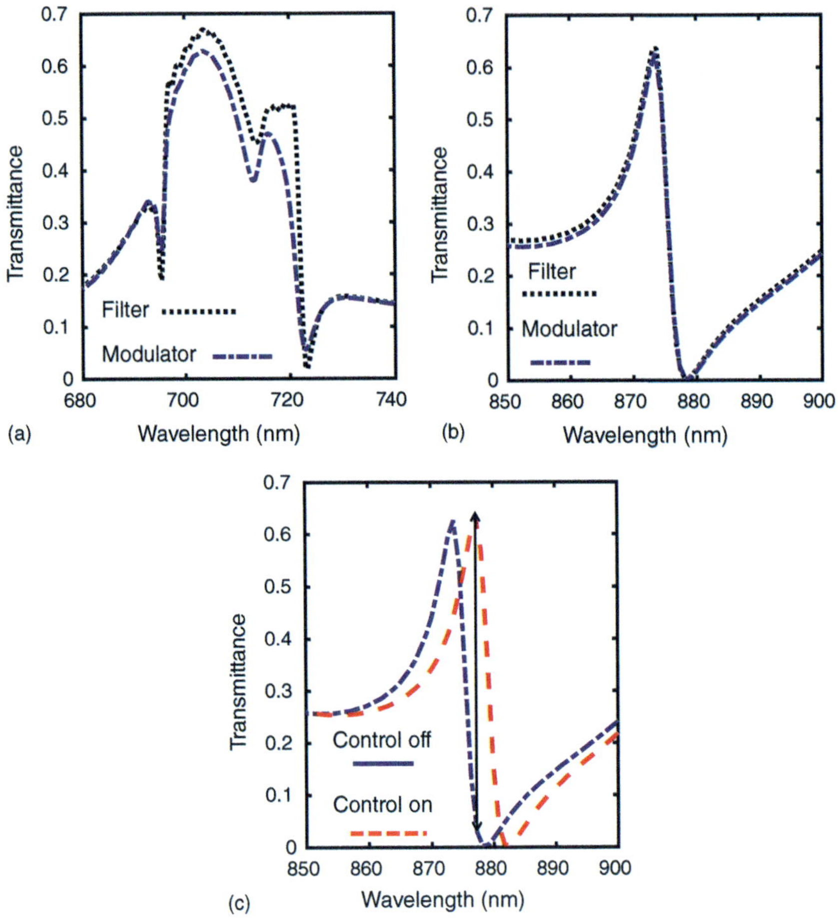

Figure 4.29 Transmission spectra of (a) the proposed modulator (from data input to data output) with only data signal source and basic filter, (b) the proposed modulator (from control input to control output) with only control signal source and basic filter, and (c) the proposed modulator with both signal sources.

the proposed amplitude modulator is that the data and control signal sources be applied, simultaneously. This case is shown in Figure 4.29c. As seen in Figure 4.29c, the transmittance value of the data wavelength (878 nm) modulates from 1.2% to 72.2% by applying the control signal source. It is because the control signal induces the Kerr effect in the nonlinear medium.

The field profile of H_z magnitude is also shown in Figure 4.30 to clarify the operating mechanism of the proposed amplitude modulator. Figure 4.30a shows that the data wavelength cannot transmit to the output port by applying only the data signal source. In addition, as expected, by applying only the control signal source, the control wavelength can propagate through the structure. As discussed, the data wavelength can pass through the structure in the presence of both data and control signal sources. Figure 4.30c shows this case.

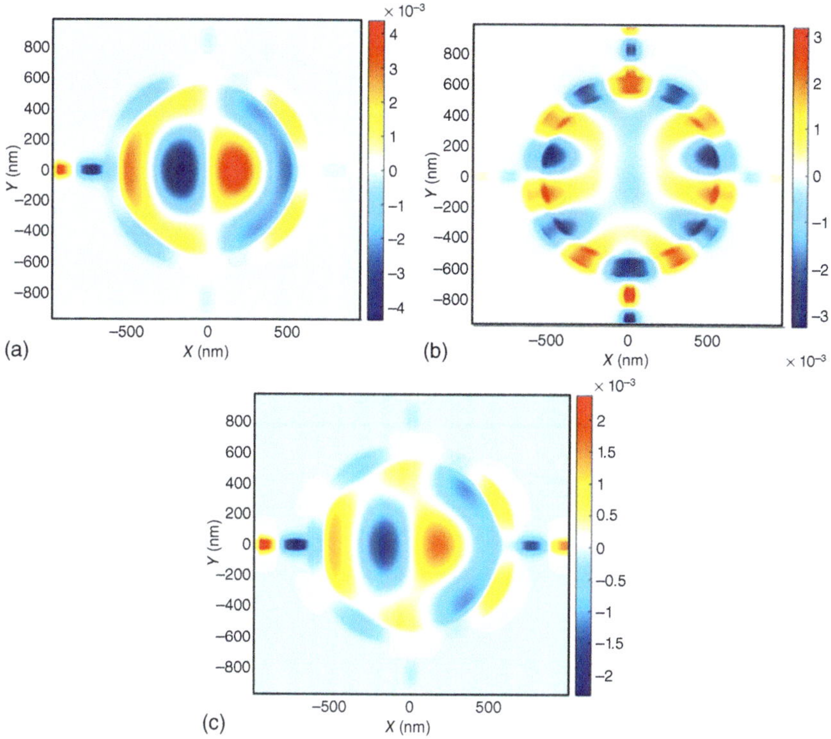

Figure 4.30 Field profile of Re(H_z) at (a) the data wavelength of 878 nm with only the data signal source, (b) the control wavelength of 702 nm with only the control signal source, and (c) the data wavelength of 878 nm with both data and control signal sources.

After designing the proposed all-optical amplitude modulator, to provide a better view of the designed modulator operation, its time-domain behavior is presented. For this purpose, the values of the input power of the data and control signal sources need to be specified. To determine the input power of the data signal source (P_{data}), a continuous wave at the wavelength of 878 nm is applied to data input port. Figure 4.31a shows the normalized transmission of the data signal versus the variations of P_{data} (input power of the data signal source). As seen in Figure 4.31a, when P_{data} increases from 0.195 to 0.78 W μm^{-1}, the maximum normalized transmission value does not change significantly, while by increasing P_{data} higher than 0.78 W μm^{-1}, the maximum normalized transmission value starts to increase. As discussed, the data wavelength cannot propagate through the data output port by applying only the data signal source. As a result, the best value for the input power of the data signal source is equal to 0.78 W μm^{-1}.

To adjust the input power of the control signal source ($P_{control}$), the data and control signal sources are applied to the input ports (data and control input ports) simultaneously. The control signal source is a continuous-wave signal at the wavelength of 702 nm. The variations of the extinction ratio for different values of $P_{control}$ are also

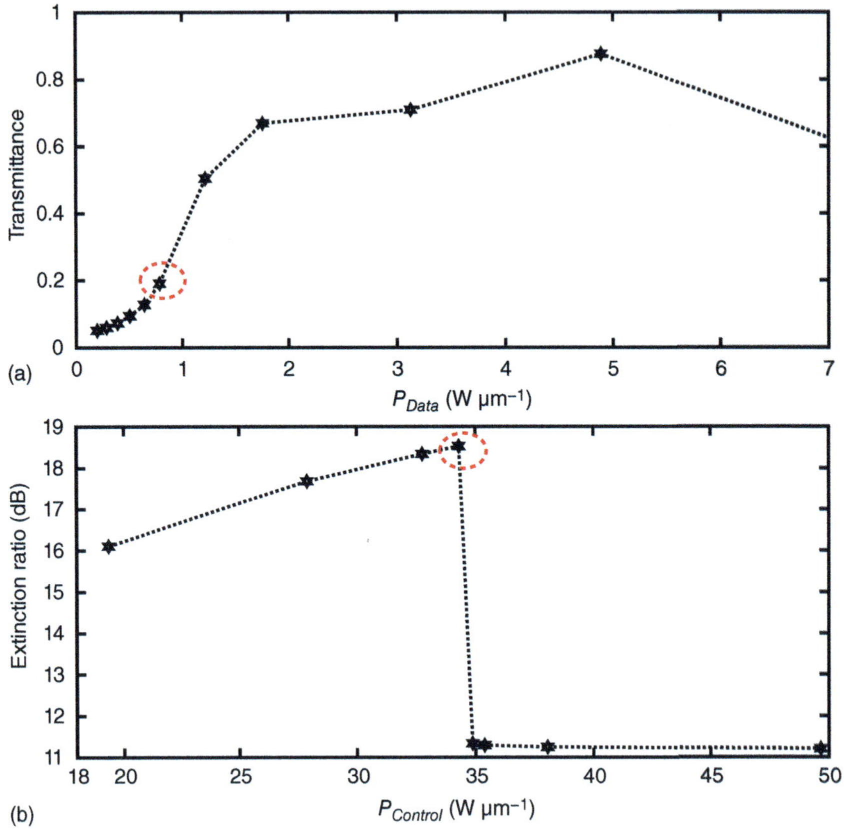

Figure 4.31 (a) The transmittance of the data wavelength for different values of P_{data} and (b) the extinction ratio of the output signal for different values of $P_{control}$.

shown in Figure 4.31b. The extinction ratio of the proposed amplitude modulator can be defined as:

$$ER = 10 \log \frac{T_{on}}{T_{off}} \tag{4.10}$$

where T_{on} and T_{off} are the transmittances in the presence and absence of the control signal source, respectively. As seen in Figure 4.31b, the extinction ratio increases initially and then decreases by increasing $P_{control}$. Therefore, $P_{control} = 34.3\ \text{W}\ \mu\text{m}^{-1}$ for the highest extinction ratio value of 18.53 dB is an appropriate choice for designing the proposed amplitude modulator.

After determining the input power of the data and control signal sources (P_{data} and $P_{control}$), the time-domain behavior of the output signal is studied. The time-domain data and control signal sources inserted into the proposed modulator are shown in Figure 4.32a and b, respectively. Figure 4.32a shows that the data signal source is a continuous-wave signal at the data wavelength of 878 nm with the selected input power of $0.78\ \text{W}\ \mu\text{m}^{-1}$. In addition, the time-domain control signal source is a continuous-wave signal at the control wavelength of 702 nm and the

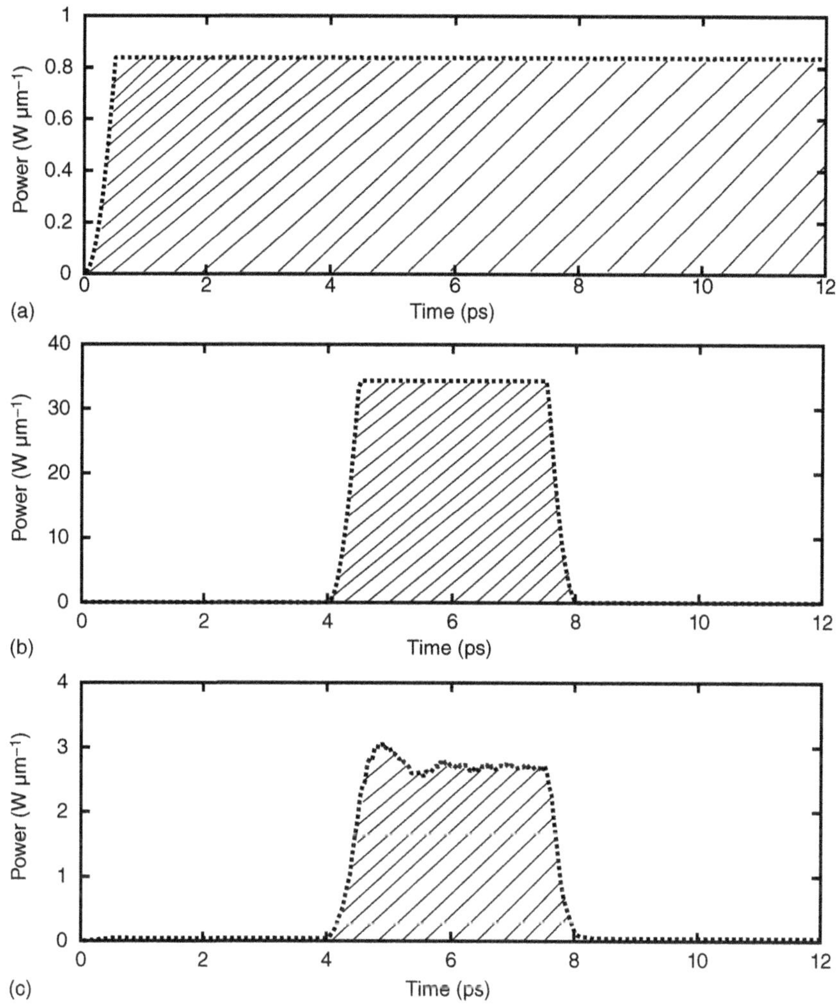

Figure 4.32 Time domain representations of (a) data signal, (b) control signal, and (c) output signal for the proposed all optical amplitude modulator (Figure 4.28).

input power of 34.3 W μm^{-1}. As seen in Figure 4.32b, this signal is "off" from 0 to 4 ps and 8 to 12 ps and it is "on" from 4.5 to 7.5 ps. The transition time of 500 fs is considered for the rising and falling edges.

The time-domain output signal derived from data and control signal sources is shown in Figure 4.32c. As seen in Figure 4.32c, at 0–4 and 8–12 ps (when only the data signal source is "on"), the output power is close to zero. When both data and control sources are "on" (in 4.5–7.5 ps), the amplitude of the data signal increases. In other words, the modulation operation occurs by applying the control signal source.

An important point for designing optical modulators is that the control signal should modulate the data signal and should not leak to the output port. For this purpose, the Fourier transform of the time-domain output signal is also obtained

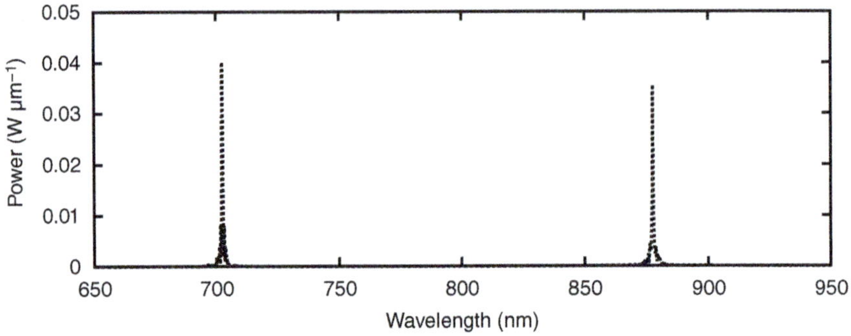

Figure 4.33 Fourier transform of the time-domain output signal for the proposed all-optical amplitude modulator (Figure 4.28).

(Figure 4.33). As seen in Figure 4.33, the control wavelength of 702 nm leaks to the output port. Consequently, the control wavelength should be filtered at the output port.

Figure 4.34 shows the designed suppression filter and its transmission spectrum. This filter is composed of five unequal stubs coupled to a MIM plasmonic waveguide. The structural parameters of the proposed suppression filter are as follows: $a_1 = 100$, $a_2 = 97$, $a_3 = 94$, $a_4 = 91$, $a_5 = 88$, $w = 60$, and $g = 100$ (all in nm).

The transmission spectrum of the suppression filter (Figure 4.34b) shows that the data wavelength can pass through the designed filter, while the control wavelength is blocked to the output port. As a result, by coupling the suppression filter at the data output port of the proposed modulator, the control wavelength cannot propagate at the output port. The operating mechanism of the suppression filter can also be explained by presenting the field profile. Figure 4.35 shows the field profile of $\mathrm{Re}(H_z)$ at both wavelengths of 702 and 878 nm. As seen, there is no light transmission between the input and output ports at the wavelength of 702 nm, while the incident light at the wavelength of 878 nm can pass through the plasmonic suppression filter.

By the combination of the designed all-optical amplitude modulator (Figure 4.28) and suppression filter (Figure 4.34a), the improved all-optical amplitude modulator is formed. Figure 4.36 shows this structure.

Here, the time-domain behavior of the improved modulator is also investigated. The continuous-wave signals for the data and control signal sources are shown in Figure 4.36a and b, respectively. As seen, such signals are similar to Figure 4.32a,b.

Figure 4.36c shows the output signal derived from these two input signals. As seen in Figure 4.36c, the amplitude modulation with a lower output power value also occurs in this case. It is because of the existence of the suppression filter at the output port.

To better analyze the results, the Fourier transform of the output time-domain signal of Figure 4.37c is presented in Figure 4.38. As seen, after adding the suppression filter, which passes the data wavelength and prohibits the propagation of the control wavelength, only the data wavelength can transmit to the output port. As a result, the mentioned problem is solved using this method (coupling the suppression filter and all-optical modulator).

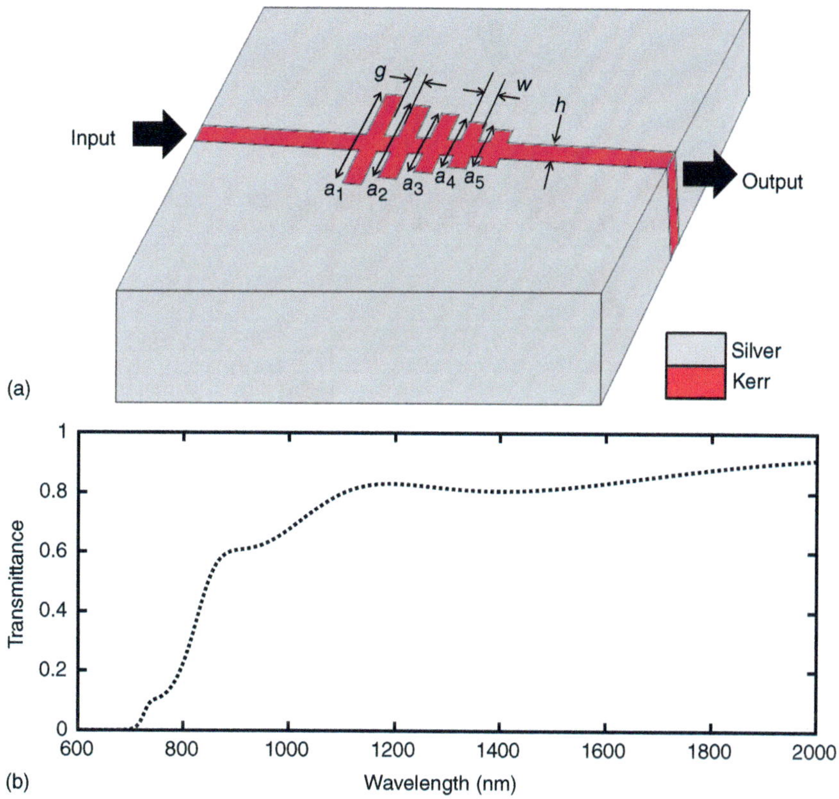

Figure 4.34 (a) Proposed structure of the suppression filter and (b) its transmission spectrum.

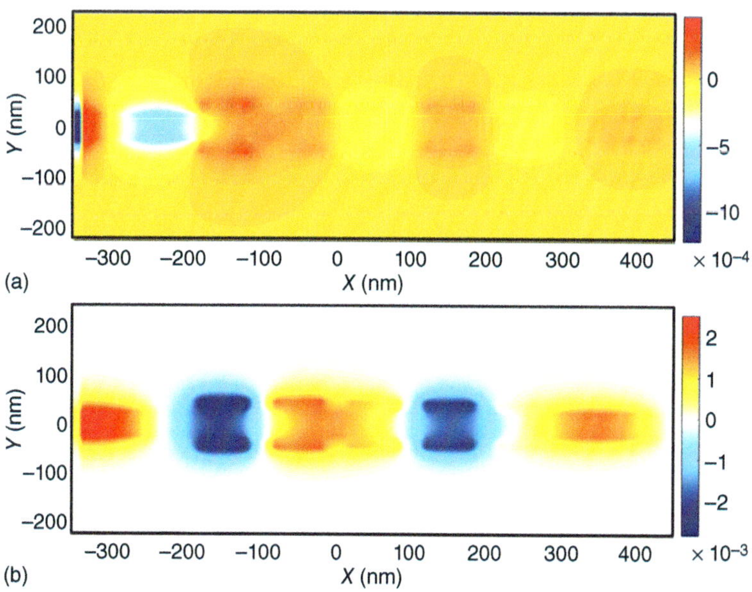

Figure 4.35 Field profile of Re(H_z) at the wavelengths of (a) 702 nm and (b) 878 nm.

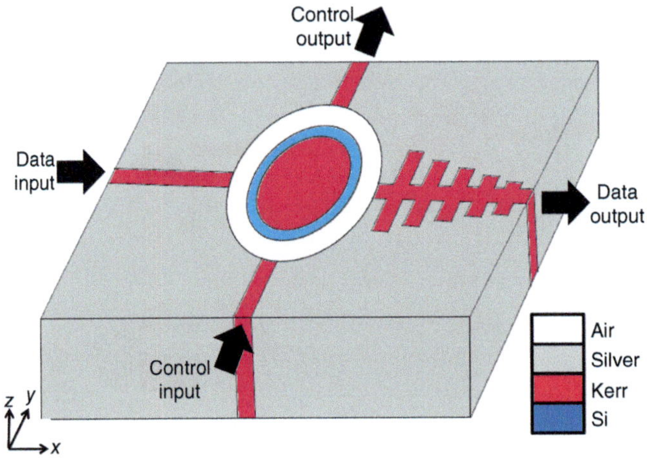

Figure 4.36 Proposed structure of the improved all-optical amplitude modulator.

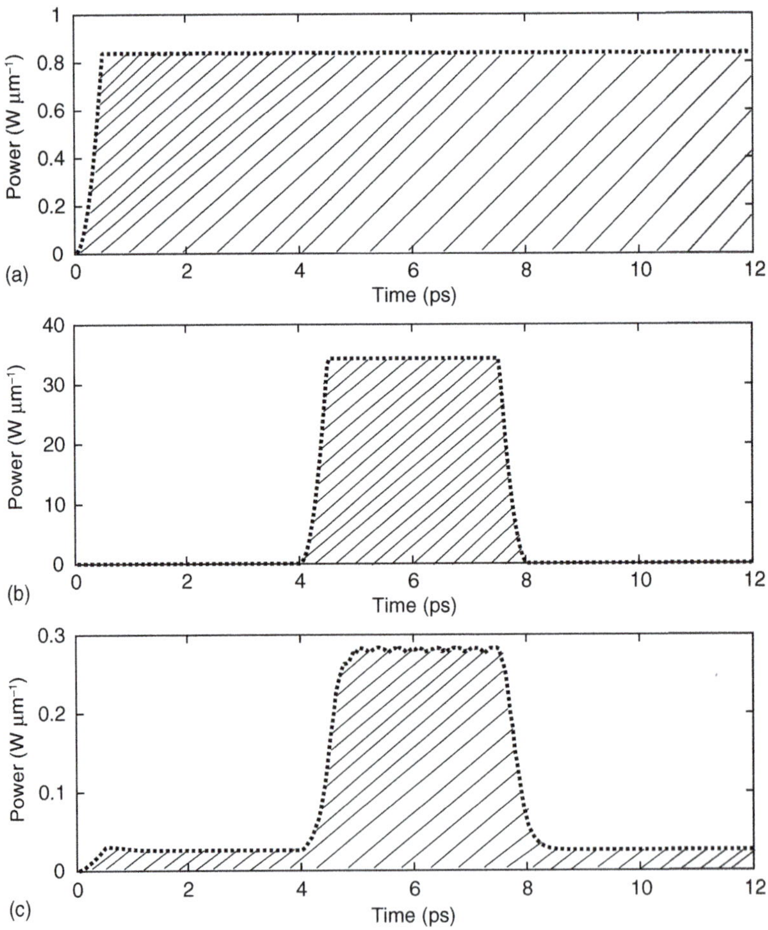

Figure 4.37 Time-domain representations of (a) data signal, (b) control signal, and (c) output signal for the proposed improved all-optical amplitude modulator (Figure 4.36).

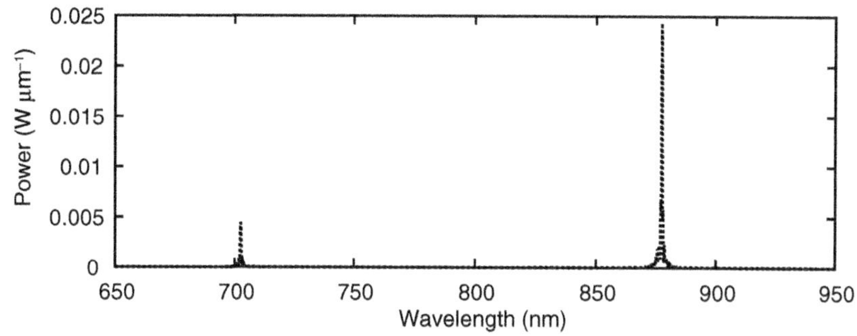

Figure 4.38 Fourier transform of the time-domain output signal for the proposed improved all-optical amplitude modulator (Figure 4.37).

Table 4.3 Summary of optical structures performance.

References	Setup	Topology	Mechanism operation	2D/3D	Isolated waveguide	Time simulation
[99]	Graphene and plasmonic	MZI	All-optical switch	3D	Yes	Yes
[101]	Plasmonic and PC	Cross waveguides	All-optical switch	2D	Yes	Yes
[110]	Plasmonic	Disk resonator and cross waveguides	All-optical amplitude modulator	2D	Yes	Yes

4.5 Summary

Since the discovery of optical structures and initial studies on the optical switches for optical sensing, there have been a lot of publications in the literature on this topic, far more than those cited here. Optical switches can be good candidates for optical sensing applications. In this chapter, some of the publications have been studied. Finally, the main properties of the discussed structures in Sections (4.2–4.4) are given in Table 4.3. Such features are the setup and topology of the optical devices, their mechanism operation, dimensions (2D/3D), and isolated waveguides. Also, whether the time-domain simulations are performed or not has been investigated. In summary, based on the mentioned properties, these structures have the potential to be used as optical sensors.

References

1 Miller, K.J., Hallman, K.A., Haglund, R.F., and Weiss, S.M. (2017). Silicon waveguide optical switch with embedded phase change material. *Opt. Express* 25 (22): 26527–26536.

2 Khani, S., Danaie, M., and Rezaei, P. (2021). Plasmonic all-optical metal–insulator–metal switches based on silver nano-rods, comprehensive theoretical analysis and design guidelines. *J. Comput. Electron.* 20 (1): 442–457.

3 Sani, M.H., Saghaei, H., Mehranpour, M.A., and Tabrizi, A.A. (2020). A novel all-optical sensor design based on a tunable resonant nanocavity in photonic crystal microstructure applicable in MEMS accelerometers. *Photonic Sensors* 11 (4): 1–15.

4 Shenai, K. (2018). The figure of merit of a semiconductor power electronics switch. *IEEE Trans. Electron Devices* 65 (10): 4216–4224.

5 Picard, J.F., Schaub, S.C., Rosenzweig, G. et al. (2019). Laser-driven semiconductor switch for generating nanosecond pulses from a megawatt gyrotron. *Appl. Phys. Lett.* 114 (16): 164102.

6 Danaie, M., Nasirifar, R., and Dideban, A. (2017). Design of adjustable T-shaped and Y-shaped photonic crystal power splitters for TM and TE polarizations. *Turk. J. Electr. Eng. Comput. Sci.* 25 (5): 4398–4408.

7 Farmani, A., Mir, A., and Irannejad, M. (2019). 2D-FDTD simulation of ultra-compact multifunctional logic gates with nonlinear photonic crystal. *J. Opt. Soc. Am. B: Opt. Phys.* 36 (4): 811–818.

8 Tavousi, A., Rakhshani, M.R., and Mansouri-Birjandi, M.A. (2018). High sensitivity label-free refractometer based biosensor applicable to glycated hemoglobin detection in human blood using all-circular photonic crystal ring resonators. *Opt. Commun.* 429: 166–174.

9 Foroughifar, A., Saghaei, H., and Veisi, E. (2021). Design and analysis of a novel four-channel optical filter using ring resonators and line defects in photonic crystal microstructure. *Opt. Quantum Electron.* 53 (2): 1–12.

10 Rakhshani, M.R. and Mansouri-Birjandi, M.A. (2013). Realization of tunable optical filter by photonic crystal ring resonators. *Optik* 124 (22): 5377–5380.

11 Rahman-Zadeh, F., Danaie, M., and Kaatuzian, H. (2019). Design of a highly sensitive photonic crystal refractive index sensor incorporating ring-shaped GaAs cavity. *Opto-Electron. Rev.* 27 (4): 369–377.

12 Aly, A.H., Zaky, Z.A., Shalaby, A.S. et al. (2020). Theoretical study of hybrid multifunctional one-dimensional photonic crystal as a flexible blood sugar sensor. *Phys. Scr.* 95 (3): 035510.

13 Okamoto, K., Tsuruda, K., Diebold, S. et al. (2017). Terahertz sensor using photonic crystal cavity and resonant tunneling diodes. *J. Infrared, Millimeter, Terahertz Waves* 38 (9): 1085–1097.

14 Danaie, M., Nasiri Far, R., and Dideban, A. (2018). Design of a high-bandwidth Y-shaped photonic crystal power splitter for TE modes. *Int. J. Opt. Photonics* 12 (1): 33–42.

15 Xu, L., Wang, Y., El-Fiky, E. et al. (2019). Compact broadband polarization beam splitter based on multimode interference coupler with internal photonic crystal for the SOI platform. *J. Lightwave Technol.* 37 (4): 1231–1240.

16 Kondo, K., Tatebe, T., Hachuda, S. et al. (2017). Fan-beam steering device using a photonic crystal slow-light waveguide with surface diffraction grating. *Opt. Lett.* 42 (23): 4990–4993.

17 Mohammadi, M., Seifouri, M., Olyaee, S., and Karamirad, M. (2021). Optimization and realization all-optical compact five-channel demultiplexer using 2D photonic crystal based hexagonal cavities. *J. Comput. Electron.* 20 (2): 984–992.

18 Tavousi, A. (2019). Wavelength-division demultiplexer based on hetero-structure octagonal-shape photonic crystal ring resonators. *Optik* 179: 1169–1179.

19 Ghomashi, M., Kaatuzian, H., and Danaie, M. (2016). Design and simulation of normally open and normally closed all-optical switches based on photonic crystal triple-waveguide directional coupler. *Opt. Quantum Electron.* 48 (1): 35.

20 Shiramin, L.A., Xie, W., Snyder, B. et al. (2017). High extinction ratio hybrid graphene-silicon photonic crystal switch. *IEEE Photonics Technol. Lett.* 30 (2): 157–160.

21 Danaie, M. and Kaatuzian, H. (2011). Improvement of power coupling in a nonlinear photonic crystal directional coupler switch. *Photonics Nanostruct. Fundam. Appl.* 9 (1): 70–81.

22 Moradi, M., Mohammadi, M., Olyaee, S., and Seifouri, M. (2021). Design and simulation of a fast all-optical modulator based on photonic crystal using ring resonators. *Silicon* 14: 1–7.

23 Furukado, Y., Abe, H., Hinakura, Y., and Baba, T. (2018). Experimental simulation of ranging action using Si photonic crystal modulator and optical antenna. *Opt. Express* 26 (14): 18222–18229.

24 Salimzadeh, A. and Alipour-Banaei, H. (2018). An all optical 8 to 3 encoder based on photonic crystal OR-gate ring resonators. *Opt. Commun.* 410: 793–798.

25 Moradi, M., Danaie, M., and Orouji, A.A. (2019). Design of all-optical XOR and XNOR logic gates based on Fano resonance in plasmonic ring resonators. *Opt. Quantum Electron.* 51 (5): 1–18.

26 Rezaee, S., Zavvari, M., and Alipour-Banaei, H. (2015). A novel optical filter based on H-shape photonic crystal ring resonators. *Optik* 126 (20): 2535–2538.

27 Scholz, S., Hess, O., and Rühle, R. (1998). Dynamic cross-waveguide optical switching with a nonlinear photonic band-gap structure. *Opt. Express* 3 (1): 28–34.

28 Liu, Z., Zhou, J., and Li, B. (2013). Photonic band gap effect on energy transfer in Tb^{3+}, Eu^{3+} CO-doped TiO_2 inverse opal films. *Mod. Phys. Lett. B* 27 (14): 1350099.

29 Khani, S., Danaie, M., and Rezaei, P. (2019). All-optical plasmonic switches based on asymmetric directional couplers incorporating Bragg gratings. *Plasmonics* 15: 1–11.

30 Mansuri, M., Mir, A., and Farmani, A. (2019). Numerical modeling of a nanostructure gas sensor based on plasmonic effect. *J. Optoelectron. Nanostruct.* 4 (2): 29–44.

31 Khani, S., Danaie, M., and Rezaei, P. (2018). Realization of single-mode plasmonic bandpass filters using improved nanodisk resonators. *Opt. Commun.* 420: 147–156.

32 Ghahari, F., Walkup, D., Gutiérrez, C. et al. (2017). An on/off Berry phase switch in circular graphene resonators. *Science* 356 (6340): 845–849.

33 Farmani, H. and Farmani, A. (2020). Graphene sensing nanostructure for exact graphene layers identification at terahertz frequency. *Physica E* 124: 114375.

34 Barnes, W.L., Dereux, A., and Ebbesen, T.W. (2003). Surface plasmon subwavelength optics. *Nature* 424 (6950): 824–830.

35 Bozhevolnyi, S.I., Volkov, V.S., Devaux, E. et al. (2006). Channel plasmon subwavelength waveguide components including interferometers and ring resonators. *Nature* 440 (7083): 508–511.

36 Gramotnev, D.K. and Bozhevolnyi, S.I. (2010). Plasmonics beyond the diffraction limit. *Nat. Photonics* 4 (2): 83–91.

37 Chen, Z. and Yu, L. (2014). Multiple Fano resonances based on different waveguide modes in a symmetry breaking plasmonic system. *IEEE Photonics J.* 6 (6): 1–8.

38 Khani, S., Danaie, M., and Rezaei, P. (2019). Tunable single-mode bandpass filter based on metal–insulator–metal plasmonic coupled U-shaped cavities. *IET Optoelectron* 13 (4): 161–171.

39 Kokabi, M., Ghorbani, S., and Moayed, S.H. (2021). Tunable bandpass plasmonic filter based on graphene as the nonlinear Kerr material. *Laser Phys.* 31 (2): 026201.

40 Pathiranage, S.P.Y., Gunapala, S.D., and Premaratne, M. (2021). Tunable plasmonic resonator using conductivity modulated Bragg reflectors. *J. Phys. Condens. Matter* 33: 245301.

41 Zhuang, H., Zhuang, J., Kong, F., and Li, K. (2020). Tunable Bragg reflector with parallel bulk Dirac semimetals at terahertz frequencies. *J. Mod. Opt.* 67 (11): 1010–1016.

42 Mansouri, M., Mir, A., Farmani, A., and Izadi, M. (2021). Numerical modeling of an integrable and tunable plasmonic pressure sensor with nanostructure grating. *Plasmonics* 16 (1): 27–36.

43 Khani, S. and Hayati, M. (2021). An ultra-high sensitive plasmonic refractive index sensor using an elliptical resonator and MIM waveguide. *Superlattices Microstruct.* 156: 106970.

44 Rakhshani, M.R. (2020). Optical refractive index sensor with two plasmonic double-square resonators for simultaneous sensing of human blood groups. *Photonics Nanostruct. Fundam. Appl.* 39: 100768.

45 Khani, S., Danaie, M., and Rezaei, P. (2018). Double and triple-wavelength plasmonic demultiplexers based on improved circular nanodisk resonators. *Opt. Eng.* 57 (10): 107102.

46 Khani, S., Farmani, A., and Mir, A. (2021). Reconfigurable and scalable 2, 4-and 6-channel plasmonics demultiplexer utilizing symmetrical rectangular resonators containing silver nano-rod defects with FDTD method. *Sci. Rep.* 11 (1): 1–13.

47 Ghosh, S. and Rahman, B.M.A. (2019). Design of on-chip hybrid plasmonic Mach–Zehnder interferometer for temperature and concentration detection of chemical solution. *Sens. Actuators, B* 279: 490–502.

48 Khani, S., Danaie, M., and Rezaei, P. (2020). Compact and low-power all-optical surface plasmon switches with isolated pump and data waveguides and a rectangular cavity containing nano-silver strips. *Superlattices Microstruct.* 141: 106481.

49 Farmani, A., Zarifkar, A., Sheikhi, M.H., and Miri, M. (2017). Design of a tunable graphene plasmonic-on-white graphene switch at infrared range. *Superlattices Microstruct.* 112: 404–414.

50 Khani, S., Danaie, M., and Rezaei, P. (2020). Realization of a plasmonic optical switch using improved nano-disk resonators with Kerr-type nonlinearity: a theoretical and numerical study on challenges and solutions. *Opt. Commun.* 477: 126359.

51 Negahdari, R., Rafiee, E., and Emami, F. (2019). Realization of all-optical plasmonic MIM split square ring resonator switch. *Opt. Quantum Electron.* 51 (7): 1–14.

52 Khani, S. and Hayati, M. (2017). Compact microstrip lowpass filter with wide stopband and sharp roll-off. *Microwave J.* 60 (11): 86–92.

53 Valizade, A., Rezaei, P., and Orouji, A.A. (2015). Design of reconfigurable active integrated microstrip antenna with switchable low-noise amplifier/power amplifier performances for wireless local area network and WiMAX applications. *IET Microwaves Antennas Propag.* 9 (9): 872–881.

54 Khani, S., Danaie, M., Rezaei, P., and Shahzadi, A. (2020). Compact ultra-wide upper stopband microstrip dual-band BPF using tapered and octagonal loop resonators. *Frequenz* 74 (1–2): 61–71.

55 Mousavi, S., Makki, S., and Siahkamari, H. (2016). Design of microstrip lowpass filter using bend configuration with excellent sharpness in transition band. *Frequenz* 70 (5-6): 237–243.

56 Pitarke, J.M., Silkin, V.M., Chulkov, E.V., and Echenique, P.M. (2006). Theory of surface plasmons and surface-plasmon polaritons. *Rep. Prog. Phys.* 70 (1): 1.

57 Jablan, M., Buljan, H., and Soljačić, M. (2009). Plasmonics in graphene at infrared frequencies. *Phys. Rev. B: Condens. Matter* 80 (24): 245435.

58 Li, S., Liu, H., Sun, Q., and Huang, N. (2016). Multi-channel terahertz wavelength division demultiplexer with defects-coupled photonic crystal waveguide. *J. Mod. Opt.* 63 (10): 955–960.

59 Zhang, L., Tang, L., Wei, W. et al. (2016). Enhanced near-infrared absorption in graphene with multilayer metal-dielectric-metal nanostructure. *Opt. Express* 24 (18): 20002–20009.

60 Dai, D., Wang, J., Chen, S. et al. (2015). Monolithically integrated 64-channel silicon hybrid demultiplexer enabling simultaneous wavelength-and mode-division-multiplexing. *Laser Photonics Rev.* 9 (3): 339–344.

61 Hamouleh-Alipour, A., Mir, A., and Farmani, A. (2020). Analytical modeling and design of a graphene metasurface sensor for thermo-optical detection of terahertz plasmons. *IEEE Sens. J.* 21: 4525–4532.

62 Milaninia, K.M., Baldo, M.A., Reina, A., and Kong, J. (2009). All graphene electromechanical switch fabricated by chemical vapor deposition. *Appl. Phys. Lett.* 95 (18): 183105.

63 Novoselov, K.S., Geim, A.K., Morozov, S.V. et al. (2005). Two-dimensional gas of massless Dirac fermions in graphene. *Nature* 438 (7065): 197–200.

64 Zhang, Y., Tan, Y.W., Stormer, H.L., and Kim, P. (2005). Experimental observation of the quantum Hall effect and Berry's phase in graphene. *Nature* 438 (7065): 201–204.

65 Koppens, F.H., Chang, D.E., and Garcia de Abajo, F.J. (2011). Graphene plasmonics: a platform for strong light–matter interactions. *Nano Lett.* 11 (8): 3370–3377.

66 Vakil, A. and Engheta, N. (2011). Transformation optics using graphene. *Science* 332 (6035): 1291–1294.

67 Aghaee, T. and Orouji, A.A. (2020). Reconfigurable multi-band, graphene-based THz absorber: circuit model approach. *Results Phys.* 16: 102855.

68 Hamzavi-Zarghani, Z., Yahaghi, A., Matekovits, L., and Farmani, A. (2019). Tunable mantle cloaking utilizing graphene metasurface for terahertz sensing applications. *Opt. Express* 27 (24): 34824–34837.

69 Aghaee, T. and Orouji, A.A. (2021). Dual-band terahertz absorber based on graphene periodic arrays of disks and ribbons: circuit model approach. *J. Comput. Electron.* 20 (1): 611–625.

70 Khosravian, E., Mashayekhi, H.R., and Farmani, A. (2020). Tunable plasmonics photodetector in near-infrared wavelengths using graphene chemical doping method. *AEU Int. J. Electron. Commun.* 127: 153472.

71 Tian, X., Luo, H., Wei, R. et al. (2018). An ultrabroadband mid-infrared pulsed optical switch employing solution-processed bismuth oxyselenide. *Adv. Mater.* 30 (31): 1801021.

72 Monfared, S.A., Seifouri, M., Hamidi, S.M., and Mohseni, S.M. (2021). Electro-optical switch based on one-dimensional graphene-plasmonic crystals. *Opt. Mater.* 115: 111051.

73 Sahoo, P.K., Sarkar, S., and Joseph, J. (2017). High sensitivity guided-mode-resonance optical sensor employing phase detection. *Sci. Rep.* 7 (1): 1–7.

74 Ismail, M., Khan, M.I., Akhtar, K. et al. (2018). Biosynthesis of silver nanoparticles: a colorimetric optical sensor for detection of hexavalent chromium and ammonia in aqueous solution. *Physica E* 103: 367–376.

75 Jiang, W.S., Xin, W., Xun, S. et al. (2017). Reduced graphene oxide-based optical sensor for detecting specific protein. *Sens. Actuators, B* 249: 142–148.

76 Rajasekar, R., Parameshwari, K., and Robinson, S. (2019). Nano-optical switch based on photonic crystal ring resonator. *Plasmonics* 14 (6): 1687–1697.

77 Ahmed, A.M. and Mehaney, A. (2019). Ultra-high sensitive 1D porous silicon photonic crystal sensor based on the coupling of Tamm/Fano resonances in the mid-infrared region. *Sci. Rep.* 9 (1): 1–9.

78 Nozhat, N. and Granpayeh, N. (2014). All-optical nonlinear plasmonic ring resonator switches. *J. Mod. Opt.* 61 (20): 1690–1695.

79 Khodadadi, M., Moshiri, S.M.M., and Nozhat, N. (2020). Theoretical analysis of a simultaneous graphene-based circular plasmonic refractive index and thickness bio-sensor. *IEEE Sens. J.* 20 (16): 9114–9123.

80 Liang, C., Niu, G., Chen, X. et al. (2019). Tunable triple-band graphene refractive index sensor with good angle-polarization tolerance. *Opt. Commun.* 436: 57–62.

81 Nejat, M. and Nozhat, N. (2020). Sensing and switching capabilities of a graphene-based perfect dual-band metamaterial absorber with analytical methods. *J. Opt. Soc. Am. B: Opt. Phys.* 37 (5): 1359–1366.

82 Li, J., Tao, J., Chen, Z.H., and Huang, X.G. (2016). All-optical controlling based on nonlinear graphene plasmonic waveguides. *Opt. Express* 24 (19): 22169–22176.

83 Luo, L., Wang, K., Ge, C. et al. (2017). Actively controllable terahertz switches with graphene-based nongroove gratings. *Photonics Res.* 5 (6): 604–611.

84 Zhang, Z., Yang, J., He, X. et al. (2018). All-optical multi-channel switching at telecommunication wavelengths based on tunable plasmon-induced transparency. *Opt. Commun.* 425: 196–203.

85 Wang, M., Zhang, M., Wang, Y. et al. (2019). Fano resonance in an asymmetric MIM waveguide structure and its application in a refractive index nanosensor. *Sensors* 19 (4): 791.

86 Zhang, Y., Kuang, Y., Zhang, Z. et al. (2019). High-sensitivity refractive index sensors based on Fano resonance in the plasmonic system of splitting ring cavity-coupled MIM waveguide with tooth cavity. *Appl. Phys. A* 125 (1): 1–5.

87 Shi, X., Ma, L., Zhang, Z. et al. (2018). Dual Fano resonance control and refractive index sensors based on a plasmonic waveguide-coupled resonator system. *Opt. Commun.* 427: 326–330.

88 Shahamat, Y., Ghaffarinejad, A., and Vahedi, M. (2020). Plasmon induced transparency and refractive index sensing in two nanocavities and double nanodisk resonators. *Optik* 202: 163618.

89 Farmani, A. (2019). Three-dimensional FDTD analysis of a nanostructured plasmonic sensor in the near-infrared range. *J. Opt. Soc. Am. B: Opt. Phys.* 36 (2): 401–407.

90 Bian, Z.Y., Liang, R.S., Zhang, Y.J. et al. (2015). Multifunctional disk device for optical switch and temperature sensor. *Chin. Phys. B* 24 (10): 107801.

91 Bazgir, M., Jalalpour, M., Zarrabi, F.B., and Arezoomand, A.S. (2020). Design of an optical switch and sensor based on a MIM coupled waveguide using a DNA composite. *J. Electron. Mater.* 49 (3): 2173–2178.

92 Shahamat, Y. and Vahedi, M. (2019). Mid-infrared plasmonically induced absorption and transparency in a Si-based structure for temperature sensing and switching applications. *Opt. Commun.* 430: 227–233.

93 Jiang, X., Chen, D., Zhang, Z. et al. (2020). Dual-channel optical switch, refractive index sensor and slow light device based on a graphene metasurface. *Opt. Express* 28 (23): 34079–34092.

94 Ghafari, S., Forouzeshfard, M.R., and Vafapour, Z. (2019). Thermo optical switching and sensing applications of an infrared metamaterial. *IEEE Sens. J.* 20 (6): 3235–3241.

95 Shahamat, Y. and Vahedi, M. (2017). Pump-tuned plasmon-induced transparency for sensing and switching applications. *Opt. Commun.* 401: 40–45.

96 Poorgholam-Khanjari, S., Razavi, Z., and Zarrabi, F.B. (2021). Reconfigurable optical rectangular particle array absorber based on metal–DNA–metal structure as a refractive index sensor and optical switch. *Opt. Commun.* 489: 126866.

97 Drachev, V.P., Chettiar, U.K., Kildishev, A.V. et al. (2008). The Ag dielectric function in plasmonic metamaterials. *Opt. Express* 16 (2): 1186–1195.

98 Falkovsky, L.A. (2008). Optical properties of graphene. *J. Phys. Conf. Ser.* 129 (1): 012004. IOP Publishing.

99 Armaghani, S., Khani, S., and Danaie, M. (2019). Design of all-optical graphene switches based on a Mach–Zehnder interferometer employing optical Kerr effect. *Superlattices Microstruct.* 135: 106244.

100 Khani, S., Danaie, M., and Rezaei, P. (2019). Design of a single-mode plasmonic bandpass filter using a hexagonal resonator coupled to graded-stub waveguides. *Plasmonics* 14 (1): 53–62.

101 Khani, S., Danaie, M., and Rezaei, P. (2020). Hybrid all-optical infrared metal–insulator–metal plasmonic switch incorporating photonic crystal bandgap structures. *Photonics Nanostruct. Fundam. Appl.* 40: 100802.

102 Ahmed, I., Khoo, E.H., Kurniawan, O., and Li, E.P. (2011). Modeling and simulation of active plasmonics with the FDTD method by using solid state and Lorentz–Drude dispersive model. *J. Opt. Soc. Am. B: Opt. Phys.* 28 (3): 352–359.

103 Haffner, C., Chelladurai, D., Fedoryshyn, Y. et al. (2018). Low-loss plasmon-assisted electro-optic modulator. *Nature* 556 (7702): 483–486.

104 Bhasker, P., Norman, J., Bowers, J., and Dagli, N. (2020). Low voltage, high optical power handling capable, bulk compound semiconductor electro-optic modulators at 1550 nm. *J. Lightwave Technol.* 38 (8): 2308–2314.

105 Li, F., Tang, T., Li, J. et al. (2020). Chiral coding metasurfaces with integrated vanadium dioxide for thermo-optic modulation of terahertz waves. *J. Alloys Compd.* 826: 154174.

106 Liao, C., Li, C., Wang, C. et al. (2019). High-speed all-optical modulator based on a polymer nanofiber Bragg grating printed by femtosecond laser. *ACS Appl. Mater. Interfaces* 12 (1): 1465–1473.

107 Gardes, F.Y., Brimont, A., Sanchis, P. et al. (2009). High-speed modulation of a compact silicon ring resonator based on a reverse-biased pn diode. *Opt. Express* 17 (24): 21986–21991.

108 Moradiani, F., Seifouri, M., Abedi, K., and Gharakhili, F.G. (2021). High extinction ratio all-optical modulator using a vanadium-dioxide integrated hybrid plasmonic waveguide. *Plasmonics* 16 (1): 189–198.

109 Sun, F., Xia, L., Nie, C. et al. (2019). An all-optical modulator based on a graphene–plasmonic slot waveguide at 1550 nm. *Appl. Phys Express* 12 (4): 042009.

110 Khani, S., Danaie, M., and Rezaei, P. (2021). Fano resonance using surface plasmon polaritons in a nano-disk resonator coupled to perpendicular waveguides for amplitude modulation applications. *Plasmonics* 16: 1–18.

5

Optical Interferometric Sensing

Hailong Wang[1,2] and Jietai Jing[1]

[1] *State Key Laboratory of Precision Spectroscopy, Joint Institute of Advanced Science and Technology, School of Physics and Electronic Science, East China Normal University, Shanghai 200062, China*
[2] *College of Optical and Electronic Technology, China Jiliang University, Hangzhou 310018, China*

5.1 Introduction

Optical interferometer is the basis for optical interferometric sensing because different kinds of sensors, such as temperature, strain, refractive index, and humidity sensors, can be prepared based on the optical interference principle. In general, traditional optical interferometers such as Mach–Zehnder interferometer (MZI) consist of two linear beam splitters (BSs), with one acting as wave splitter and the other as wave combiner, while their corresponding phase sensitivity is always bounded by standard quantum limit (SQL) due to Heisenberg uncertainty principle. In order to surpass this limit, in 1986, Yurke et al. proposed a so-called SU(1,1) interferometer, which can be constructed by replacing the linear beam splitters in MZI with nonlinear parametric processes, and compared with MZI, its corresponding phase sensitivity can reach the ultimate quantum limit of phase measurement, that is Heisenberg limit (HL) [1]. Due to the fact that maximum fringe intensity of the output beams in SU(1,1) interferometer depends quadratically on the intensity of the phase-sensing field, an SU(1,1) interferometer can also be known as nonlinear interferometer. The proposal of nonlinear interferometer has attracted extensive attention; e.g. Plick et al. proposed the injection of a coherent state to boost the sensitivity of nonlinear interferometer [2].

More than 20 years later, in 2011, Jing et al. experimentally realized this nonlinear interferometer with four-wave mixing (FWM) processes acting as nonlinear beam splitters to split and recombine an incoming optical field [3]. Amplification process is actively involved in the interferometer, leading to enhanced sensitivity for phase measurement compared to MZI. Benefiting from the major advantages of nonlinear interferometer, it can realize phase-sensitivity enhancement, entanglement enhancement, quantum-noise cancellation (QNC), high-sensitivity interferometric surface–plasmon–resonance (SPR) sensing, and so on. These facts indicate a nonlinear interferometer can find potential applications in quantum metrology.

Optical Imaging and Sensing: Materials, Devices, and Applications, First Edition.
Edited by Jiang Wu and Hao Xu.

To give the comprehensive descriptions about nonlinear interferometric sensing, this chapter is organized as follows: in Section 5.2, we will discuss the above nonlinear interferometer, including experimental implementation of phase-locking, enhancement of phase sensitivity, experimental realization of entanglement enhancement and QNC; in Section 5.3, we will describe other types of nonlinear interferometers, including a nonlinear Sagnac interferometer (NSI), a hybrid interferometer consisting of a nonlinear FWM process and a linear beam splitter, a phase-sensitive FWM process acting as a nonlinear beam splitter, and interference-induced quantum-squeezing enhancement so on; in Section 5.4, a nonlinear interferometric SPR sensor has been theoretically proposed and demonstrated to have its sensing advantages by using sensing parameters such as degree of intensity-difference squeezing (IDS), estimation precision, and signal–noise ratio; in Section 5.5, we will give a summary and outlook of this chapter.

5.2 Nonlinear Interferometer

As discussed above, nonlinear interferometer can be constructed by replacing the linear beam splitters in MZI with nonlinear FWM processes; its corresponding conceptual sketch is shown in Figure 5.1. As shown in Figure 5.1, nonlinear interferometer consists of two FWM processes. The first FWM (FWM_1) process amplifies the input "signal" beam and meanwhile produces a conjugate "idler" beam. Thus, the first FWM serves as an equivalent beam splitter and splits the input beam into two. The second FWM (FWM_2) process serves as a beam combiner, which mixes the amplified signal beam and the idler beam. The action of a FWM process is well written by

$$A_s^{out} = GA_s^{in} + gA_i^{in*}; \quad A_i^{out} = GA_i^{in} + gA_s^{in*} \tag{5.1}$$

where s and i stand for signal and idler beams, respectively. G is the amplitude gain of the FWM process and $G^2 - g^2 = 1$.

Therefore, if there is no idler beam input at FWM_1, the beams after FWM_1 are

$$A_{s1} = G_1 A_{s0}; \quad A_{i1} = g_1 A_{s0}^* \tag{5.2}$$

Here, A_{s0} is the input signal beam. Note that when amplitude gain is large enough, i.e. $G_1 \approx g_1 \geq 1$, we have a nearly equal splitting of A_{s0}, but the splitting beams are amplified from the input beam. Assuming the idler beam is subjected to a phase shift of φ, we obtain the outputs of FWM_2 from Eqs. (5.1) and (5.2) as

$$A_{s2} = G(\varphi)A_{s0}; \quad A_{i2} = g(\varphi)A_{s0}^* \tag{5.3}$$

with

$$G(\varphi) = G_1 G_2 + g_1^* g_2 e^{-i\varphi}; \quad g(\varphi) = G_1^* g_2 + G_2 g_1 e^{i\varphi} \tag{5.4}$$

In the case when both FWM processes have the same gain: $G_1 = G_2 = G_0$, $g_1 = g_2 = g_0$, we have

$$G(\varphi) = 1 + {g_0}^2(1 + e^{-i\varphi}); \quad g(\varphi) = G_0 g_0(1 + e^{i\varphi}) \tag{5.5}$$

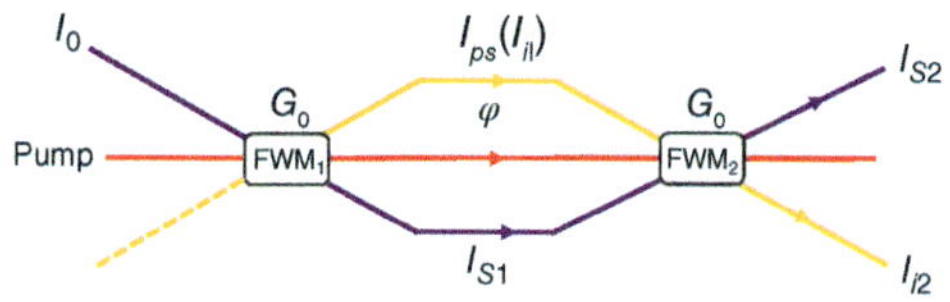

Figure 5.1 A nonlinear interferometer with two FWM processes as two equivalent beam splitters. I_0 is input signal beam; $I_{ps}(I_{i1})$ and I_{s1} are the amplified idler and signal beams, respectively; G_0 is amplitude gain for FWM_1 and FWM_2 processes, respectively; I_{s2} and I_{i2} are the output signal and idler beams from nonlinear interferometer, respectively.

From Eq. (5.2), we find the intensity of the phase-sensing beam inside the nonlinear interferometer as

$$I_{i1} = A_{i1}{}^2 = g_0{}^2 I_0 \tag{5.6}$$

with $I_0 = A_{s0}{}^2$ as the input intensity. The intensities of the output beams can also be easily obtained as

$$I_{s2} = I_0 \left[1 + 2G_0{}^2 g_0{}^2(1 + \cos\varphi)\right], \quad I_{i2} = 2I_0 G_0{}^2 g_0{}^2(1 + \cos\varphi) \tag{5.7}$$

So both the output beams show interference fringes when the phase is scanned. From Eq. (5.7), we find the maximum output intensity at the idler output to be

$$I_{i2}^M = 4G_0{}^2 g_0{}^2 I_0 = 4G_0{}^2 I_{ps} = \frac{4(I_{ps} + I_0)I_{ps}}{I_0} \tag{5.8}$$

Here $I_{ps} = I_{i1}$ is the intensity of phase-sensing beam. For a large gain and a fixed input I_0, we have $I_{ps} \geq I_0$ and $I_{i2}^M \propto I_{ps}^2$. Thus, Eq. (5.8) is the main feature of this nonlinear interferometer, which is obviously different from the traditional linear interferometer (MZI). After the above basic principle about nonlinear interferometers is discussed, we will discuss the experimental implementation of phase-locking, quantum enhancement of phase sensitivity, enhancement of entanglement, QNC, and so on in detail in Sections 5.2.1–5.2.3.

5.2.1 Experimental Implementation of Phase Locking

In this section, phase-locking of nonlinear interferometer will be experimentally implemented [4]. Phase-locking techniques are widely used in many fields, including electronics and optics, such as frequency demodulators and frequency synthesizers. As the fields of fundamental quantum optics and quantum information science develop, phase-locking technique has become an indispensable experimental tool for the generation, manipulation, and detection of quantum states. For example, it has been utilized to stabilize the relative phase between multiple squeezed states to prepare multipartite entanglement and cluster states. It can also manipulate the behavior of quantum squeezing and entanglement in both single-mode and two-mode phase-sensitive FWM processes based on rubidium vapor. Phase-locking techniques are also a critical part of balanced homodyne detection systems, which can be used to measure the quadrature variance of a quantum state. In addition, phase-locking technique is required for the use of an interferometer in real applications. For example, gravitational wave detectors operate close to a dark fringe, which not only cancels the common mode noise of the system but also maximizes the signal-to-shot noise ratio.

Our group has experimentally implemented a nonlinear interferometer and investigated its nonlinear character and visibility dependence on system parameters. More importantly, we studied the quantum nature of its performance and demonstrated a 4 dB improvement in signal-to-noise ratio (*SNR*) for the phase measurement that results from an increase in signal and the cancellation of quantum noise. In this case, nonlinear interferometer is operated on its dark fringe caused by destructive interference via the control of the internal phase. On the other hand, the interferometer can also be operated on its bright fringe by modifying the internal phase. As a result, the quantum correlations generated in the first FWM stage can be further enhanced by the second stage. Consequently, different physical phenomena can be observed at different phase-locking points of our nonlinear interferometer. From this point of view, the ability to phase lock this nonlinear interferometer is important for its use in real applications. Thus, here we investigate the physical implementation of phase-locking in such a nonlinear interferometer.

There are several methods for phase-locking: beat note locking (BNL), coherent modulation locking (CML), and quantum noise locking (QNL). CML and BNL can be mainly used to lock the relative phase of bright optical fields, such as coherent states and squeezed coherent states, whereas the QNL can lock the relative phase between a squeezed vacuum state and a coherent state. Here we investigate the phase-locking of nonlinear interferometers using two different methods, i.e. BNL and CML.

As shown in Figure 5.2, entangled twin beams, called probe and conjugate, are generated in the first cell by the FWM process. The second cell plays the role of the second beam splitter in an interferometer and leads to interference by mixing the probe, conjugate, and $pump_2$. The relative phase between these three beams needs to be stabilized in order to lock the interferometer.

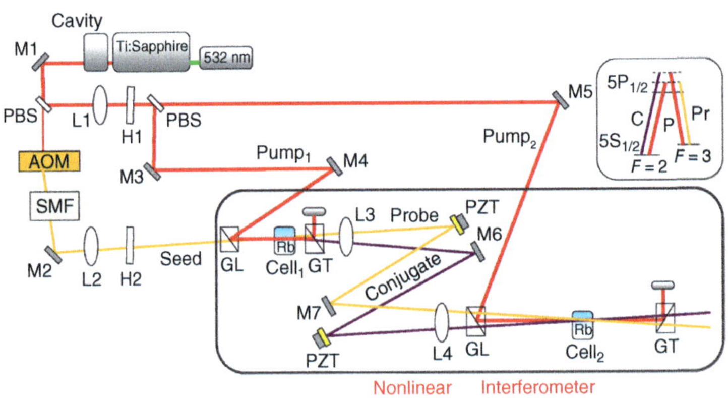

Figure 5.2 Experimental setup for nonlinear interferometer. PBS, polarizing beam splitter; AOM, acousto-optic modulator; SMF, single-mode fiber; L1–4, lens; H1–2, half-wave plates; M1–7, mirrors; GL, Glan laser polarizer; $Cell_1$ and $Cell_2$, atomic vapor cells; GT, Glan–Thompson polarizer; and PZT, piezo-electric transducer.

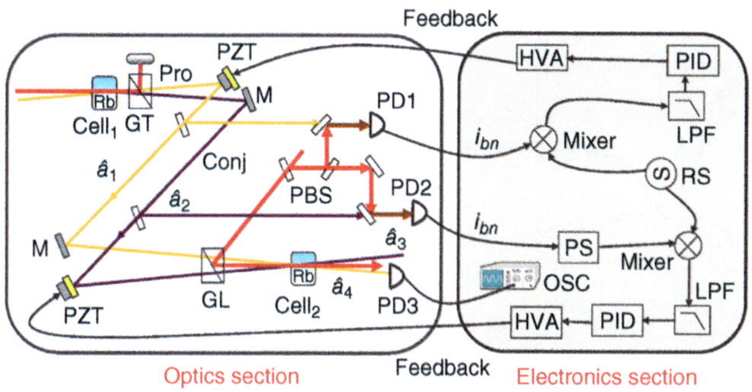

Figure 5.3 Detailed experimental layout for phase-locking nonlinear interferometer using BNL method. In the optics section: PD1, PD2, broad bandwidth photodiodes; PD3, monitoring photodiode. In the electronics section: OSC, oscilloscope; Mixer, frequency mixer; PS, phase shifter; RS, radio frequency source; LPF, low-pass filter; PID, proportional-integral-derivative controller; and HVA, high-voltage amplifier.

To implement phase-locking, it is straightforward to set up two sets of phase-locking systems, one for probe and pump$_2$, and the other for conjugate and pump$_2$. This is the basic idea of BNL method. As shown in Figure 5.3, we first pick off a small portion of each beam inside the interferometer. Then, we overlap the portion of the probe or conjugate that we picked off with pump$_2$ using a beam splitter. In this way, with high-bandwidth detectors, two beat note signals at 3.04 GHz (one for the probe and one for the conjugate) are obtained and are given as

$$i_{bn} = \Gamma\cos[(\omega_{pro(conj)} - \omega_{pump2})t + (\phi_{pro(conj)} - \phi_2)] \tag{5.9}$$

where Γ depends on the field amplitudes of the probe or conjugate and pump$_2$ beams at the position of the photodiode. $\omega_{pro(conj)} - \omega_{pump2}$ is equal to ± 3.04 GHz, which is determined by the hyperfine splitting of the two ground states of Rb^{85}. $\phi_{pro(conj)}$ is the phase of the probe (conjugate) beam, and ϕ_2 is the phase of pump$_2$. What we care about in phase-locking is the phase difference term $\phi_{pro(conj)} - \phi_2$. To obtain this term, we mix the beat note signal with an electronic local oscillator, which has the same frequency of 3.04 GHz. After the mixing process, we use two low-pass filters (LPFs) to filter out the high-frequency terms at about 6.08 GHz. This results in a low-frequency signal that depends on the phase difference between the probe or conjugate and pump$_2$. This gives us the error signal of the BNL method

$$\zeta = \Gamma\cos(\phi_{pro(conj)} - \phi_2) \tag{5.10}$$

which has zero crossings at

$$\phi_{pro(conj)} = \phi_2 \pm \frac{\pi}{2} \tag{5.11}$$

In order to lock the phase difference between pump$_2$ and the probe or conjugate, we introduce a phase shifter (PS) in one of the beat note signals before the frequency mixer (Mixer). The phase-locking results for BNL method are shown in Figure 5.4. In Figure 5.4a, the black curve (trace A) shows the interference fringes when we scan the two piezo-electric transducers (PZTs) shown in Figure 5.3. This signal is used to

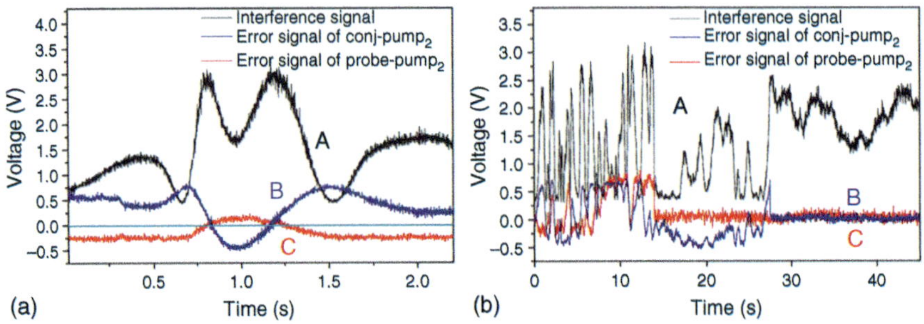

Figure 5.4 The phase-locking results for BNL method. (a) Signals when scanning the phase of the interferometer. (b) Signals when the interferometer is not scanned. Trace A: interference signal used for monitoring; Trace B (C): error signal used for locking the conjugate and pump$_2$ (probe and pump$_2$).

monitor the stability of the phase-locking method due to its phase-sensitive nature. The blue curve (trace B) is the error signal of the phase lock loop for the conjugate and pump$_2$, whereas the red curve (trace C) is the error signal of the phase lock loop for the probe and pump$_2$. We then stop scanning the two PZTs and enable the phase lock loop by feeding back the error signal to the PZT (its bandwidth of several tens of kHz determines the locking bandwidth) through a proportional-integral-derivative controller (PID) and a high-voltage amplifier (HVA), as also shown in Figure 5.3. The result is shown in Figure 5.4b. The three curves (traces A, B, and C) that we obtain without scanning correspond to the same curves with scanning in Figure 5.4a. From 0 to 14 seconds, both phase-locking systems operate without feedback. The fluctuations of the two error signals and the interference signal are large. After 14 seconds, we enable only one locking system (probe-pump$_2$). As we can see, the fluctuations of the interference signal have become smaller. We enable both locking systems at 28 seconds; however, this does not significantly improve the locking quality. Figure 5.4b shows that BNL method does not phase lock nonlinear interferometer successfully. This is due mainly to the fact that the beat note signals are derived at the positions of photodiodes PD1 and PD2 instead of at the center of cell$_2$, where the interference occurs. That is to say, BNL method is trying to use a "virtual" locking position (PD1 and PD2) instead of the real locking position (the center of cell$_2$), which is not part of the phase-locking loop. In this case, only when the optical paths after the polarization beam splitter (PBS) used to pick off the beams to obtain the beat note signal are stable enough, would the BNL method result in a stable lock. However, these optical paths cannot be made short enough to ensure the required stability in our current experimental setup because of the optical elements of the nonlinear interferometer. In principle, the phase-locking quality of BNL method could be further improved by redesigning nonlinear interferometer in a compact configuration. However, it is important to note that this technique cannot directly stabilize the FWM process in cell$_2$. To overcome this limitation of BNL method, other more suitable techniques are needed to achieve stable operation of nonlinear interferometer. In order to do this, it is necessary to directly stabilize the FWM process. This involves phase-locking three beams at different frequencies

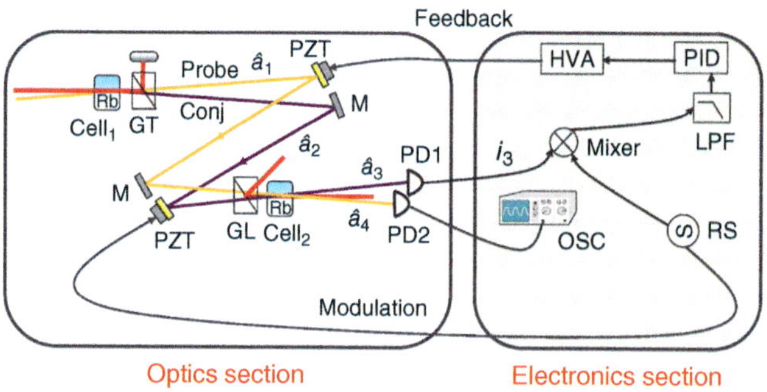

Figure 5.5 Detailed experimental layout for phase-locking nonlinear interferometer using CML method.

with only one phase-locking loop. The CML method allows us to do so and achieve much better stability for nonlinear interferometers, as discussed below. In addition, compared with BNL method, CML method, as shown in Figure 5.5, does not require any extra optical elements for picking off beams inside the interferometer. In this way, it avoids insertion losses that introduce vacuum noise and deteriorate the quantum correlations between the two arms of the interferometer. As shown in Figure 5.5, we first detect the optical field $\hat{a}_3$ using photodiode PD1, whose output can be expressed after making a linear approximation as

$$i_3(t) = |G_2|^2\overline{a}_2{}^2 + |g_2|^2\overline{a}_1{}^2 + 2\overline{a}_1\overline{a}_2G_2g_2\cos\theta \tag{5.12}$$

where G_2 is the gain of cell$_2$ and g_2 satisfies the relationship $|G_2|^2 - |g_2|^2 = 1$. $\overline{a}_1$ and $\overline{a}_2$ are the mean values of the optical fields $\hat{a}_1$ and $\hat{a}_2$, and θ is the relative phase between the three beams.

To obtain the error signal for CML method, it is necessary to introduce a relative phase modulation between the input fields. We include this in the phase difference of the input fields as

$$\theta = 2\phi_2 - \phi_{pro} - \phi_{conj} = \theta_0 + \theta_1 \sin\Omega t \tag{5.13}$$

the dependence of θ on the phases of the probe, conjugate, and pump beams is determined by the FWM process. θ_0 is the average phase of the phase difference between the three beams, θ_1 is the modulation depth, and Ω is the modulation frequency. For our case, this modulation depth is very small ($\theta_1 \ll \theta_0$), so we can make the approximation

$$e^{i\theta_1 \sin\Omega t} \simeq J_0(\theta_1) + J_1(\theta_1)e^{i\Omega t} - J_1(\theta_1)e^{-i\Omega t} \tag{5.14}$$

where $J_0(\theta_1)$ and $J_1(\theta_1)$ are the zero and first-order Bessel functions. We substitute this approximation into Eq. (5.12) to obtain

$$i_3(t) = |G_2|^2\overline{a}_2{}^2 + |g_2|^2\overline{a}_1{}^2 + 2\overline{a}_1\overline{a}_2G_2g_2[J_0(\theta_1)\cos\theta_0 - 2J_1(\theta_1)\sin\theta_0\sin\Omega t] \tag{5.15}$$

we then mix the above photocurrent with an electronic local oscillator, which has the same frequency of Ω with a Mixer and use an LPF to filter the high-frequency

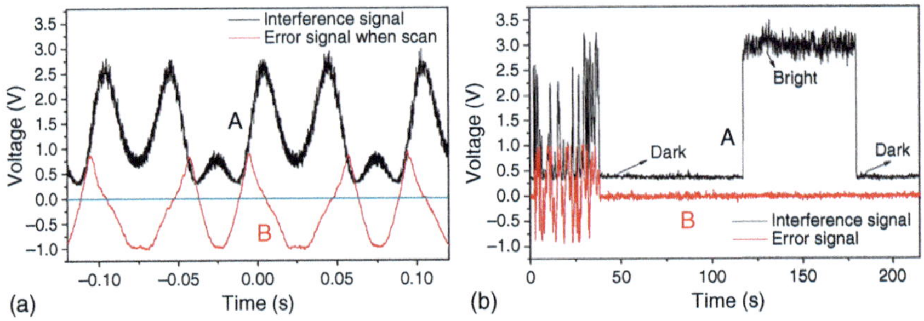

Figure 5.6 The phase-locking results for CML method. (a) Signals when scanning the phase of the interferometer. (b) Signals when the interferometer is not scanned. Trace A: interference signal used for monitoring; Trace B: error signal for the CML method.

components. In this way, the error signal reduces to

$$\zeta = -2\overline{a}_1\overline{a}_2 G_2 g_2 J_1(\theta_1) \sin \theta_0 \tag{5.16}$$

From Eq. (5.16), we can see that the error signal of CML method is proportional to $J_1(\theta_1)$, the gain of cell$_2$, and the mean value of fields $\hat{a}_1$ and $\hat{a}_2$. At the same time, it has zero crossings at $\theta_0 = 0, \pi$, which correspond to the bright and dark fringes of nonlinear interferometer, respectively. Consequently, the nonlinear interferometer can be locked to the bright and dark fringes by using CML method. The phase-locking results for CML method are shown in Figure 5.6. In Figure 5.6a, the black curve (trace A) shows the interference fringe when we scan one of the two PZTs shown in Figure 5.5. The red curve (trace B) shows the error signal of the phase-lock loop for nonlinear interferometer. As we can see, the zero crossings of the error signal almost perfectly correspond to either the bright or dark fringes of the interference signal. This makes it possible for CML method to provide an accurate and stable lock. We then stop scanning the PZT and enable the phase-lock loop by feeding back the error signal to the PZT through a PID and a HVA, as shown in Figure 5.5. The results are shown in Figure 5.6b, where trace A and trace B correspond to the same curves as in Figure 5.6a but without scanning. When we first let the system operate without feedback for about 38 seconds, the acquired interference signal and error signal showed large fluctuations, indicating large-phase fluctuations inside the interferometer. We then enable the feedback system for about 78 seconds, and nonlinear interferometer is locked to the dark fringe. In this way, the large fluctuations of the interference signal and the error signal are largely suppressed, showing very good phase stability. Then, we lock the interferometer to the bright fringe simply by switching the sign of the proportional term in the PID.

We can obtain an estimate for the phase stability of the interferometer when locked by using the error signal when the system is locked as well as the slope around the locking points (zero crossings). With this analysis, we can estimate that for CML method, phase stability is $\Delta\theta{\sim}0.05$ rad at the dark fringe and $\Delta\theta{\sim}0.10$ rad at the bright fringe. The level of stability obtained at the dark fringe will enable, under certain conditions, the study of the SU(1,1) interferometer well within the operational phase region where it is possible to obtain an enhancement in phase estimation below standard quantum limit (SQL) or shot noise limit (SNL). A similar analysis can be done for BNL method; however, in this case, the locking stability does not provide a direct measure of the stability of the interferometer, as this method does not provide a direct stabilization of the FWM process in $cell_2$.

The fluctuations of the interference signal when it is locked to the bright fringe seem noisier than the ones for the dark fringe. This results in part from the locking stability, as described above, and from the different levels of noise added by the nonlinear interferometer at its different operating points. The output states of nonlinear interferometer when locked to the bright fringe are thermal states with significant excess noise due to the amplification from the cascaded FWM processes. However, the output states are coherent when the interferometer is locked to the dark fringe due to QNC. As we can see, the stability of the phase lock with CML method is long enough for quantum correlation and entanglement measurements at both dark and bright fringes.

In conclusion, we have locked the internal phase of a nonlinear interferometer to a level that will make it possible to study its operation for phase estimation in the sub-shot noise regime. We studied two types of phase-locking techniques, i.e. BNL and CML. Compared with the BNL method, the CML method can achieve much better phase-locking stability. The CML method used in our current work could also be applied to more complicated processes that require phase-locking multiple beams with different frequencies. For example, we can use CML technique to lock the relative phase between multiple beams of a phase-sensitive N-wave mixing process or an N-port nonlinear interferometer. This will make it possible to control the relative phase of a complex nonlinear system. Our results pave the way for real applications of a nonlinear interferometer in precision measurement and quantum manipulation, for example, phase control in phase sensitive amplifiers (PSAs) and quantum-enhanced real-time phase tracking.

5.2.2 Quantum Enhancement of Phase Sensitivity

Although we have experimentally implemented the phase-locking of nonlinear interferometer in Section 5.2.1, in this section we use a more general method to measure phase sensitivity [5]. Phase sensitivity as another important physical quantity, is used to character the uncertainty in the internal phase estimation of an interferometer. Figure 5.7a and 5.7b show the phase-sensitivity measurement scheme of MZI and bright-seeded SU(1,1) interferometer. The equal photon number (Ns) inside them can provide a fair comparison premise. Due to the amplification nature of the first parametric process, the seed intensity of MZI should be increased to balance the internal photon number from the two different interferometers.

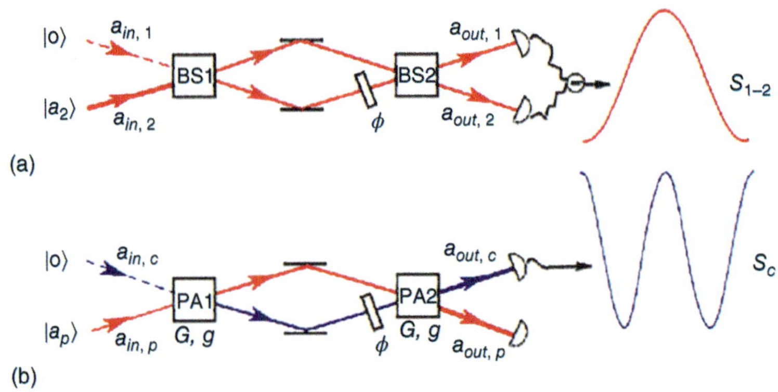

Figure 5.7 The two types of interferometer. (a) MZI and (b) bright-seeded SU(1,1) interferometer. G is the amplitude gain of the parametric process, ϕ is the internal phase difference, S_{1-2} is the interference fringe obtained by subtraction from the two arms in MZI, and S_c is the interference fringe of the conjugate arm in bright-seeded SU(1,1) interferometer.

Firstly, the input-output relation of the bright-seeded SU(1,1) interferometer can be expressed as

$$\begin{aligned}\hat{a}_{out,p} &= (G^2 + g^2e^{-i\phi})\hat{a}_{in,p} + Gg(1 + e^{-i\phi})\hat{a}^{\dagger}_{in,c} \\ \hat{a}_{out,c} &= Gg(1 + e^{i\phi})\hat{a}^{\dagger}_{in,p} + (G^2e^{i\phi} + g^2)\hat{a}_{in,c}\end{aligned} \tag{5.17}$$

in which p and c represent the probe and conjugate beams, G is the amplitude gain and has the relation of $g^2 = G^2 - 1$, and ϕ is the internal phase variation in the conjugate beam path. The phase sensitivity $\Delta\phi$ of the interferometer can be expressed as

$$\Delta\phi^2 = \frac{\langle(\Delta\hat{O})^2\rangle}{|\partial\langle\hat{O}\rangle/\partial\phi|^2} \tag{5.18}$$

where $\hat{O}$ stands for some measurement combinations of the output fields. Due to the importance role of the measurement method and thus the different strategies may lead to different phase sensitivities. For a coherent state $|\alpha_p\rangle$ injection of the probe port (bright-seeded) and a vacuum state $|0\rangle$ injection of the conjugate port, the interference fringe of the conjugate field is chosen as the measurement strategy. As shown in Figure 5.7b, $\hat{O}_c = \langle\hat{N}_{out,c}\rangle + \Delta\hat{N}_{out,c}$, where $\hat{N}$ represents the photon number operator of the fields. Here $\langle\hat{N}_{out,c}\rangle$ and $\Delta\hat{N}^2_{out,c}$ denote the average intensity and intensity noise of the conjugate field, respectively. On the basis of this measurement strategy, the interference fringe slope corresponding to the denominator in Eq. (5.18) is amplified by the second PA process. Then, based on Eq. (5.18), the phase sensitivity of such a conjugate-port direct measurement strategy can be expressed by

$$\Delta\phi_c = \frac{\sqrt{4G^4g^4(\cos\phi + 1)^2(2N + 1) + 2G^2g^2(\cos\phi + 1)(N + 1)}}{2G^2g^2|\sin\phi|(N + 1)} \tag{5.19}$$

in which $N = \left\langle \hat{a}^{\dagger}_{in,p}\hat{a}_{in,p} \right\rangle$. For the bright-seeded SU(1,1) interferometer ($N \gg 1$), the Eq. (5.19) can be further approximated as

$$\Delta\phi_c = \frac{\sqrt{[8G^4g^4(\cos\phi + 1)^2 + 2G^2g^2(\cos\phi + 1)](G^2 + g^2)}}{2G^2g^2|\sin\phi|\sqrt{N_s}} \tag{5.20}$$

in which $N_s = (G^2 + g^2)\left\langle \hat{a}^{\dagger}_{in,p}\hat{a}_{in,p} \right\rangle$ is the total internal photon number inside the interferometer. To corroborate the quantum enhancement of phase sensitivity brought by the bright-seeded SU(1,1) interferometer, its phase sensitivity should be compared with regard to SQL. The SQL is normally measured by direct subtraction of the photocurrents from the two output fields in MZI. In this way, the phase sensitivity of MZI is given by $\Delta\phi_{MZI} = 1/(|\sin\phi|\sqrt{N_s})$, where $N_s = \left\langle \hat{a}^{\dagger}_{in,2}\hat{a}_{in,2} \right\rangle$. The corresponding SQL is the minimum value of $\Delta\phi_{MZI}$, which is given by $\Delta\phi_{SQL} = 1/\sqrt{N_s}$. From Eq. (5.20), we can see that the phase sensitivity for bright-seeded SU(1,1) interferometer reaches its minimum when $\phi = \pi$. It is given by

$$\Delta\phi_{c,min} = \sqrt{\frac{G^2 + g^2}{4G^2g^2}}\frac{1}{\sqrt{N_s}} \tag{5.21}$$

It can be seen that the bright-seeded SU(1,1) interferometer will beat the SQL for any $G^2 > 1.32$ in the dark fringe. Such quantum enhancement of the phase sensitivity increases with the gain increase. Such SQL beating is due to the factor $\sqrt{4G^2g^2/(G^2 + g^2)}$, and the scaling of such a bright-seeded SU(1,1) interferometer is still $1/\sqrt{N_s}$. Secondly, the detailed experimental arrangement for measuring phase sensitivity of interference fringes S_c and S_{1-2} is shown in Figure 5.7 [5]. Here, i_1 and i_2 represent the photocurrent signals from the two detectors, i_3 (i_5) is the dc component of i_1 (i_2), and i_4 is the subtraction result of *rf* components of i_1 and i_2 for the detection of S_{1-2}. For the detection of S_c, i_4 is the *rf* component of i_1. An oscilloscope with a 28 ms acquisition time and a 50 MHz sampling rate records i_3, i_4, and i_5. To experimentally implement the above theoretical proposal, the detailed experimental arrangement is shown in Figure 5.8.

Figure 5.9 shows a group of typical experimental data when we keep N_s equal for the MZI and the bright-seeded SU(1,1) interferometer. Here we set the gain values of both cells at $G^2 = 2.28$. Figure 5.9a shows i_3–i_5 and i_4 for the MZI, which are displayed in dark red (trace i) and light red (trace ii), respectively. The corresponding data for the bright-seeded SU(1,1) interferometer are depicted in Figure 5.9b, where i_3 and i_4 are displayed in dark blue (trace i) and light blue (trace ii), respectively. In order to obtain the phase sensitivity of the two interferometers, we must obtain both the intensity noise term (numerator) and the slope term (denominator) in Eq. (5.18). First, we divide the 1.4×10^6 points in traces i and ii from Figure 5.9a,b into 250 groups along the time axis. Then, to calculate the intensity noise of the interferometer, fast Fourier transform (FFT) is used to implement the spectrum analysis for each group in trace ii. Considering the quantum correlation from the

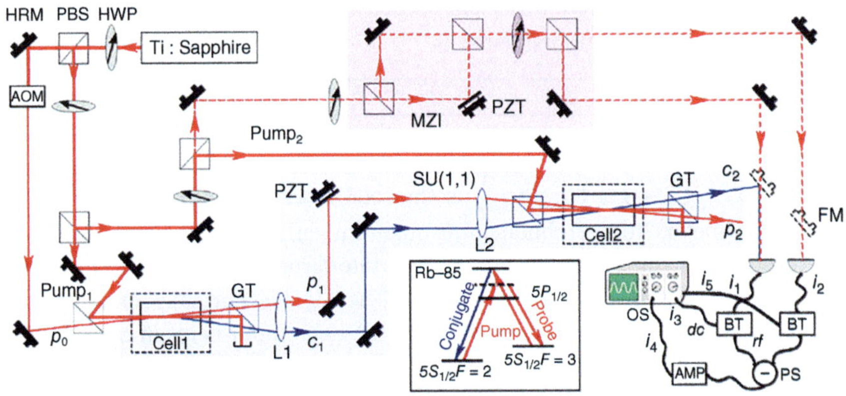

Figure 5.8 Detailed experimental arrangement. AMP, amplifier; AOM, acousto-optic modulator; FM, flip mirror; GT, Glan–Thompson polarizer; HRM, high reflectivity mirror; HWP, half-wave plate; L1–2, lens; OS, oscilloscope.

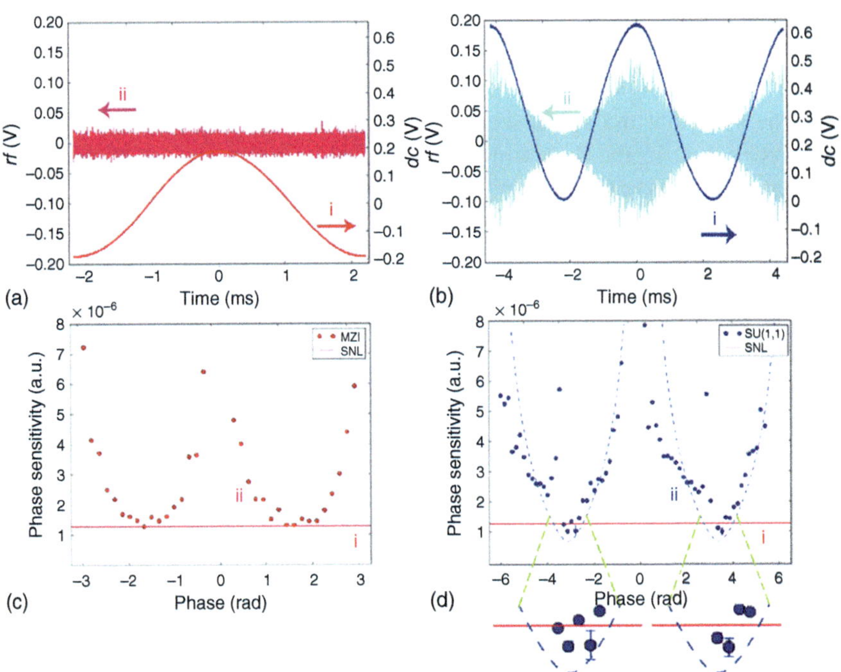

Figure 5.9 Typical experimental data and corresponding phase sensitivity for MZI and bright-seeded SU(1,1) interferometer. (a) Traces i and ii represent the *dc* and *rf* components respectively of the interference fringe for MZI. (b) Traces i and ii represent the *dc* and *rf* components respectively of the interference fringe for the bright-seeded SU(1,1) interferometer. (c) The phase sensitivity of the MZI (trace ii) versus the phase; trace i denotes the corresponding SQL. (d) The phase sensitivity of the bright-seeded SU(1,1) interferometer (trace ii) versus the phase. The dashed blue lines are the corresponding theoretical predictions. Trace i denotes the corresponding SQL.

FWM process has optimal values in the frequency range between 1 and 1.5 MHz, we treat the average value of the FFT spectrum obtained in this range as the intensity noise of each group. Since the interference fringe in our experiment is acquired by scanning the phase ϕ in time inside the interferometer, one phase point corresponds to one time point. In this sense, the time-dependent data are actually the phase-dependent data. The corresponding interference fringe slope of each group can be obtained from the average value of the above-mentioned group in trace i and the corresponding phase of each group. Finally, we use Eq. (5.18) to calculate the phase sensitivity of each phase point for the two interferometers, as shown in Figure 5.9c,d. As shown in Figure 5.9c, the red curve (trace ii) gives the phase sensitivity versus the phase inside the MZI. The red line (trace i) gives the corresponding SQL of the MZI. The corresponding data for the bright-seeded SU(1,1) interferometer are shown in Figure 5.9d, where the blue curve (trace ii) gives the phase sensitivity of the bright-seeded SU(1,1) interferometer. This curve clearly shows the phase-sensitivity enhancement of the bright-seeded SU(1,1) interferometer compared with the SQL in the dark fringe due to direct intensity detection. The enhancement factor is about 1.28, which is calculated by comparing the best phase sensitivity in traces i and ii as shown in Figure 5.9d. The corresponding theoretical prediction is also shown in Figure 5.9d (dashed blue lines), which gives an enhancement factor of 1.81. As shown in Figure 5.9d, the experimental result and the corresponding theoretical prediction have very similar trends, and the slight difference may be the result of the loss from atomic absorption and imperfect detection. To show the phase-sensitivity scaling, we perform a series of measurements as shown in Figure 5.9 with different N_s by changing the power of the seeded beams for both the two interferometers. The phase-sensitivity scaling of the SQL (trace i) and that of the bright-seeded SU(1,1) interferometer (trace ii) are plotted in red and blue curves respectively in Figure 5.10a on a $10\log_{10}$–$10\log_{10}$ scale. As shown in Figure 5.10a, the sensitivity of the bright-seeded SU(1,1) interferometer (trace ii) increases with increasing power of the seeded beam. Trace i is about 1.15 ± 0.16 dB above the phase-sensitivity scaling of the bright-seeded SU(1,1) interferometer (trace ii); that is, the bright-seeded SU(1,1) interferometer has beaten the SQL by an amount of about 1.15 dB. The corresponding theoretical predicted curves are shown as traces i and ii in Figure 5.10b. The experimental enhancement factor (1.15 dB) compared with the SQL is less than the theoretical prediction (2.6 dB). The main reason is that our theory does not include the losses existing in the system, which can affect the performance of the interferometer to some extent.

In this section, by using direct intensity detection, we experimentally demonstrate the quantum enhancement of the phase sensitivity for the bright-seeded SU(1,1) interferometer with reference to SQL. It is this direct detection scheme that ensures the quantum enhancement in real-time and broadens its applications in quantum metrology, such as a nonlinear interferometer based on quantum-enhanced real-time phase tracking and detection of an N-port interferometer with quantum-correlated outputs.

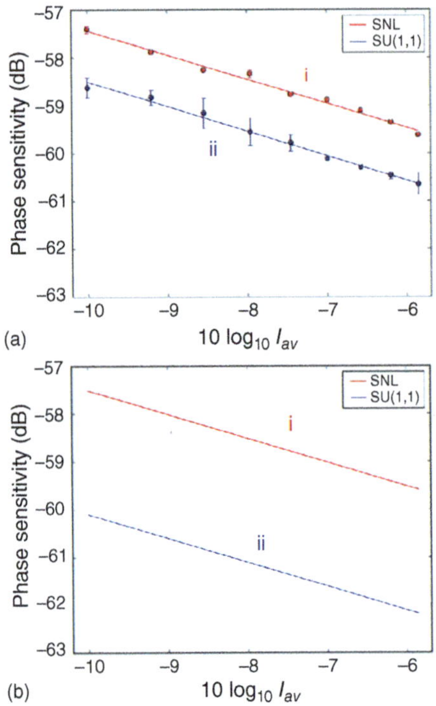

Figure 5.10 The phase-sensitivity scaling of the bright-seeded SU(1,1) interferometer and the SQL. (a) Traces i and ii denote the phase sensitivity of SQL and the bright-seeded SU(1,1) interferometer respectively versus the average intensity (I_{av}) of the fields inside the interferometers. (b) The theoretical prediction corresponding to (a).

5.2.3 Enhancement of Entanglement and Quantum Noise Cancellation

After phase sensitivity of nonlinear interferometer has been experimentally measured in Section 5.2.2, based on similar experimental setup, enhancement of entanglement [6] and QNC [7] will be discussed in this section. Here we will take enhancement of entanglement as the first example, i.e. more than 3 dB enhancement of joint quadrature squeezing compared with one of the single FWM process is observed. By varying the pump power of the cascaded system, we also show how the gain affects the entanglement enhancement. Theoretical predictions with the considerations of losses in the experiment process are also investigated and a similar trend in the low-gain regime can be found with the experimental results. The scheme of cascaded FWM processes is shown in Figure 5.11. For two vacuum fields, respectively, named as "probe" ($\hat{a}_{in,p}$) and "conjugate" ($\hat{a}_{in,c}$) entering into the first FWM process, the input–output relation for a single FWM process can be given by

$$\hat{A}_p = G\hat{a}_{in,p} + g\hat{a}^{\dagger}_{in,c}, \hat{A}_c = G\hat{a}_{in,c} + g\hat{a}^{\dagger}_{in,p} \tag{5.22}$$

where $\hat{A}_p$ and $\hat{A}_c$ are the probe and conjugate fields after the first FWM process, G is the amplitude gain, and we have the relation $g^2 = G^2 - 1$. To describe the quantum entanglement between two fields such as $\hat{a}$ and $\hat{b}$, joint quadrature operators $\hat{X}_{-,ab} = \hat{X}_a - \hat{X}_b$ and $\hat{Y}_{+,ab} = \hat{Y}_a + \hat{Y}_b$ are defined, where $\hat{X}_{a(b)} = \hat{a}^{\dagger}(\hat{b}^{\dagger}) + \hat{a}(\hat{b})$ and $\hat{Y}_{a(b)} = i\hat{a}^{\dagger}(\hat{b}^{\dagger}) - i\hat{a}(\hat{b})$ are the quadrature amplitude phase of the corresponding fields. For the fields of $\hat{A}_p$ and $\hat{A}_c$, the corresponding SQL is the noise of the joint

Figure 5.11 Scheme of cascaded FWM processes for entanglement enhancement.

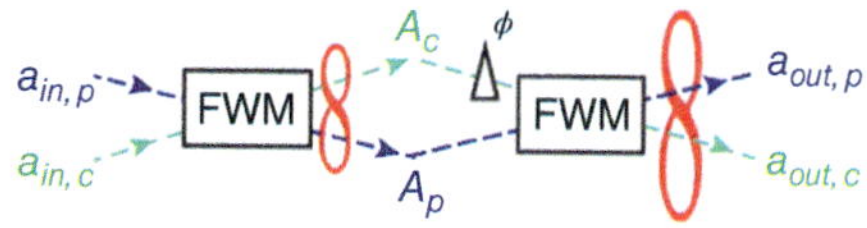

quadratures of two vacuum states $\left(\left\langle \Delta\hat{X}^2_{-,vv}\right\rangle\ and\ \left\langle \Delta\hat{Y}^2_{+,vv}\right\rangle\right)$, which gives an upper bound for the entanglement criterion. For a single FWM process, the noise of the joint quadratures can be calculated and given by

$$\left\langle \Delta\hat{X}^2_{-,pc}\right\rangle = \left\langle \Delta\hat{Y}^2_{+,pc}\right\rangle = 2(G-g)^2 \tag{5.23}$$

the value of Eq. (5.23) is smaller than 2 (the corresponding SQL) for any $G^2 > 1$. That is to say the twin beams ($\hat{A}_p$ and $\hat{A}_c$) output from the single FWM process have the properties of inseparability or entanglement, which has been demonstrated in the previous experiment. In order to obtain a larger degree of entanglement, the cascaded FWM processes are used by seeding the two entangled fields $\hat{A}_p$ and $\hat{A}_c$ into the second FWM process, as shown in Figure 5.11. With the control of the internal phase (ϕ) between $\hat{A}_p$ and $\hat{A}_c$, the output fields ($\hat{a}_{out,p}$ and $\hat{a}_{out,c}$) after the second FWM process can be written as

$$\hat{a}_{out,p} = G\hat{A}_p + g\hat{A}_c^{\dagger}e^{-i\phi}, \hat{a}_{out,c} = G\hat{A}_c e^{i\phi} + g\hat{A}_p^{\dagger} \tag{5.24}$$

Similar to the single FWM process, the noise of the joint quadratures of cascaded FWM processes can be calculated and given by

$$\left\langle \Delta\hat{X}^2_{-,pc}\right\rangle = \left\langle \Delta\hat{Y}^2_{+,pc}\right\rangle = 2 - 4Gg(G-g)^2(1+\cos\phi) \tag{5.25}$$

From Eq. (5.25), it can be found that for the scheme of the cascaded FWM processes, the noise of the joint quadratures is phase-dependent. In addition, the minimal value, which happen at $\phi = 2k\pi$, where k is an integer, of the $\left\langle \Delta\hat{X}^2_{-,pc}\right\rangle$ and $\left\langle \Delta\hat{Y}^2_{+,pc}\right\rangle$ are both $2 - 8Gg(G-g)^2$, which is smaller than the one of the single FWM process in Eq. (5.23) for any $G^2 > 1$. In other words, a larger degree of entanglement can be obtained through the cascaded FWM processes, with reference to the single FWM process.

The detailed experimental layout is similar to that in Figure 5.8, and the typical experimental results are shown in Figure 5.12. As shown in Figure 5.12, the noise $\left\langle \Delta\hat{X}^2_{-,pc}\right\rangle$ of the joint quadrature $\hat{X}_{-,pc}$ for the scheme of the cascaded FWM processes is shown by the minimum value of trace i. To show the entanglement enhancement of the scheme of the cascaded FWM processes, the noise (the minimum value of trace ii) of the joint quadrature of single FWM process is also measured by blocking the pump beam of the second FWM process in the experiment process. Trace iii gives the corresponding SQL for the two schemes of single and cascaded FWM processes when we block the two pumps of the cascaded FWM processes. As shown in Figure 5.12, it is clear that both the two minimal values of both traces i and ii are below trace iii, which is sufficient to claim the inseparability or entanglement between the probe and conjugate beam output from both single and cascaded FWM processes. By comparing the traces i and iii in Figure 5.12, the maximal quadrature

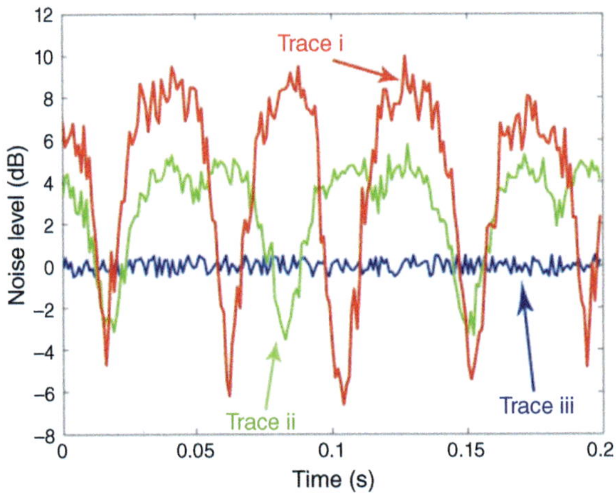

Figure 5.12 Group of typical experimental data. See the details in the main text.

squeezing of −6.6 dB can be obtained by the scheme of the cascaded FWM processes, which is about 2.8 dB better than the one of single FWM process. That is, a higher degree of entanglement is experimentally obtained by the scheme of cascaded FWM processes compared with the single FWM process.

In order to experimentally study the maximal quadrature squeezing that can be obtained by the scheme of the cascaded FWM processes, we make a series of measurements, as shown in Figure 5.13, with different G^2 by changing the power of the pump beams of the FWM processes. For simplicity, we always keep both FWM processes with the same gains. In each measurement, the minimal values of traces i and ii in Figure 5.12 and the mean value of trace iii are picked up, which gives us the quadrature squeezing of the single and cascaded FWM processes versus the gain, as shown by traces i and ii in Figure 5.13a. In Figure 5.13a, it can be found that the entanglement enhancement of the cascaded FWM processes can be obtained for any $1.1 < G^2 < 3.1$ and the maximal degree of about 3.1 dB entanglement enhancement happens at $G^2 = 2.1$. Moreover, a maximal joint quadrature squeezing of -6.8 ± 0.4 dB is obtained at the same gain point. However, by comparing our experimental results with the theoretical predictions from Eqs. (5.23) and (5.25), as shown by traces i and ii in Figure 5.13b, we find that the experimental results are far from achieving the theoretical predictions. This discrepancy can be explained by the effect of losses and pump beam scattering, such as the conical emission from the FWM process.

It is well known that entanglement is fragile and can be easily degraded by the loss of systems. For the entanglement based on the cascaded FWM processes, the losses, such as the absorptions of the atomic vapor cells and the unavoidable losses in light path, will introduce extra vacuum noise into the entangled beams, which is also amplified by the successive FWM processes. As a result, the entanglement between the probe and conjugate beams will be experimentally degraded and cannot reach the theoretical predictions. The detailed theoretical descriptions can be seen in Ref. [6], in which the losses are considered. In summary, our experiment

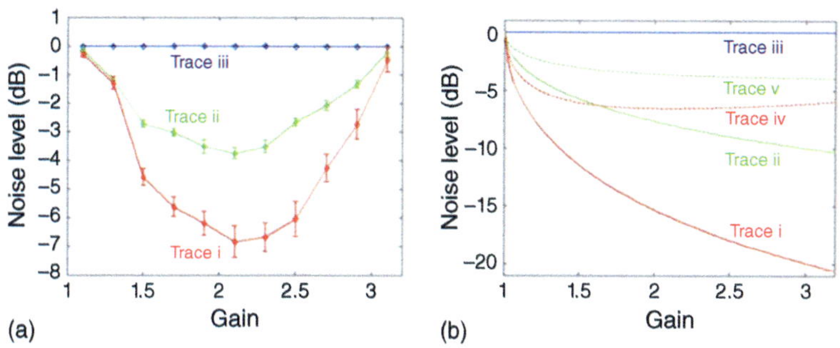

Figure 5.13 Experimental and theoretical results of the quadrature squeezing versus the gain for both schemes of single and cascaded FWM processes. See the details in the main text.

provides a scheme for entanglement enhancement by phase manipulation. In this way, our scheme with strongly enhanced entanglement may help to detect these physical quantities with high sensitivities better than SQL. In addition, our scheme provides a method to overcome the gain limitation in generating continuous variable entanglement from a single FWM process.

Next, as the second example, QNC is an effective method to control the noise of the quantum system, which reduces or even eliminates the noise of the quantum system by utilizing destructive interference in the quantum system [7]. However, QNC can be extremely dependent on the losses inside the system. Along this line, we experimentally and theoretically study how the losses can affect the QNC in nonlinear interferometer. We find that losses in the different arms inside nonlinear interferometer can have different effects on QNC in the output fields from nonlinear interferometer. And QNC in nonlinear interferometer can almost be insensitive to the losses in some cases.

The typical experimental results, under the condition of keeping the gains of both two FWM processes at $G^2 = 2$, are depicted in Figure 5.14. As shown in Figure 5.14, traces i and ii denote the noise levels of $a_{out,p}$ and $a_{out,c}$, respectively, when the phase is scanned. The amplified vacuum noise levels on the output probe field coming from the first and second FWM processes are also shown as traces iii and iv in Figure 5.14, respectively, which almost overlap with the corresponding traces of conjugate fields (not shown in Figure 5.14 for clarity). Traces v and vi show the vacuum noise levels double-amplified by the two FWM processes when we block the probe and conjugate fields inside nonlinear interferometer, respectively. Trace vii is SQL when we block all the pump beams in nonlinear interferometer. As shown in Figure 5.14, traces i and ii can be lower than traces iii and iv and close to trace vii when the phase is scanned, that is to say QNC happens. However, it can also be demonstrated that the noise levels of $a_{out,p}$ and a $a_{out,c}$ cannot be reduced back to SQL as the minimum values of traces i and ii are 1.24 ± 0.07 dB and 1.96 ± 0.15 dB higher than the mean value of trace vii. It is due to the losses inside the quantum system, e.g. the absorptions of the atomic vapor cells. Actually, the losses will introduce extra vacuum noise, which

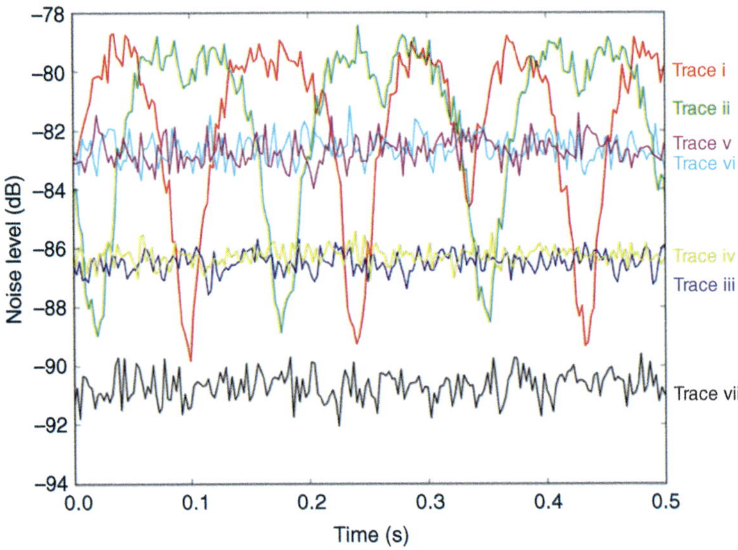

Figure 5.14 Typical experimental data. Traces i and ii, quantum noise levels of the output fields from nonlinear interferometer when the phase is scanned. Traces iii and iv, quantum noise levels of the vacuum fields amplified by the first and second FWM processes, respectively. Traces v and vi, quantum noise levels of the vacuum noise levels double amplified by the two FWM processes when we block the probe and conjugate fields, respectively. Trace vii, SQL.

is uncorrelated with A_p or A_c but is also amplified by the second FWM process. As a result, the deterioration to the quantum correlation between A_p and A_c cannot cancel completely the quantum noise of the two fields $a_{out,\,p}$ and $a_{out,\,c}$. On the other hand, it seems that QNC in $a_{out,\,p}$ is more effective than the one in $a_{out,\,c}$, which may depend on the different absorption levels of the Rb cells for the probe and conjugate beams, which are 21% and 10% in the current experiment, respectively. In order to experimentally study how the losses in the probe and conjugate beams can differently impact QNC happened on $a_{out,\,p}$ and $a_{out,\,c}$, a variable attenuation is introduced on the probe and conjugate fields inside the interferometer. First, we introduce an attenuation on the probe arm to study how the loss can affect QNC in $a_{out,\,p}$ and $a_{out,\,c}$. To theoretically explain the lossy mechanism in the system, the fictitious beam splitters with the reflectivities of η and L are used to simulate the losses. Then, the probe ($\hat{A}'_p$) and conjugate ($\hat{A}'_c$) fields after the first FWM process can be expressed as

$$\hat{A}'_p = \sqrt{1-\eta_a}\left(\sqrt{1-L_p}\hat{A}_p e^{i\phi} + \sqrt{L_p}\hat{\nu}_L\right) + \sqrt{\eta_a}\hat{\nu}_a \tag{5.26}$$

$$\hat{A}'_c = \sqrt{1-\eta_b}\hat{A}_c + \sqrt{\eta_b}\hat{\nu}_b \tag{5.27}$$

in which η_a and η_b are the absorption losses of the two Rb cells for the probe and conjugate fields, and L_p represents the attenuation in the probe field. $\hat{\nu}_L$, $\hat{\nu}_a$, and $\hat{\nu}_b$ are vacuum-state operators. Then, the quantum noise levels of the two fields $a_{out,\,p}$

and $a_{out,\,c}$ can be given by

$$\left\langle \Delta\hat{X}^2_{p,p\downarrow} \right\rangle = G^4(1-L_p)(1-\eta_a) + g^4(1-\eta_b) + G^2g^2(1-\eta_b) + G^2g^2(1-L_p)(1-\eta_a) + g^2\eta_b + G^2L_p(1-\eta_a) + G^2\eta_b + 4G^2g^2\sqrt{1-L_p}\sqrt{1-\eta_a}\sqrt{1-\eta_b}\cos\phi \tag{5.28}$$

$$\left\langle \Delta\hat{X}^2_{c,p\downarrow} \right\rangle = G^4(1-\eta_b) + g^4(1-L_p)(1-\eta_a) + G^2g^2(1-\eta_b) + G^2g^2(1-L_p)(1-\eta_a) + G^2\eta_b + g^2L_p(1-\eta_a) + g^2\eta_a + 4G^2g^2\sqrt{1-L_p}\sqrt{1-\eta_a}\sqrt{1-\eta_b}\cos\phi \tag{5.29}$$

in which $X_{p,\,p\downarrow}$ ($X_{c,\,p\downarrow}$) is the quadrature-phase amplitude of the probe (conjugate) field, as attenuation is applied on the probe field.

Similarly, the quantum noise levels ($\langle \Delta\hat{X}^2_{p,c\downarrow} \rangle$ and $\langle \Delta\hat{X}^2_{c,c\downarrow} \rangle$) of the two fields $a_{out,\,p}$ and $a_{out,\,c}$, when the attenuation (L_c) is applied on the conjugate field can be given by

$$\left\langle \Delta\hat{X}^2_{p,c\downarrow} \right\rangle = G^4(1-\eta_a) + g^4(1-L_c)(1-\eta_b) + G^2g^2(1-\eta_a) + G^2g^2(1-L_c)(1-\eta_b) + g^2\eta_b + G^2L_c(1-\eta_b) + G^2\eta_a + 4G^2g^2\sqrt{1-L_c}\sqrt{1-\eta_a}\sqrt{1-\eta_b}\cos\phi \tag{5.30}$$

$$\left\langle \Delta\hat{X}^2_{c,c\downarrow} \right\rangle = G^4(1-L_c)(1-\eta_b) + g^4(1-\eta_a) + G^2g^2(1-\eta_a) + G^2g^2(1-L_c)(1-\eta_b) + G^2\eta_b + G^2L_c(1-\eta_b) + g^2\eta_a + 4G^2g^2\sqrt{1-L_c}\sqrt{1-\eta_a}\sqrt{1-\eta_b}\cos\phi \tag{5.31}$$

in which $X_{p,\,c\downarrow}$ ($X_{c,\,c\downarrow}$) is the quadrature-phase amplitude of the probe (conjugate) field as the attenuation is applied on the conjugate arm.

Figure 5.15 shows the experimental data and theoretical simulations about the effect of losses on QNC. The vertical axis is the excess noise, which can be obtained by subtracting the vacuum noise level from the minimum quantum noise level of the output fields. The horizontal axis is the attenuation level. The effects of the losses on QNC of the probe field when attenuation is applied on the conjugate ($p, c\downarrow$) and probe ($p, p\downarrow$) fields are shown as traces i and ii in Figure 5.15a, respectively. The solid and dashed lines are the experimental results and theoretical results, respectively, in which the absorptions of the two Rb cells for probe (η_a) and conjugate (η_b) fields are 21% and 10%. The corresponding results for the effect of losses on QNC of the conjugate field when attenuation is applied on the probe ($c, p\downarrow$) and conjugate ($c, c\downarrow$) fields are shown as traces iii and iv in Figure 5.15b, respectively. As shown in Figure 5.15, the experimental and theoretical results have similar trends, and the difference can be explained by the other losses from the system, such as propagation

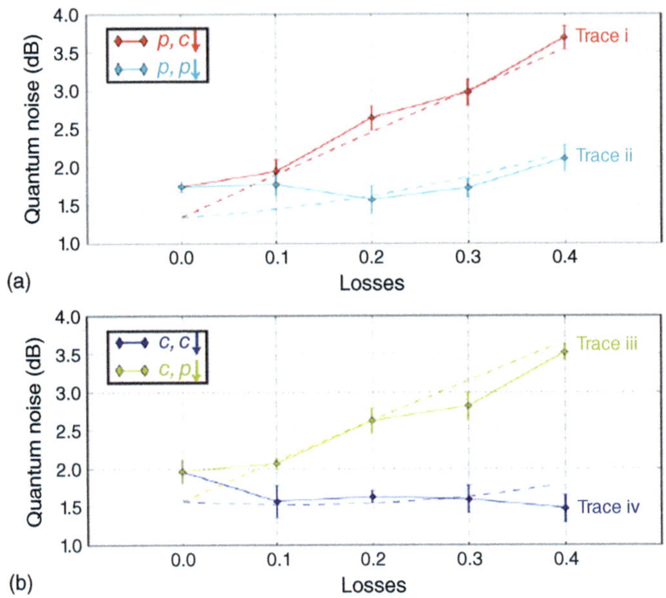

Figure 5.15 Experimental and theoretical data for the effect of losses on QNC inside nonlinear interferometer. Trace i (iii) labeled as *p*, *c*↓ (*c*, *p*↓) is the loss-sensitive case in probe (conjugate) field by introducing the loss on the conjugate (probe) field; trace ii (iv) labeled as *p*, *p*↓(*c*, *c*↓) is the loss-insensitive case in probe (conjugate) field by introducing the loss on the probe (conjugate) field. The dashed lines are the corresponding theoretical simulations.

losses from the optical path and the imperfect detection from the detectors. It can be found that the larger the losses are introduced, the more difficult QNC can be realized in $a_{out,p}$ and $a_{out,c}$. It is mainly because the losses will introduce the vacuum noise, which is completely random and uncorrelated with the internal fields (A_p, A_c) inside nonlinear interferometer. As a result, the vacuum noise amplified by the second FWM process cannot be cancelled completely and deteriorates QNC in the two fields $a_{out,p}$ and $a_{out,c}$ to a certain extent. By comparing the curves as shown in Figure 5.15, the most interesting thing is that QNC in the cases of *p*, *p*↓ and *c*, *c*↓ is less sensitive to the losses than the cases of *p*, *c*↓ and *c*, *p*↓. Especially for the case of *c*, *c*↓, it can be found that QNC is almost insensitive to the losses in the case of *c*, *c*↓ as shown in Figure 5.15b. This can be simply explained by the fact that the losses inside the interferometer can not only degrade the quantum correlations between the internal probe and conjugate fields but also reduce their thermal noise levels. The former makes the performance of QNC worse, nevertheless the latter makes it better. For some special cases, such as *p*, *p*↓ and *c*, *c*↓, the constraint between the degradation of the quantum correlations and the reduction of the thermal noise levels of the twin beams output from the interferometer makes QNC insensitive to the losses inside the interferometer. If the cases of *p*, *p*↓ and *c*, *c*↓ are called the "parallel cases," whereas *p*, *c*↓ and *c*, *p*↓ are the so-called "cross cases," then we have our conclusion that in nonlinear interferometer, QNC in the parallel cases is less sensitive to the losses than the one in the cross cases, which is quite different from the assumption that losses in either field inside the quantum system are catastrophic to the quantum characteristics of the output fields. Herein, we have completed the basic descriptions of nonlinear interferometers; other types of nonlinear interferometer will be discussed in Section 5.3, which will undoubtedly broaden and enrich our knowledge about nonlinear interference from various nonlinear interferometers.

5.3 Other Types of Nonlinear Interferometers

Nonlinear interferometer discussed in Section 5.2 comprises two cascaded FWM processes, that is, two linear beam splitters in MZI are replaced with two nonlinear FWM process. Then phase-locking technique, high phase sensitivity, enhancement of entanglement, and QNC have also been discussed. However, many other nonlinear interferometers can also be obtained by replacing the linear elements in basic interferometer with nonlinear FWM processes. For example, a NSI can be realized by replacing the beam splitter in the traditional Sagnac interferometer (TSI) with a FWM process [8], and a hybrid interferometer can also be constructed by using a FWM process as a wave splitter and beam splitter as a wave combiner [9]. Therefore, in Sections 5.3.1 and 5.3.2, these two nonlinear interferometers will be discussed in detail.

5.3.1 Nonlinear Sagnac Interferometer

When two electromagnetic waves travel oppositely along a closed circular path with rotation, there will be a phase difference between these two counter-propagating waves. This phenomenon is called the Sagnac effect, which has been widely used in the high precision measurement of rotation, such as in fiber-optic gyroscopes and global navigation systems. A TSI can be simply viewed as a combination of a circular ring and a 50:50 beam splitter (BS), as shown in Figure 5.16a. As a beam is seeded into the TSI, it can be calculated that the phase difference (ϕ) between the two output beams from the interferometer is

$$\phi = \frac{8\pi NA\Omega}{c\lambda} \tag{5.32}$$

where N is the number of the turns, A the effective area of the TSI, Ω the angular velocity of the rotation, c the speed of light in vacuum, and λ the wavelength of the light.

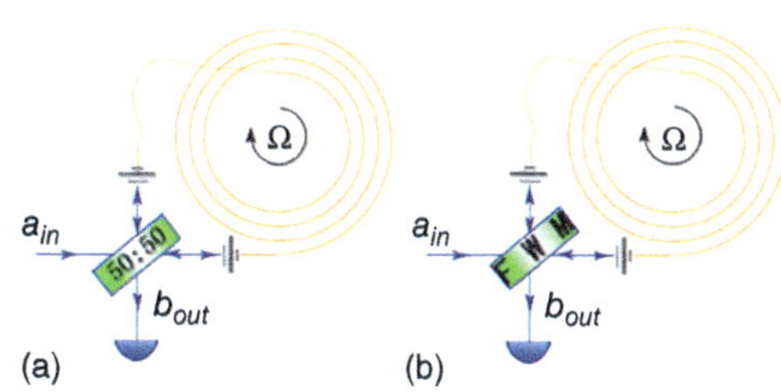

Figure 5.16 The schemes of the TSI and NSI. (a) The TSI; (b) the NSI.

Angular velocity sensitivity, the uncertainty in the estimation of the angular velocity of the Sagnac interferometer, is one of the most important physical quantities characterizing the performance of a Sagnac interferometer. In recent years, substantial researches have been done to improve the angular velocity sensitivity of the Sagnac interferometer. Quantum input, such as the Fock state, has been proposed to improve the measurement precision of the Sagnac interferometer. Atom-based Sagnac interferometer, which uses the wave character of atoms instead of light, has found its advantages over the optical Sagnac interferometer.

Here, a new NSI that can be constructed by replacing the BS in the TSI with a FWM process [8]. Such a NSI has better angular velocity sensitivity than the one of TSI. SQL can be beaten, and HL can even be reached for the ideal case by NSI. We also study the effect of the losses on the angular velocity sensitivity of NSI and find that the optimal angular velocity, where the best angular velocity sensitivity can be obtained, of NSI may be dependent on the losses of the interferometer.

Figure 5.16 shows the schemes of TSI and NSI. It can be found that the BS in the TSI is replaced by the FWM process in NSI. Consider coherent state a_{in} entering into both of TSI and NSI, while vacuum states b_{in} (not marked in Figure 5.16) enter into the other ports. For TSI, the output fields (a_{BS} and b_{BS}) from the 50:50 BS can be given by

$$\begin{pmatrix} \hat{a}_{BS} \\ \hat{b}_{BS} \end{pmatrix} = U_{BS} \begin{pmatrix} \hat{a}_{in} \\ \hat{b}_{in} \end{pmatrix} = \frac{1}{\sqrt{2}} \begin{pmatrix} 1 & 1 \\ 1 & -1 \end{pmatrix} \begin{pmatrix} \hat{a}_{in} \\ \hat{b}_{in} \end{pmatrix} \tag{5.33}$$

where U_{BS} is the transformational matrix of the 50:50 BS. As a_{BS} and b_{BS} pass through the circular ring, there will be a phase difference ϕ between them based on the Sagnac effect as mentioned above. This process can be written as

$$\begin{pmatrix} \hat{a}_{SG} \\ \hat{b}_{SG} \end{pmatrix} = T_{SG} \begin{pmatrix} \hat{a}_{BS} \\ \hat{b}_{BS} \end{pmatrix} = \begin{pmatrix} 1 & 0 \\ 0 & e^{i\phi} \end{pmatrix} \begin{pmatrix} \hat{a}_{BS} \\ \hat{b}_{BS} \end{pmatrix} \tag{5.34}$$

where T_{SG} is the transformational matrix of the Sagnac effect, a_{SG} and b_{SG} the fields after the a_{BS} and b_{BS} pass through the circular ring, and $\phi = 8\pi NA\Omega/(c\lambda)$ from Eq. (5.32). Then the output fields (a_{out} and b_{out}) from TSI can be given by

$$\begin{pmatrix} \hat{a}_{out} \\ \hat{b}_{out} \end{pmatrix} = U_{BS} T_{SG} U_{BS} \begin{pmatrix} \hat{a}_{in} \\ \hat{b}_{in} \end{pmatrix} = \begin{pmatrix} e^{i\phi} \cos \frac{\phi}{2} \hat{a}_{in} - ie^{i\phi} \sin \frac{\phi}{2} \hat{b}_{in} \\ -ie^{i\phi} \sin \frac{\phi}{2} \hat{a}_{in} + e^{i\phi} \cos \frac{\phi}{2} \hat{b}_{in} \end{pmatrix} \tag{5.35}$$

Angular velocity sensitivity $\Delta\Omega$, the minimal angular velocity difference that can be detected, is one of the most important quantities characterizing the performance of a Sagnac interferometer, which can be defined by the error propagation formula

$$\Delta\Omega^2 = \frac{\langle(\Delta\hat{N})^2\rangle}{|\partial\langle\hat{N}\rangle/\partial\Omega|^2} \tag{5.36}$$

where N consists of some measurement combinations of the output fields. For the direct measurement of the b_{out}, $\hat{N} = \hat{N}_b = \hat{b}^\dagger_{out}\hat{b}_{out}$ denotes the photon number operator of the corresponding field. For the coherent state a_{in} and vacuum state b_{in}

entering into the TSI, the intensity noise $\langle(\Delta\hat{N}_b)^2\rangle$ and the slope $\partial\langle\hat{N}_b\rangle/\partial\Omega$ of the interference fringe of b_{out} can be calculated from Eq. (5.35) and given by

$$\left\langle \Delta\hat{N}_b^2 \right\rangle = \sin^2\frac{\beta\Omega}{2}\left\langle a_{in}^{\dagger}a_{in}\right\rangle \tag{5.37}$$

$$\frac{\partial\hat{N}_b}{\partial\Omega} = \beta\sqrt{\frac{2}{(1+\cos\beta\Omega)N_s}} \tag{5.38}$$

where $\beta = 8\pi NA/(c\lambda)$. Then the angular velocity sensitivity $\Delta\Omega_{TSI}$ of TSI can be easily derived from Eqs. (5.36)–(5.38) and written as

$$\Delta\Omega_{TSI} = \frac{1}{\beta}\sqrt{\frac{2}{(1+\cos\beta\Omega)N_s}} \tag{5.39}$$

where $N_s = \hat{a}_{in}^{\dagger}\hat{a}_{in}$ is the total intensity of the fields inside TSI. For the NSI, as shown in Figure 5.16b, the output fields (a_{FWM} and b_{FWM}) from the FWM process can be written as

$$\begin{pmatrix}\hat{a}_{FWM}\\ \hat{b}_{FWM}^{\dagger}\end{pmatrix} = U_{PA}\begin{pmatrix}\hat{a}_{in}\\ \hat{b}_{in}^{\dagger}\end{pmatrix} = \begin{pmatrix}G & g\\ g & G\end{pmatrix}\begin{pmatrix}\hat{a}_{in}\\ \hat{b}_{in}^{\dagger}\end{pmatrix} \tag{5.40}$$

where U_{PA} is the transformational matrix of the FWM process, G the amplification gain of the FWM process, and we have the relation $G^2 - g^2 = 1$. Similar to the TSI, the output fields from the NSI can be derived from Eqs. (5.34) and (5.40) and written as

$$\begin{pmatrix}\hat{a}_{out}\\ \hat{b}_{out}\end{pmatrix} = \begin{pmatrix}(G^2 + g^2e^{-i\phi})\hat{a}_{in} + Gg(1+e^{-i\phi})\hat{b}_{in}^{\dagger}\\ Gg(1+e^{i\phi})\hat{a}_{in}^{\dagger} + (G^2e^{i\phi} + g^2)\hat{b}_{in}\end{pmatrix} \tag{5.41}$$

For the coherent state a_{in} and vacuum state b_{in} entering into NSI, the intensity noise and the slope of the interference fringe of the b_{out} can be calculated from Eq. (5.41) and given by

$$\begin{aligned}\left\langle \Delta\hat{N}_b^2 \right\rangle &= 4G^2g^2(1+\cos\beta\Omega)^2\left(2\left\langle \hat{a}_{in}^{\dagger}\hat{a}_{in}\right\rangle + 1\right)\\ &\quad + 2G^2g^2(1+\cos\beta\Omega)\left(\left\langle \hat{a}_{in}^{\dagger}\hat{a}_{in}\right\rangle + 1\right)\end{aligned} \tag{5.42}$$

$$\frac{\partial\langle\hat{N}\rangle}{\partial\Omega} = -2\beta G^2g^2\sin\beta\Omega\left(\left\langle \hat{a}_{in}^{\dagger}\hat{a}_{in}\right\rangle + 1\right) \tag{5.43}$$

By neglecting the small items due to $\left\langle \hat{a}_{in}^{\dagger}\hat{a}_{in}\right\rangle \gg 1$ in Eqs. (5.42) and (5.43), the angular velocity sensitivity $\Delta\Omega_{NSI}$ of NSI can be easily derived from Eqs. (5.36), (5.42), and (5.43), and we find that

$$\Delta\Omega_{NSI} = \frac{1}{\beta}\sqrt{\frac{4G^2g^2(G^2+g^2)(1-\cos\beta\Omega) + G^2 + g^2}{2G^2g^2N_s(1+\cos\beta\Omega)}} \tag{5.44}$$

where $N_s = (G^2 + g^2)\left\langle \hat{a}_{in}^{\dagger}\hat{a}_{in}\right\rangle$ is the total intensity of the fields inside the NSI.

Figure 5.17 shows the comparison of the angular velocity sensitivity versus the angular velocity between the TSI and NSI with $N_s = 10^4$. It is convenient to nondimensionalize $\Delta\Omega$ with the substitution of $\beta = 1$. The black line in Figure 5.17 is the angular velocity sensitivity of TSI, whereas the green and blue lines are the

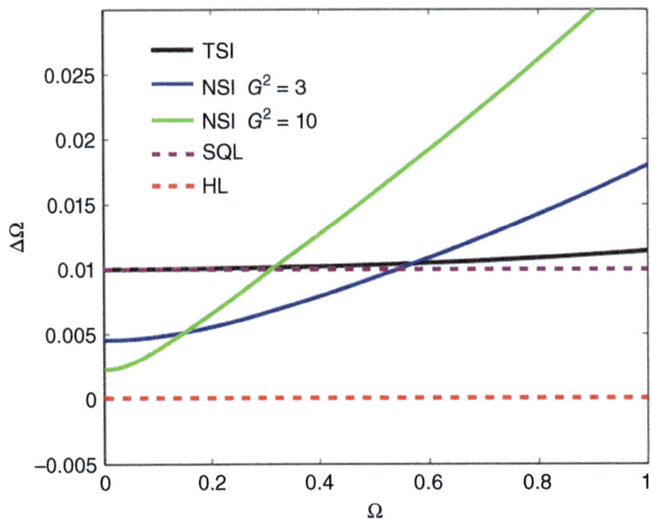

Figure 5.17 The comparison of the angular velocity sensitivity versus the angular velocity between TSI and NSI, where it is assumed that the total intensity of the fields inside both two interferometers is $N_s = 10^4$. The black line is the angular velocity sensitivity of the TSI. The green and blue lines are the corresponding ones of the NSI with $G^2 = 3$ and $G^2 = 10$. The dash lines colored by purple and red are the corresponding SQL and HL, which are characterized by the angular velocity sensitivity scalings of $1/\sqrt{N_s}$ and $1/N_s$, respectively. Here it should be noted that the value of the HL ($\Delta\Omega = 1/N_s = 10^{-4}$) is too small to be distinguished from the zero line in this figure.

corresponding ones of NSI with $G^2 = 3$ and $G^2 = 10$. The dashed lines colored purple and red are the corresponding SQL and HL, which are characterized by the angular velocity sensitivity scalings of $1/\sqrt{N_s}$ and $1/N_s$, respectively. As shown in Figure 5.17, the angular velocity sensitivity of both TSI and NSI depends on the angular velocity. In other words, both TSI and NSI cannot work well at all angular velocities. It is because the angular velocity sensitivity characterized by the error propagation formula, as shown by Eq. (5.36), is determined by both the intensity noise and the intensity interference slope. When the angular velocity deviates from the best sensitivity point, the increasing intensity noise will dominate and lead to the degradation of the angular velocity sensitivity for both TSI and NSI. Therefore, it is natural to ask whether the dependence of the angular velocity sensitivity on the angular velocity for both TSI and NSI is a consequence of the intensity measurement being utilized or the properties of the interferometers themselves in general. In order to further study the fundamental properties of TSI and NSI, quantum Fisher information (Q) should be considered, which gives the bound $\left(\Delta\Omega \geq \frac{1}{\beta}\frac{1}{\sqrt{Q}}\right)$ of the estimation uncertainty of the angular velocity sensitivity. The Q for TSI (Q_{TSI}) and NSI (Q_{NSI}) can be calculated and expressed as $Q_{TSI} = 2N_s$ and $Q_{NSI} = 8G^2g^2 2N_s/(G^2 + g^2)$, respectively, from which it can be found that both the Q_{TSI} and Q_{NSI} do not depend on the angular velocity. Therefore, the dependence of the angular velocity sensitivity on the value of the underlying angular velocity being estimated for both TSI and NSI is merely due to the intensity measurement being utilized. In order to obtain the best angular velocity sensitivity, the angular velocity has to be manipulated. And the optimal angular velocity Ω_{opt}, where the best angular velocity sensitivity can be obtained, happens for both of the two interferometers at the dark interference fringe $\Omega = 2k\pi$, where k is an integer. From Eqs. (5.39) and (5.44), it can be calculated that with the optimal angular velocity $\Omega_{opt} = 2k\pi$, the ratio of the $\Delta\Omega_{TSI}$ to $\Delta\Omega_{NSI}$ is $\sqrt{\frac{4G^2g^2}{G^2+g^2}}$, which is larger than 1 for any $G^2 > 1$, that is to say the best angular velocity sensitivity of NSI can always be better than the one of the TSI. For SQL, it can be found that SQL cannot be beaten by TSI at all angular velocities, while for NSI, it can be beaten in some regions near the dark interference fringe $\Omega = 2k\pi$. By comparing the green and blue lines in Figure 5.17, it can be found that the best angular velocity sensitivity of the NSI decreases with the increasing of the gain (G) of the FWM process. Therefore, it can be inferred that the HL can even be reached by NSI if G is large enough.

To further study the performance of the NSI, a particular situation is considered where there is no input into the NSI, i.e. $\left\langle \hat{a}_{in}^{\dagger}\hat{a}_{in} \right\rangle = 0$ in Eqs. (5.42) and (5.43). Then the angular velocity sensitivity of NSI in Eq. (5.44) becomes

$$\Delta\Omega_{NSI} = \frac{1}{\beta}\sqrt{\frac{N_s(N_s+2)(1-\cos\beta\Omega)+2}{N_s(N_s+2)(1+\cos\beta\Omega)}} \tag{5.45}$$

where $N_s = 2g^2$ is the total intensity of the fields inside NSI. Figure 5.18 shows the angular velocity sensitivity scalings of the NSI for the case of no input into the interferometer. The green, blue, brown, and black lines represent the angular velocity

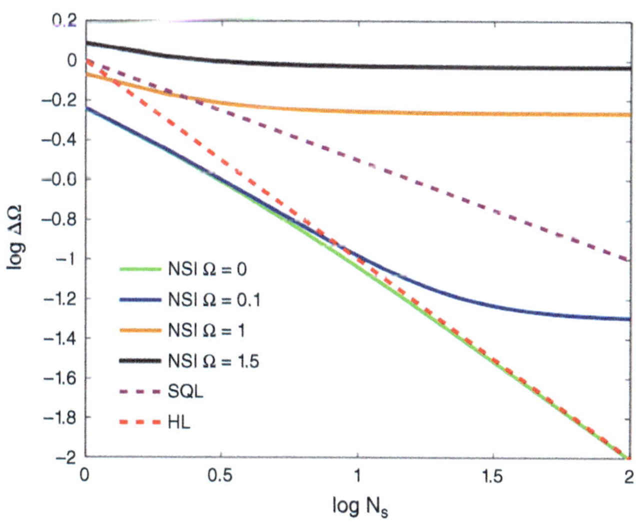

Figure 5.18 The angular velocity sensitivity scalings of NSI for the case of no input into the interferometer. The green, blue, brown, and black lines are the angular velocity sensitivity of NSI versus N_s with different Ω. The dashed lines colored purple and red are the corresponding SQL and HL, respectively.

sensitivity of the NSI versus N_s with different Ω. The dashed lines colored by purple and red are the corresponding SQL and HL. As shown in Figure 5.18, in the case of no input into the interferometer, the performance of NSI is also dependent on the angular velocity, which is similar to the case of seeding coherent state as shown in Figure 5.17. And the best angular velocity sensitivity of NSI also happens at $\Omega = 2k\pi$. More interestingly, when $\Omega = 2k\pi$, the N_s terms will only exist in the denominator of Eq. (5.45), which leads to the characteristic of an asymptote, as shown by the green line in Figure 5.18. Otherwise, when $\Omega \neq 0$ and N_s is large enough, Eq. (5.45) will only depend on the angular velocity Ω and approach a certain value for a given angular velocity Ω, as shown by blue, brown, and black lines in Figure 5.18. In this sense, we could name this phenomenon as saturation effect of the angular velocity sensitivity.

It is well known that loss is one of the most important factors that can extremely affect the performance of an interferometer. Also, the angular velocity sensitivity of NSI can be strongly dependent on the losses of the interferometer. To interpret the losses of the NSI, fictitious beam splitters with reflectivity of η are modeled on the input fields (a_{in} and b_{in}), inside the NSI (a_{FWM} and b_{FWM}), and on the output fields (b_{out}) of the interferometer. For one beam splitter, the input–output relation can be expressed as

$$\hat{a}' = \sqrt{1-\eta}\hat{a} + \sqrt{\eta}\hat{v} \tag{5.46}$$

where $\hat{a}$ is the input field, $\hat{a}'$ the output field, and v the vacuum field.

Similar to the study of the NSI with ideal case, we start by investigating the effect of the losses on the angular velocity sensitivity of the NSI with the entering of the coherent state a_{in} and vacuum state b_{in}. From Eqs. (5.34), (5.36), (5.40), and (5.46), the angular velocity sensitivity of the NSI can be calculated and given by

$$\Delta\Omega_{NSI} = \frac{1}{\beta}\sqrt{\frac{(G^2+g^2)[4(1-\eta_2)(1-\eta_3)G^2g^2(1-\cos\beta\Omega) + 1 - \eta_2\eta_3 + 2g^2\eta_2(1-\eta_3)]}{2G^2g^2N_s(1-\eta_1)(1-\eta_2)(1-\eta_3)(1+\cos\beta\Omega)}} \tag{5.47}$$

where $N_s = (G^2+g^2)\left\langle \hat{a}_{in}^{\dagger}\hat{a}_{in}\right\rangle$ is the total intensity of the fields inside the NSI, η_1, η_2, and η_3 stands for the losses on the input fields, inside the NSI, and on the output fields of the NSI, respectively.

Since the functional form of Eq. (5.47) is similar to the ones of Eqs. (5.44) and (5.45), it can be easily inferred that the angular velocity sensitivity of the NSI will become better with a larger gain (G) of the FWM process, and the best angular velocity sensitivity will be obtained at $\Omega = 2k\pi$, which has been discussed above. Therefore, for simplicity, the effect of the losses on the angular velocity sensitivity of NSI is studied on the assumption of $G^2 = 3$ and $\Omega = 2k\pi$. Figure 5.19 shows the effect of the losses on the angular velocity sensitivity of NSI with the entering of the coherent state a_{in} and vacuum state b_{in}. The green, yellow, blue, and black lines are the angular velocity sensitivity scalings of NSI with different η_1, η_2, and η_3. The dashed lines colored purple and red are the corresponding SQL and HL, respectively. By comparing the green and yellow lines in Figure 5.19, it can be found that the losses on the input ($\eta_1 = 0.1$) and output ($\eta_3 = 0.1$) fields of the NSI,

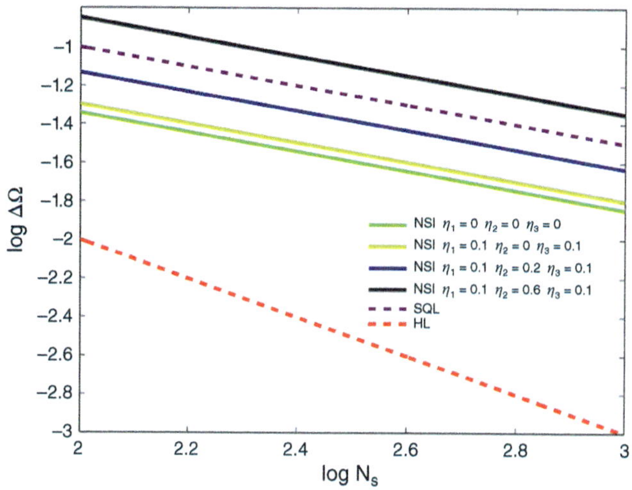

Figure 5.19 The effect of the losses on the angular velocity sensitivity of NSI for the case of seeding coherent state where $\Omega = 2k\pi$. The green, yellow, blue, and black lines are the angular velocity sensitivity scalings of NSI with different η_1, η_2, and η_3. The dashed lines colored purple and red are the corresponding SQL and HL, respectively.

such as the detection inefficiency, do affect the angular velocity sensitivity of NSI. Here we also study the loss inside NSI, which can be caused by atomic absorptions and unavoidable losses in the optical fiber. As shown by yellow, blue, and black lines, the angular velocity sensitivity of NSI becomes worse with the increasing loss η_2 inside the interferometer. However, all three types of losses cannot change the functional form of the angular velocity sensitivity of the NSI, which has a characteristic of $1/\sqrt{N_s}$. For the SQL, it can be beaten by NSI over a wide range of losses (for example, $\eta_1 = 0.1$, $\eta_2 < 0.43$, and $\eta_3 = 0.1$).

Next, we study how the losses can affect the angular velocity sensitivity of NSI for the case of no input into the interferometer. Similar to the above cases, for the case of no input into the interferometer, the angular velocity sensitivity of NSI with the consideration of the losses can be given by

$$\Delta\Omega_{NSI} = \frac{1}{\beta}\sqrt{\begin{aligned}&\frac{1-\cos\beta\Omega}{1+\cos\beta\Omega} + \frac{2(N_s\eta_2 + 1 - N_s\eta_2\eta_3)}{N_s(N_s+2)(1-\eta_2)(1-\eta_3)(1+\cos\beta\Omega)}\\ &+ \frac{\eta_2(N_s\eta_2 - N_s\eta_2\eta_3 + 2 - 2\eta_3 + 2\eta_2\eta_3)}{N_s(N_s+2)^2(1-\eta_2)^2(1-\eta_3)\sin^2\beta\Omega}\end{aligned}} \tag{5.48}$$

where $N_s = 2g^2$ is the total intensity of the fields inside NSI. Here it should be pointed out that, in the case of no input into the interferometer, the loss on the input field η_1 does not affect the angular velocity sensitivity of NSI due to the fact that the vacuum state cannot be affected by any losses. From Eq. (5.48), it can be found that for the case of no input into the interferometer, the functional form of the angular velocity sensitivity of NSI has been changed by the introduction of the losses, which is different from the case of seeding coherent state in Eq. (5.47). Figure 5.20 shows the effect of the losses on the angular velocity sensitivity of NSI with no input into the

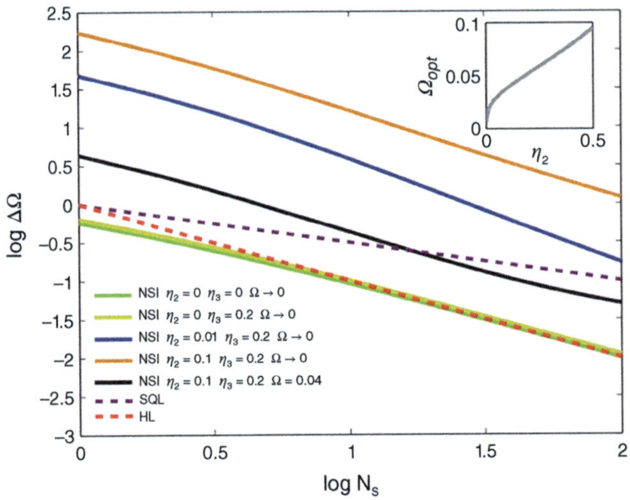

Figure 5.20 The effect of the losses on the angular velocity sensitivity of NSI with no input into the interferometer. The green (yellow) line is the angular velocity sensitivity of NSI with $\Omega \to 2k\pi$ and $\eta_2 = 0$ for the case of $\eta_3 = 0$ ($\eta_3 = 0.2$). Here it should be noted that green line is too close to the yellow line, such that it is hard to distinguish the green line from the yellow one. The blue (brown) line is the corresponding one with $\eta_3 = 0.2$ and $\Omega \to 2k\pi$ for the case of $\eta_2 = 0.01$ ($\eta_2 = 0.1$). The black line is the corresponding one of NSI with $\Omega = 0.04$ for the case of $\eta_2 = 0.1$ and $\eta_3 = 0.2$. The dashed lines colored purple and red are the corresponding SQL and HL. The inset of this figure shows the optimal angular velocity versus different η_2.

interferometer. The green (yellow) line is the angular velocity sensitivity of NSI with $\Omega \to 2k\pi$ and $\eta_2 = 0$ for the case of $\eta_3 = 0$ ($\eta_3 = 0.2$). The blue (brown) line is the corresponding one with $\eta_3 = 0.2$ and $\Omega \to 2k\pi$ for the case of $\eta_2 = 0.01$ ($\eta_2 = 0.1$). The black line is the corresponding one of NSI, with $\Omega = 0.04$ for the case of $\eta_2 = 0.1$ and $\eta_3 = 0.2$. The dashed lines colored purple and red are the corresponding SQL and HL, respectively. From the green and yellow lines in Figure 5.20, it can be found that the loss on the output field of the interferometer η_3 can affect the angular velocity sensitivity of NSI. By comparing the green, blue, and brown lines in Figure 5.20, it is clear that the angular velocity sensitivity of NSI is much more sensitive to the loss η_2 than the case of seeding coherent state as shown in Figure 5.19. This is because for $\Omega \to 2k\pi$, the third term of Eq. (5.48) becomes significantly large with $\eta_2 > 0$.

As a result, NSI is extremely sensitive to the loss (η_2) inside the system, and HL cannot be reached by NSI with the introduction of η_2. By comparing the black and blue lines in Figure 5.20, the most interesting thing is that for the cases of $\eta_2 = 0.1$, $\eta_3 = 0.2$, and $\Omega = 0.04$, the angular velocity sensitivity of the NSI can be better than the one with $\eta_2 = 0.01$, $\eta_3 = 0.2$, and $\Omega \to 2k\pi$, which is different from all the cases mentioned above where the optimal angular velocity is $\Omega_{opt} = 2k\pi$. It seems that with the introduction of the losses, a new optimal angular velocity Ω_{opt} is required to be searched in order to obtain the best angular velocity sensitivity of the NSI. The inset of Figure 5.20 shows the optimal angular velocity Ω_{opt} versus different η_2. It can be found that in order to obtain the best angular velocity sensitivity of the NSI, the angular velocity is required to be far away from $\Omega = 2k\pi$ with the increasing η_2. From this point of view, the manipulation of the angular velocity can be used to compensate the effect of the loss (η_2) in such a NSI. Since losses are always unavoidable in the real system, our results here are valuable for the real applications of the NSI.

In summary, in this work we propose a new NSI by replacing the BS in TSI with a FWM process. Such a NSI has better angular velocity sensitivity than the one of TSI. The SQL can be beaten, and the HL can even be reached in the ideal case by NSI. The effect of the losses inside the interferometer on the angular velocity sensitivity of NSI is also studied, and we find that the optimal angular velocity of NSI may be dependent on the losses of the interferometer. Such a NSI may find its potential applications in quantum metrology.

5.3.2 Hybrid Interferometer with a Nonlinear FWM Process and a Linear Beam-splitter

Section 5.3.1 discussed a NSI, and a hybrid interferometer composed of a nonlinear FWM process and a linear beam splitter will be discussed in the present section [9]. Here, nonlinear FWM process is a frequency-degenerate four-wave mixing (DFWM) process, in which one only needs to split a small portion of the pump beam using a linear beam splitter to obtain the seed beam. On the contrary, the seed beam of the NDFWM is obtained by double passing a small portion of pump beam through a GHz-level acoustic–optic modulator (AOM). This is determined by the double-Λ energy-level configuration of their NDFWM process. Therefore, the DFWM can be realized more easily than NDFWM due to its frequency degeneracy. As shown in Figure 5.21a, a weak-signal beam with the same frequency as the pump is crossed with the strong pump beam in the center of the Rb-85 vapor cell at a small angle. During this process, the seed (signal) field is amplified, and a phase conjugate (idler) field is generated in the meantime. On account of the phase-matching condition, the pump-idler angle is identical with the pump-signal angle on the other side of the pump. Due to energy conservation, the idler beam has the same frequency as both the signal beam and pump beam. This is different from the non-degenerate four-wave mixing (NDFWM) as shown in Figure 5.21b, where the idler beam has a 6.08 GHz frequency shift from the signal beam.

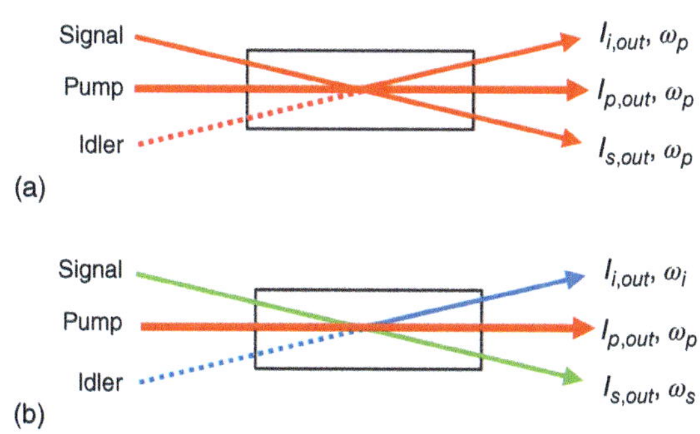

Figure 5.21 Comparison between the scheme of DFWM and the scheme of NDFWM. (a) The scheme of DFWM, where all the three output beams have the same frequency. (b) The scheme of NDFWM, where all the three output beams have different frequencies. $I_{s,\,out}$, $I_{i,\,out}$, $I_{p,\,out}$, the intensities of the output beams; ω_s, ω_i, ω_p, the frequency of the output beams.

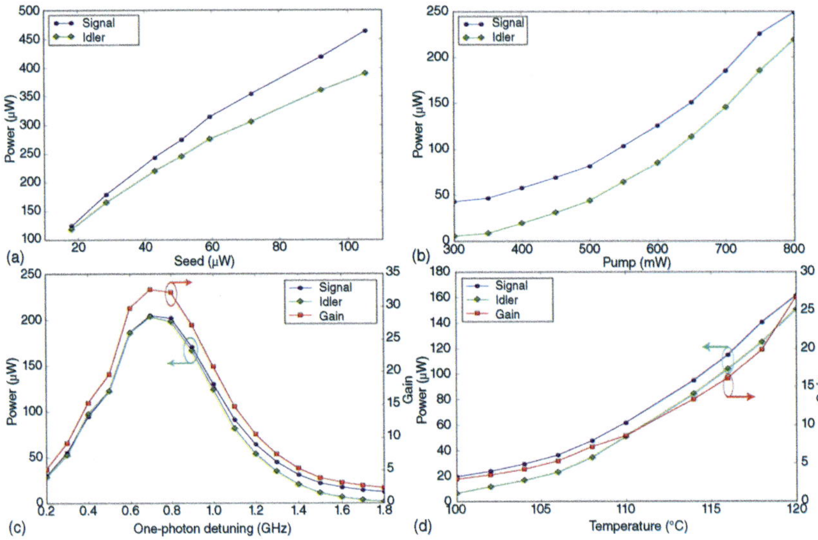

Figure 5.22 The relationships between the power of two output ports of DFWM and various system parameters. (a) The relationship between the power of two output ports (signal beam, blue dot and idler beam, green diamond) and the seed power. (b) The relationship between the power of two output ports (signal beam, blue dot and idler beam, green diamond) and the pump power. (c) The power of two output ports (signal beam, blue dot and idler beam, green diamond) as a function of the one-photon detuning. (d) The power of two output ports (signal beam, blue dot and idler beam, green diamond) as a function of the temperature of the Rb-85 vapor cell. Both (c) and (d) figures include the traces for the gain of the system (red square).

In order to prove that the above-mentioned DFWM can be used as a PA for constructing the hybrid interferometer, we study the relationships between the power of the two output ports and different system parameters, such as the seed power, the pump power, the one-photon detuning, and the temperature of the Rb-85 vapor cell. Firstly, we set the pump power at 900 mW, maintain the temperature of Rb-85 vapor cell at 120 °C by using a digital temperature controller, set the one-photon detuning at 0.9 GHz, and change the seed power from 18 to 105 μW. As shown in Figure 5.22a, we find that the output power of the signal and idler beams approximately have linear dependences on the seed power. And the difference between the power of signal and idler beams is approximately equal to the input seed power. Secondly, while keeping the seed power at 50 μW, the temperature of Rb-85 vapor cell at 120 °C, the one-photon detuning at 0.9 GHz, and varying the pump power from 300 to 800 mW, we record the power of the signal and idler beams at the output. As shown in Figure 5.22b, the signal and idler power of two output ports increase with the increasing pump power. The power difference between the signal and idler beams almost keeps constant and is approximately equal to the input seed power. Thirdly, we fix the seed power at 5 μW, the pump power at 1 W, the temperature of Rb-85 vapor cell at 120 °C, and scan the one-photon detuning from 0.2 to 1.8 GHz. The result is shown in Figure 5.22c. When the one-photon detuning is 0.7 GHz, the system gain reaches its maximum of about 32. The system gain is determined by calculating the ratio between the output signal power and the input seed power. Finally, we vary the temperature of the Rb-85 vapor cell from 100 to 120 °C and set the seed power at 5 μW, the pump power at 1 W, and the one-photon detuning at 0.9 GHz to study the influence of the temperature. As shown in Figure 5.22d, it clearly shows that the power of both the two output ports increases as the temperature increases. Based on all the experimental results as shown in Figure 5.22, we can see that DFWM can work as a PA, and, therefore, it can be used to construct the hybrid interferometer involving both the nonlinear PA and linear beam splitter. When the wave splitter is completed, a linear beam splitter will be used as the wave combiner to interfere with the signal and idler beams with the identical frequency. As interference visibility is a direct measure of how perfectly the two interfering waves can cancel due to destructive interference and gives the degree of coherence between the waves, it is worth studying the dependence of interference visibilities on the different system parameters, such as the gain of the PA, the one-photon detuning, and the temperature of the Rb-85 vapor cell. Firstly, in order to explore the relationship between the interference visibility and the gain of the PA, we vary the pump power from 300 to 1100 mW, which ranges the gain from 1.5 to 26. In this case, we keep the frequency of the pump beam unchanged at blue, detuned about 0.9 GHz from the D_1 line Rb-85 ($5S_{1/2}, F = 2 \rightarrow 5P_{1/2}$), which is called one-photon detuning (Δ). The seed beam has the same frequency as the pump, and its power is fixed at about 5 μW. The temperature of Rb-85 vapor cell is kept at 120 °C. As shown in Figure 5.23a, the blue dot line represents the relationship between the interference visibility and the gain for D_1 output port, and the green diamond line is for D_2 output port. Both of the interference visibilities of the D_1 and D_2 output ports increase as the gain increases. The two visibilities are almost identical and can be above 90% for a gain larger than about 12. Secondly, we investigate how the interference visibility depends on the one-photon detuning of the PA. In order to study this effect, we set the seed power at 5 μW, the pump power at 1 W, the temperature of Rb-85 vapor cell at 120 °C, and scan the one-photon detuning from 0.2 to 1.8 GHz by changing the frequency of the pump beam. As shown in Figure 5.23b, the blue dot line represents the relationship between the interference visibility and the one-photon detuning for D_1 output port, and the green diamond line is for the D_2 output port. The red square line represents the gain of the PA. We find that the interference visibilities of D_1 and D_2 output ports maintain above 90% for the range of the one-photon detuning from 0.7 to 1.2 GHz. And when the one-photon detuning is 0.9 GHz, the interference visibility reaches its maximum of about 95%. Thirdly, we fix the seed power at 5 μW, the pump power at 1 W, the one-photon detuning at 0.9 GHz, and vary the temperature of the Rb-85 vapor cell from 100 to 120 °C. The result is shown in Figure 5.23c. The interference visibility of D_1 and D_2 output ports increases with the increasing temperature. Within the range of the temperature from 115 to 120 °C, the interference visibilities for both output ports are all about 90%.

The frequency degeneracy of our hybrid interferometer ensures that the signal and idler fields experience same levels of self-focusing and optical losses. Therefore, the visibility of our hybrid interferometer will not be affected by these two

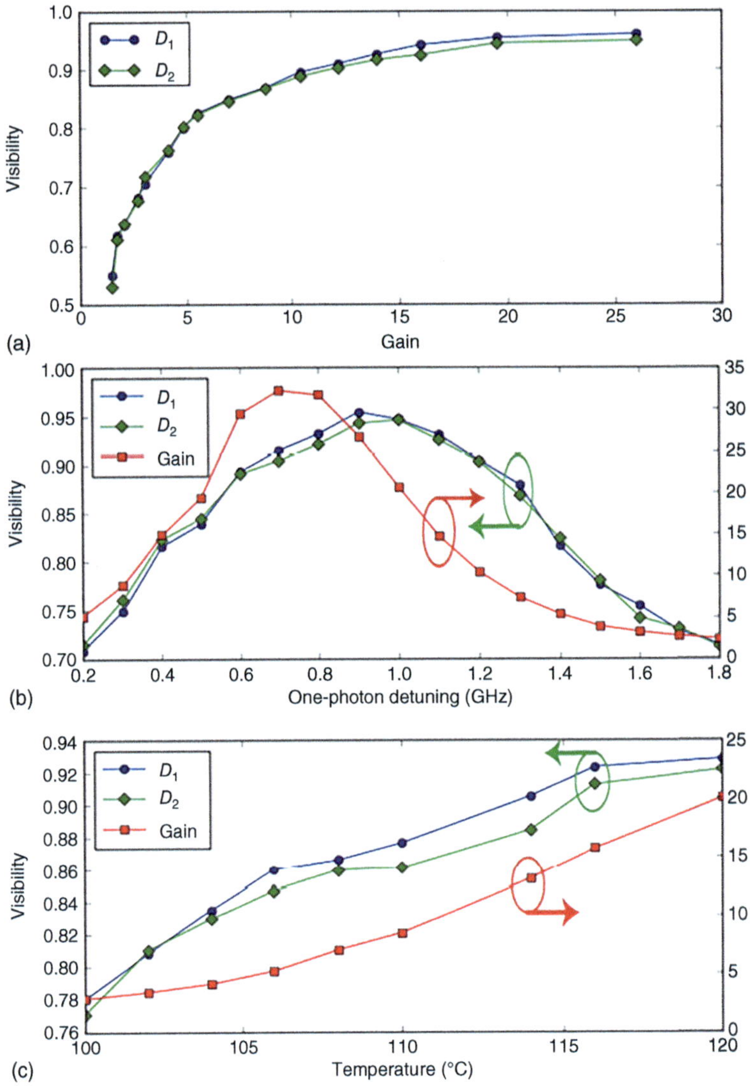

Figure 5.23 The dependence of the interference visibilities on various system parameters. (a) The interference visibilities of the two output ports as a function of the gain of the PA. (b) The relationship between the interference visibilities of the two output ports and the one-photon detuning. (c) The relationship between the interference visibilities of the two output ports and the temperature of the Rb-85 vapor cell. Both (b) and (c) figures include the traces for the gain of the PA.

experimental imperfections. However, the visibility of nonlinear interferometer can be affected by these two experimental imperfections due to the fact that the signal and idler fields at different frequencies experience different levels of self-focusing and optical losses. Thus, the hybrid interferometer is more suitable for finding potential application in precision measurement when amplification, robustness, and conciseness are required.

5.3.3 Experimental Implementation of a Phase-Sensitive Parametric Amplifier

The previous mentioned interference phenomena involve the twice utilization of a FWM process; can the interference phenomenon happen when a FWM process is utilized only once? This section will give the answer to this question [10]. FWM process, as another type of nonlinear beam splitter, can combine two incoming optical fields in coherent state with different frequencies. This is actually the proposal for two-mode PSA. As shown in Figure 5.24a, a normal beam splitter combines two input beams of the same frequency and induces interference phenomena at the two output ports. Considering the schematic shown in Figure 5.24b, two beams with different frequencies are injected into a nonlinear beam splitter, which is based on PA process. The input–output relationship of a PA is well described by

$$\begin{aligned}\hat{a}_{out} &= \sqrt{G}\hat{a}_{in} + e^{i\phi}\sqrt{G-1}\hat{b}_{in}^{\dagger}\\ \hat{b}_{out}^{\dagger} &= e^{-i\phi}\sqrt{G-1}\hat{a}_{in} + \sqrt{G}\hat{b}_{in}^{\dagger}\end{aligned} \tag{5.49}$$

where G is the intensity gain, and $\hat{a}$ ($\hat{b}$) indicates the signal (idler) field. Here, we assume that the idler beam is subject to a phase shift of ϕ, while the phases of the signal and pump beams are fixed.

When both input ports are seeded by coherent fields, the intensities of the output fields are given by

$$\begin{aligned}I_{s,out} &= GI_a + (G-1)I_b + 2\sqrt{G(G-1)I_aI_b}\cos\phi\\ I_{i,out} &= GI_b + (G-1)I_a + 2\sqrt{G(G-1)I_aI_b}\cos\phi\end{aligned} \tag{5.50}$$

where I_a and I_b indicate the intensities of the input fields. Here, we have neglected the spontaneous emission term $G-1$, since both input beams (in μW order, corresponding to photon number of 4×10^{12} within one second) are bright in the experiment. From Eq. (5.50), one can easily find that the output fields can

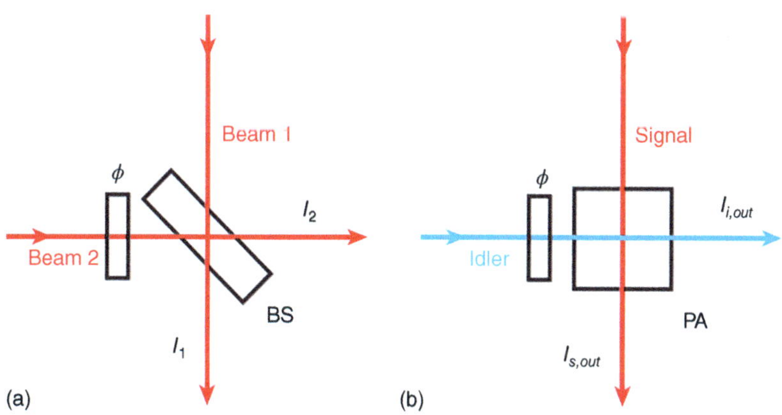

Figure 5.24 (a) A normal beam splitter combining two beams. (b) A nonlinear beam splitter based on a PA. BS, beam splitter; PA, parametric amplifier; ϕ, the phase shift of beam 2 or idler beam; I_1; I_2; $I_{s,out}$; and $I_{i,out}$, the intensities of the output beams.

be amplified or de-amplified under varying phases, leading to the interference phenomena at the two output ports. The interference visibilities for both output ports are calculated

$$V_s = \frac{2\sqrt{\beta}\sqrt{G(G-1)}}{\beta G + G - 1}$$
$$V_i = \frac{2\sqrt{\beta}\sqrt{G(G-1)}}{\beta G + G - \beta} \tag{5.51}$$

where $\beta = I_a/I_b$. The interference phenomenon can be explained as two individual PAs seeded only by the signal or idler field interfering with each other. Each PA amplifies or generates its own signal and idler beams, respectively. The two output signal beams of the same frequency are well overlapped under our current symmetric geometry and so are the two output idler beams. Thus, these two output signal (idler) beams will interfere with each other, leading to the amplification or de-amplification of the beams. From this point of view, our nonlinear beam splitter induces interference phenomena at each output port.

The experimental layout is shown in Figure 5.25. The nonlinear beam splitter is based on a double-Λ configuration FWM process in a Rb-85 vapor cell (as shown in Figure 5.25b). We use an external cavity diode laser (ECDL) and a tapered amplifier (TA) as our laser system. The linewidth of the ECDL is 100 kHz, and the frequency is blue-detuned about 0.8 GHz to the $5S_{1/2}, F = 2 \rightarrow 5P_{1/2}$ transition. A PBS is used to split the beam into two. One is sent to the TA to generate the pump beam, which is vertically polarized. The other one is passed through an AOM 1. A weak-signal beam tuned 3.04 GHz to the red side of the pump is derived. Afterwards, the zero-order beam passed through the first AOM is sent to the second AOM. In the same way, a weak idler beam tuned 3.04 GHz to the blue side of the pump is produced. Both signal and idler beams are horizontally polarized. All these beams are combined with a Glan laser polarizer and crossed at the center of the Rb vapor cell, whose

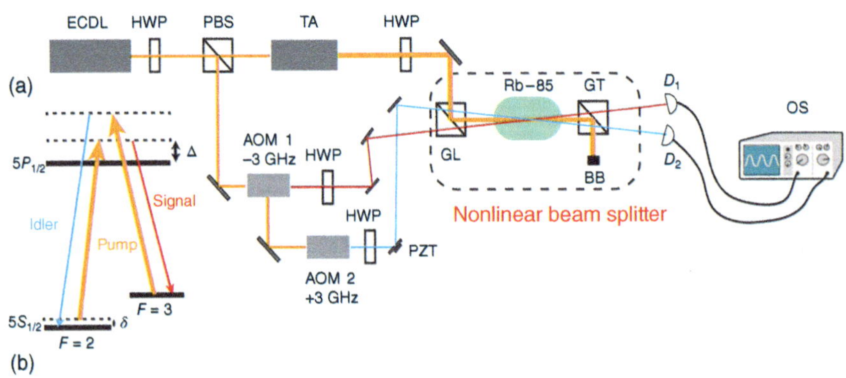

Figure 5.25 Experimental layout. (a) ECDL, external cavity diode laser; HWP, half-wave plate; PBS, polarization beam splitter; TA, tapered amplifier; AOM, acousto-optic modulator; PZT, piezo-electric transducer; GL, Glan laser polarizer; GT, Glan–Thompson polarizer; BB, beam blocker; D1, D2, photodetectors; OS, oscilloscope. (b) Double-Λ scheme of the D1 line of Rb-85. Δ, one-photon detuning; δ, two-photon detuning.

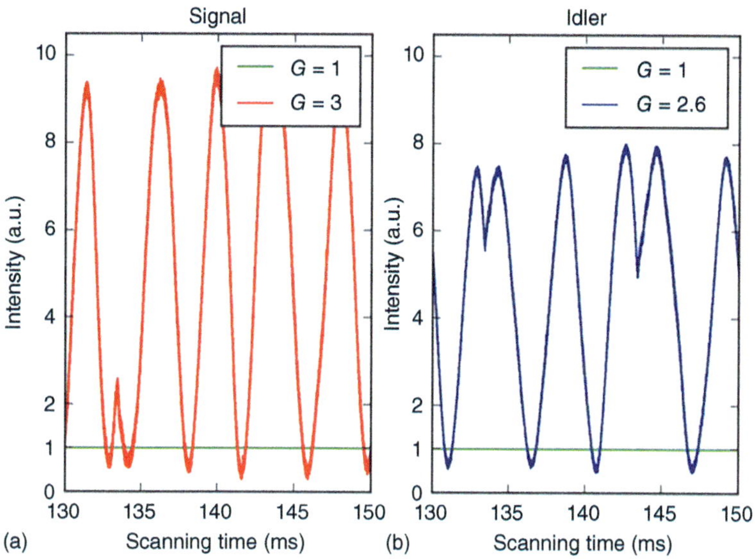

Figure 5.26 Interference fringes at the output ports of the nonlinear beam splitter. (a) Signal output. (b) Idler output. The green lines, the output of signal/idler without the pump beam.

temperature is stabilized at 116 °C. The beam waist of the signal (idler) beam is 200 (215) μm, and 500 μm for the pump beam. A Glan–Thompson polarizer with an extinction ratio of $10^5 : 1$ is placed after the cell to filter out the residual pump beam. The output signal and idler beams are sent to two photodetectors (D_1 and D_2).

As shown in Figure 5.26, the typical interference fringes of the two outputs are obtained by scanning the PZT mirror on the input idler beam path. The interference visibilities of the signal and idler output ports are 91.6% and 90.0%, respectively. It should be pointed out that the data on visibility of the signal and idler beams are not measured at the same time. The two interference fringes are modulated in sync, in principle. The output fields can be amplified (above the straight green lines) or de-amplified (below the straight green lines) by this nonlinear beam splitter depending on the value of phase ϕ. The measured intensities of the two output beams when the pump beam is blocked are defined as units and denoted by the straight green lines in Figure 5.26. The interference pattern intensities of both output beams are normalized to the two straight green lines, respectively.

In order to investigate the influence of intensity ratio β between the two input beams on the visibilities of the two output ports, we keep the idler beam power (10 μW) unchanged and vary the signal beam power from 2.5 to 40 μW. At the same time, we change the intensity gain of the system by varying the pump beam power. The other parameters are set as follows: $\delta = 4$ MHz and $\Delta = 0.8$ GHz. The experimental results (b) and theoretical predictions (a) agree very well, as shown in Figure 5.27. For the experimental results, the maximal visibilities for both the signal and idler beams are limited by absorption losses and imperfect spatial mode matching

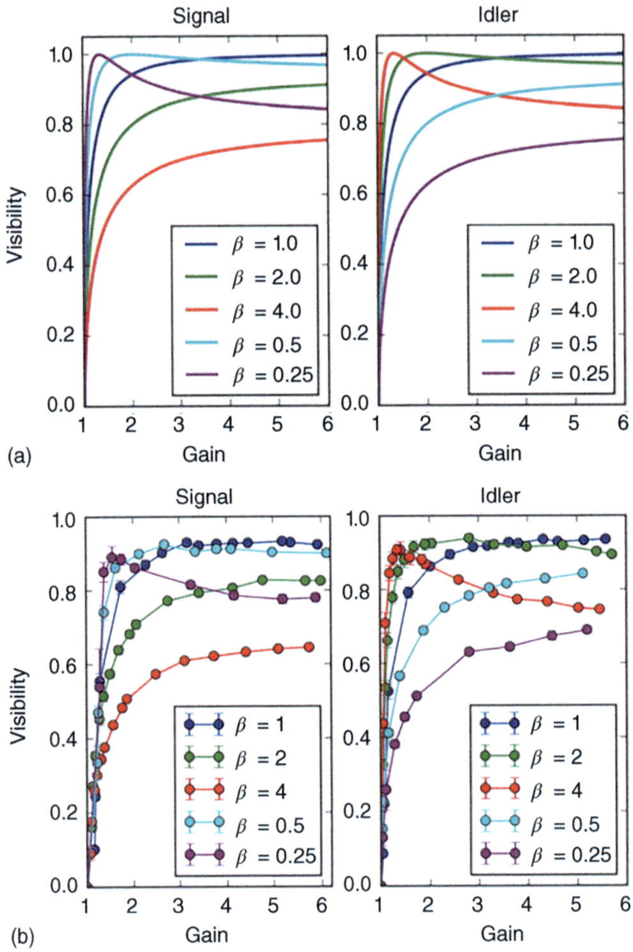

Figure 5.27 The intensity ratio and intensity-gain dependence of the two output port visibilities. (a) the theoretical simulations from Eq. (5.51), (b) the experimental results for the signal and idler outputs.

between the input beams at the center of the vapor cell. At large-intensity gain regime, the better intensity balancing of the two input beams gives better visibilities.

The one-photon detuning and two-photon detuning of the system can affect the behavior of the nonlinear beam splitter. Therefore, the visibilities also will be influenced by these two parameters. To study this effect, we fix the signal and idler beam powers at 20 μW. First, we set the two-photon detuning at 4 MHz and scan the one-photon detuning from 0.4 to 2.0 GHz. At the same time, we record the intensity gains and visibilities of two output ports. The results are shown in Figure 5.28a. The red diamond trace is for the signal visibility as a function of one-photon detuning, and the blue diamond trace is for the idler. The green and yellow dot traces are for the intensity gains of the signal and idler beams, respectively. Within the range from 0.5 to 1.0 GHz, the visibilities for both channels are all above 90%. Both of the

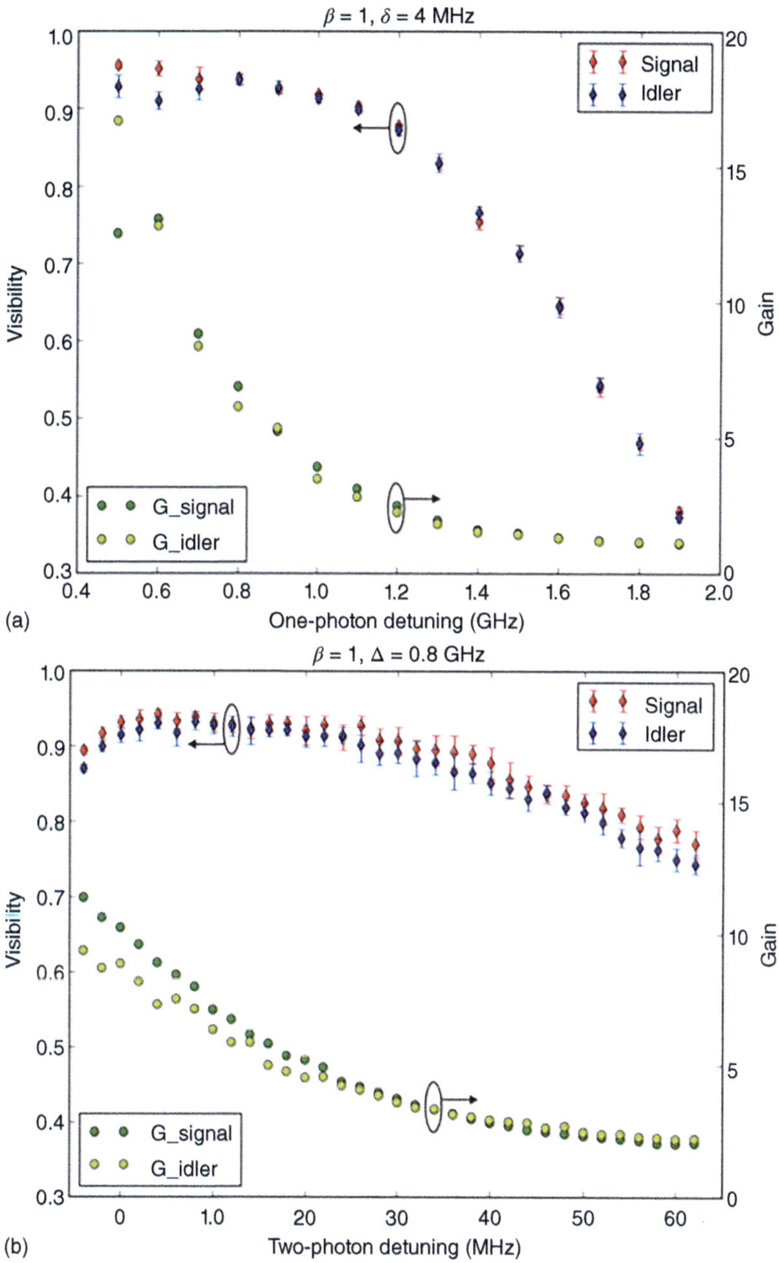

Figure 5.28 (a) The visibilities of the signal and idler fringe as a function of one-photon detuning. (b) The visibilities of the signal and idler fringe as a function of two-photon detuning. The intensity gains for both beams are also depicted in both figures.

intensity gains decrease as the detuning increases. Second, we set the one-photon detuning at 0.8 GHz and scan the two-photon detuning from −5 to 65 MHz. We vary the frequencies of both signal and idler beams symmetrically to change the two-photon detuning. The results are shown in Figure 5.28b. We find that the visibilities keep above 90% within a wide range from 0 to 30 MHz. The best visibility for both output ports is achieved at about 4 MHz.

It must be pointed out that the nonlinear beam splitter described here is different from both stages of our previous SU(1,1) interferometer. In our previous SU(1,1) interferometer, the first PA works in phase-insensitive configuration, amplifying the signal beam and generating a new idler beam. The signal and idler beams are quantum-correlated and very noisy after this amplification (well above their corresponding SQLs). The second PA is injected with signal, idler, and pump beams and thus becomes phase-sensitive. The gain of the second PA will depend on the phase of the three beams. Under amplification condition, quantum correlation shared between the two output ports of the PSA stage will be enhanced. Under de-amplification condition, the second PA can recover the signal and idler beams back to the initial states (coherent state and vacuum). Instead of being seeded by two noisy signal and idler beams, the current system described here is seeded by two coherent states. When the system operates at amplification condition, it will generate two output beams with IDS, and at de-amplification condition, these two outputs will show intensity sum squeezing as predicted by our theoretical study. We believe it is worth studying how the sensitivity of the interferometer changes if we apply the system studied here to the first stage of the SU(1,1) interferometer.

5.3.4 Interference-Induced Quantum-Squeezing Enhancement

After classical behaviors of nonlinear beam splitter are analyzed in Section 5.3.3, in this section, quantum-squeezing enhancement induced by interference will be discussed [11]. Following Eq. (5.49), the *IDS* degree between the signal and idler beams from Figure 5.25 with respect to SQL can be expressed as:

$$IDS = \frac{\mathrm{Var}(\hat{N}_{a,out} - \hat{N}_{b,out})_{PSA}}{\langle \hat{N}_{a,out}\rangle + \langle \hat{N}_{b,out}\rangle}$$
$$= \frac{1}{(2G-1) + \dfrac{2(G-1)}{\langle \hat{N}_{a,in}\rangle + \langle \hat{N}_{b,in}\rangle} + 4\sqrt{G(G-1)}\dfrac{\sqrt{\eta}}{1+\eta}\cos\phi} \tag{5.52}$$

where $\langle \hat{N}_{a,in}\rangle$ ($\langle \hat{N}_{a,out}\rangle$) and $\langle \hat{N}_{b,in}\rangle$ ($\langle \hat{N}_{b,out}\rangle$) represent the average input (output) photon number of the signal and idler fields, respectively. η is the input signal–idler power ratio. $\phi = 2\phi_c - \phi_a - \phi_b$, where ϕ_a and ϕ_b are the phase of the signal and idler fields, respectively. For convenience, we define ϕ as the phase of the PSA. In our experiment, the two input beams have powers of about 10 μW, i.e. $\langle \hat{N}_{a,in}\rangle = \langle \hat{N}_{b,in}\rangle \approx 4\times 10^5$ per unit coherence time (about 10 ns) of the FWM process, allowing the "bright beam" approximation $\langle \hat{N}_{a,in}\rangle + \langle \hat{N}_{b,in}\rangle \gg 1$ to be made. Therefore, for the two-beam PSA in our experiment, the term

$(2G-1)/\langle\hat{N}_{a,in}\rangle + \langle\hat{N}_{b,in}\rangle$ in the denominator of *IDS* can undoubtedly be ignored. Taking this into account, from Eq. (5.52), the *IDS* of a two-beam PSA can be given by

$$IDS \approx \frac{1}{(2G-1)+4\sqrt{G(G-1)}\frac{\sqrt{\eta}}{1+\eta}\cos(\phi)} \tag{5.53}$$

When we seed a vacuum field to the signal port ($\eta = 0$), the *IDS* degree becomes $1/(2G-1)$, which gives the quantum squeezing of the phase-insensitive amplifier (PIA). Compared with this, one can find that there is a phase-sensitive term in the denominator of Eq. (5.53). It is this phase-sensitive term, which brings the possibility of *IDS* manipulation for two-beam PSA. One can also see that the quantum-squeezing enhancement is dependent not only on the gain G but also on the input signal–idler power ratio η and the phase of PSA ϕ. When we simultaneously seed coherent fields to the signal port and idler port ($\eta \neq 0, \infty$), the *IDS* between signal and idler beams will be enhanced for any $\cos(\phi) > 0$. When $\cos(\phi) = 1$ and $\eta = 1$, such enhancement of quantum squeezing reaches its maximum, which can approach to 3 dB with $G \gg 1$ in principle. In fact, such enhancement is already 2.96 dB for $G = 3$. This clearly shows that the intrinsic interference nature of the two-beam PSA enables its ability to enhance the *IDS*.

Figure 5.29 shows the typical measured results of the two output ports of our PSA. For the phase-locking point $\phi = 0$, the corresponding intensity-difference noise power (IDNP) spectrum of our two-beam PSA is shown in Figure 5.29a. The trace A (green trace) is the normalized IDNP spectrum for the range of 0–2 MHz, while the trace B (blue trace) is the corresponding SQL. The black straight line at 0 dB, which corresponds to the mean value of data points on trace B, is taken as a reference. As we can see from Figure 5.29a, the IDNP of the PSA is 10.13 ± 0.21 dB below the SQL. In order to verify the theoretically predicted *IDS* enhancement induced by the interference of the PSA, we measure the noise power spectrum of the PIA under the same experimental conditions just by blocking the seeding signal beam or idler beam. The results of the PIA are shown in Figure 5.29b (blocking the seeding idler beam) and Figure 5.29c (blocking the seeding signal beam). As shown in Figure 5.29b, the *IDS* of signal-seeded PIA is 8.97 ± 0.24 dB. For the idler-seeded PIA, the *IDS* is 8.76 ± 0.26 dB, as shown in Figure 5.29c. It means that by changing the normal PIA process to a two-beam PSA process, the *IDS* of the system is enhanced from 8.97 ± 0.24 or 8.76 ± 0.26 to 10.13 ± 0.21 dB. In order to verify that the *IDS* enhancement is induced by the interference of the PSA from another aspect, we slightly walk off the seeding idler beam by an amount of about 3.5 mrad along the yellow arrow direction, as shown in Figure 5.29d. Such slight change makes the interference fringe of the PSA shown in Figure 5.26 disappear (become a straight line) while keeping the *IDS* of two single-seeded PIAs unchanged. The measured *IDS* of such signal–idler-seeded PIA is 8.86 ± 0.38 dB, which is almost equal to the squeezing level of the signal-seeded or idler-seeded PIA, clearly showing the decisive role of its intrinsic interference nature for quantum-squeezing enhancement in the two-beam PSA.

As mentioned above, we can see that the *IDS* enhancement of the two-beam PSA is dependent not only on the gain G but also on the input signal–idler power ratio

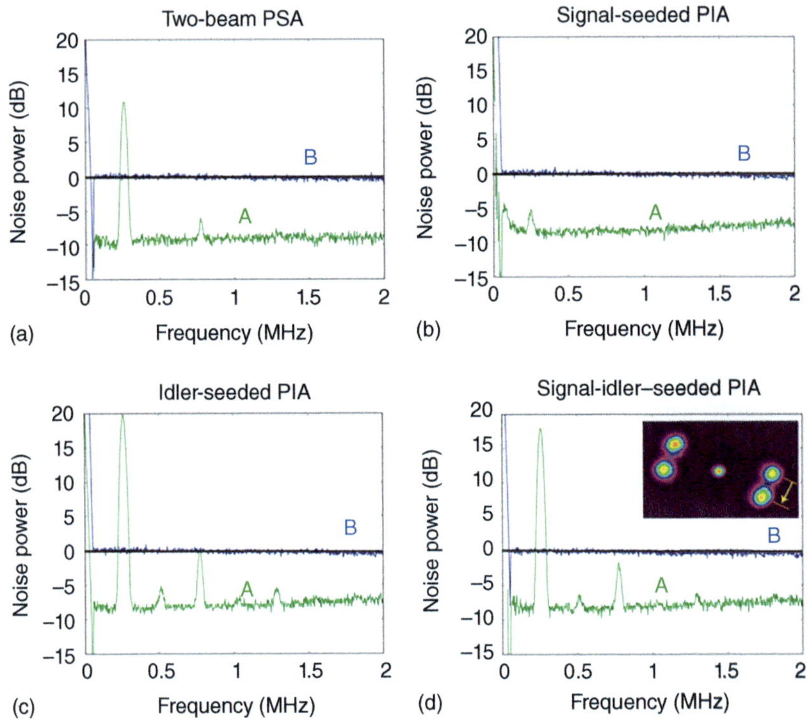

Figure 5.29 Observation of *IDS* enhancement. (a) Normalized IDNP spectrum of twin beams generated from two-beam PSA process when the phase of the PSA is locked at $\phi = 0$ (A) and the corresponding SQL (B). (b) Normalized IDNP spectrum of the twin beams generated from signal-seeded PIA process (A) and the corresponding SQL (B). (c) Normalized IDNP spectrum of the twin beams generated from idler-seeded PIA process (A) and the corresponding SQL (B). (d) Normalized IDNP spectrum of the two twin beams generated from signal-idler–seeded PIA process (A) and the corresponding SQL (B). The inset shows the output beams of the signal-idler–seeded PIA process in the far field captured by charge coupled device (CCD) camera.

η and the phase of the PSA ϕ. Thus, it is worthwhile to study how they influence the performance of the PSA. First, in order to explore the effect of gain on the *IDS* enhancement, we vary the pump power from 80 to 350 mW, which ranges the gain from 3 to 16, as shown in Figure 5.30a. In this case, we lock the phase of the PSA at 0 and keep $\eta \approx 1$. The green line at 0 dB is the normalized SQL. The theoretically predicted IDNP (red curve for PSA and blue curve for PIA) and the experimental results (red dot curve for PSA, yellow dot curve for signal-seeded PIA, and blue dot curve for idler-seeded PIA) have similar trends. The differences can be explained by including the imperfections of the system, such as the losses from atomic absorption and imperfect detection. The *IDS* of signal-seeded PIA is almost the same as the one of idler-seeded PIA. The degree of *IDS* of the two-beam PSA is always better than the one of PIA, clearly showing the squeezing enhancement. The corresponding interference visibility of PSA versus the gain is given as black triangle curve.

Second, we investigate how the *IDS* enhancement of PSA depends on the input signal–idler power ratio η. In order to study this effect, we set the gain of system at 16,

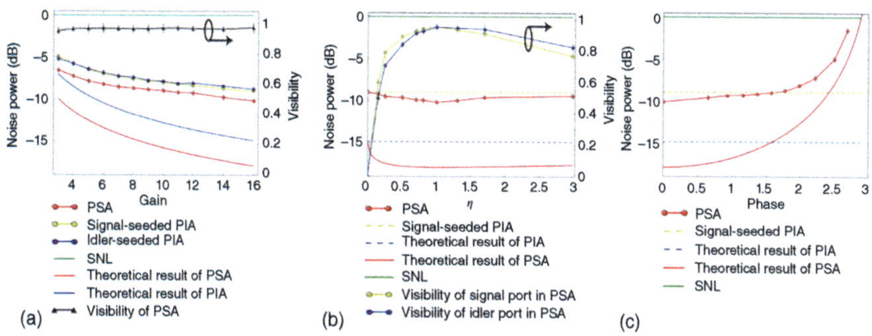

Figure 5.30 The dependence of *IDS* enhancement on various system parameters. (a) The IDNP of the two-beam PSA and PIA as a function of the gain of the system. (b) The relationship between the IDNP of the two-beam PSA and the input signal–idler power ratio η. (c) The relationship between the IDNP of the two-beam PSA and the phase. Both (a) and (b) figures include the visibility for the PSA. The curves circled by ellipses with a right-facing arrow correspond to the axis of "Visibility." The other curves and lines without ellipses correspond to the axis of "Noise power."

the phase of the PSA at 0 and scan the ratio η from 0 to 3. As shown in Figure 5.30b, the red dot curve gives the measured IDNP of PSA versus η. The corresponding theoretical prediction is shown as red curve. The yellow dashed line is the measured IDNP of signal-seeded PIA. The corresponding theoretical prediction is shown as blue dashed line. The normalized SQL is the green line at 0 dB. We find that the experimental results have the similar trend as the theoretical simulation. We can see that once the interference appears, quantum-squeezing enhancement will occur. This clearly proves that the *IDS* enhancement of the two-beam PSA is induced by its intrinsic interference nature. The *IDS* enhancement reaches its maximal value at $\eta \approx 1$, which agrees with the theoretical simulation. The corresponding measured visibility of signal (idler) port in PSA is shown as yellow (blue) dot curve.

Third, in order to explore how the phase of PSA ϕ influences the squeezing enhancement performance of the PSA, we fix the gain G of system at 16, the input signal–idler power ratio $\eta \approx 1$, and change the phase-locking point of the PSA from 0 to 2.69 in radian. The result is shown in Figure 5.30c. The red dot curve is the IDNP dependence relation on the phase of the PSA. The corresponding theoretical simulation is shown as a red curve. The yellow dashed line is the measured IDNP of signal-seeded PIA. The corresponding theoretical simulation is shown as a blue dashed line. The normalized SQL is the green line at 0 dB. We find that the *IDS* enhancement appears when $\phi < \pi/2$. The maximal squeezing level of intensity difference is obtained when $\phi = 0$. These results are consistent with the theoretical simulations. The present results clearly show that *IDS* manipulation is largely dependent on the phase point of the interference fringe, indicating that the physical mechanism of *IDS* manipulation is the intrinsic interference nature of the two-beam PSA. These results clearly show that the physical mechanism inducing the *IDS* enhancement of the two-beam PSA is its intrinsic interference nature. These results also prove that our method is an efficient way to enhance quantum squeezing, which can find potential applications in improving the fidelity of quantum information processing and the precision of quantum metrology.

5.4 Nonlinear Interferometric SPR Sensing

The Section 5.2 discussed the basic configurations of different nonlinear interferometers; here, in the present section, nonlinear interferometric SPR sensor-combining nonlinear interferometers and conventional SPR sensors will be discussed [12]. Before nonlinear interferometric SPR sensors are discussed, the operational principle for the conventional SPR sensors should be exhibited first, as shown in Figure 5.31a, an attenuated total reflection prism setup based on the Kretschmann configuration consists of three layers: prism, gold film, and sample. When the surface plasmon wave is excited, the reflection coefficient r_{spr} of the reflected light from the Kretschmann configuration can be expressed as

$$r_{spr} = \frac{r_{pg} + r_{gs}e^{2ik_{gz}d}}{1 + r_{pg}r_{gs}e^{2ik_{gz}d}} \tag{5.54}$$

where d is the thickness of gold film, and the reflection coefficient between the l-th and m-th layers r_{lm} can be given by

$$r_{lm} = \frac{k_{lz}\varepsilon_m - k_{mz}\varepsilon_l}{k_{lz}\varepsilon_m + k_{mz}\varepsilon_l} \tag{5.55}$$

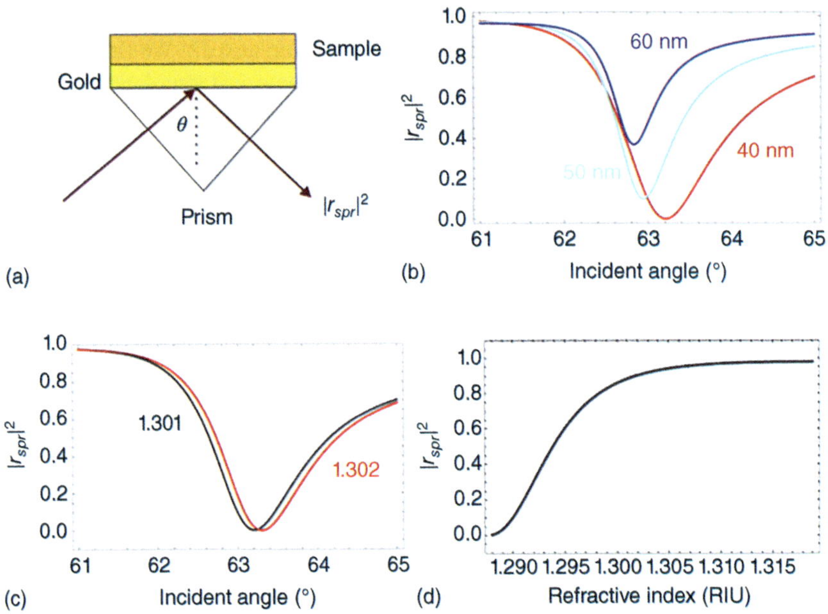

Figure 5.31 (a) An attenuated total reflection prism setup based on the Kretschmann configuration for the conventional SPR sensors. (b) The reflected light intensity as a function of the incident angle θ for the three different thicknesses of gold film: $d = 40$ nm (red curve), $d = 50$ nm (cyan curve), and $d = 60$ nm (blue curve). Here the refractive index is set to equal to 1.301. (c) The intensity of the reflected light as a function of the incident angle θ for the two different samples: $n_s = 1.301$ (black curve) and $n_s = 1.302$ (red curve). Here the thickness of gold film is set to 40 nm. (d) The reflected light intensity as a function of refractive index of the sample in the range of $1.288 \leq n_s \leq 1.319$. Here the incident angle and thickness of gold film are set to 62° and 40 nm, respectively. Other details can be seen in the main text.

where $l, m = (p$ [prism], g [gold], and s [sample]), k_{lz} is the normal component of the wave vector in the gold film layer and can be written as

$$k_{lz} = \frac{2\pi}{\lambda}\sqrt{\varepsilon_l - \varepsilon_p \sin\theta^2} \tag{5.56}$$

with the incident angle θ, the permittivities of prism ε_p and gold film ε_g for a 795 nm incident light are 1.5112 and $(0.181 + i5.126)^2$, respectively. To choose an optimal thickness of the gold film, the intensity of the reflected light as a function of the incident angle θ for the three different thicknesses of gold film: $d = 40$ nm (red curve), $d = 50$ nm (cyan curve), and $d = 60$ nm (blue curve) is depicted in Figure 5.31b. The reflection coefficient $|r_{spr}|^2$ for $d = 40$ nm shows the highest attenuation contrast compared with the other two thicknesses, and thus d is set to 40 nm in the following analysis. With a fixed thickness of gold film, the intensity of the reflected light as a function of the incident angle θ for the two different samples, $n_s = 1.301$ (black curve) and $n_s = 1.302$ (red curve), is shown in Figure 5.31c. From Figure 5.31c, it can be seen that a change of the refractive index can be identified by observing a shift of the resonance angle θ, or alternatively, the change of the reflected light intensity is observed for a given incident angle $\theta = 62°$ in the refractive index range of $1.288 \leq ns \leq 1.319$ (see Figure 5.31d), which also indicates that there is a point-to-point relationship between the reflected intensity and refractive index. Therefore, Figure 5.31c and d shows angle demodulation technique and intensity demodulation technique, respectively, for the conventional SPR sensors, because the information of refractive index of the sample can be identified by observing the change of resonant angle and reflected intensity, respectively, which constitute the basic operational principle of the conventional SPR sensors.

The Section 5.4 described the basic operational principle of conventional SPR sensors; nonlinear interferometric SPR sensors based on a nonlinear interferometer will be discussed in this section. As shown in Figure 5.32a, the amplified seed beam from FWM_1 process with the power gain G_1 interacts with the sample via the conventional SPR sensors, the reflected beam and idler beam are both seeded into the FWM_2 process with the power gain G_2, and the two output beams are analyzed using different evaluation parameters. Different from the corresponding case in Figure 5.32a, the output beam $\hat{a}_1$ from nonlinear interferometer is directly illuminated to the Kretschmann configuration and then also analyzed with the

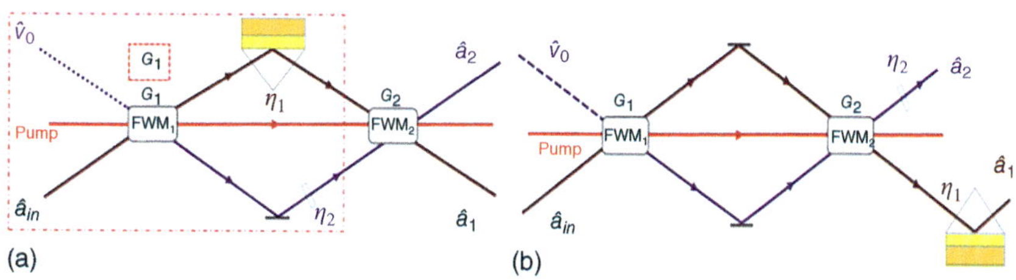

Figure 5.32 Nonlinear interferometric SPR sensors. (a) I-type nonlinear interferometric SPR sensor; (b) II-type nonlinear interferometric SPR sensor. $\hat{a}_{in}$ is coherent input beam, $\hat{v}_0$ is vacuum input, $\hat{a}_1$ and $\hat{a}_2$ are the two output beams. G_i is the power gain for FWM$_i$ process, η_i is the attenuation for the beams $\hat{a}_i$ ($i = 1, 2$). For the sake of comparison, a quantum SPR sensor based on a single FWM process with the power gain G_s is also shown in the red dashed box in (a).

help of the other beam $\hat{a}_2$. The former and latter schemes are called I-type and II-type nonlinear interferometric SPR sensors, respectively. For the simplicity of the following discussions, the reflection coefficient $|r_{spr}|^2$ can be replaced with η_1; meanwhile, η_2 represents the attenuation for the other beam $\hat{a}_2$, which can be easily realized by inserting variable neutral density filter in the corresponding optical path. To obtain the quantitative description of nonlinear interferometric SPR sensors, the input–output relation of Figure 5.32a can be expressed as

$$\begin{aligned}
\hat{a}_2 &= \left(\sqrt{G_2 g_1 \eta_2} e^{i\varphi} + \sqrt{G_1 g_2 \eta_1}\right) \hat{a}_{in}^{\dagger} + \left(\sqrt{G_1 G_2 \eta_2} e^{i\varphi} + \sqrt{g_1 g_2 \eta_1}\right) \hat{v}_0 \\
&\quad + \sqrt{G_2(1-\eta_2)} e^{i\varphi} \hat{v}_1 + \sqrt{g_2(1-\eta_1)} \hat{v}_2^{\dagger} \\
\hat{a}_1 &= \left(\sqrt{G_1 G_2 \eta_1} e^{i\varphi} + \sqrt{g_1 g_2 \eta_2}\right) \hat{a}_{in} + \left(\sqrt{G_2 g_1 \eta_1} e^{i\varphi} + \sqrt{G_1 g_2 \eta_2}\right) \hat{v}_0^{\dagger} \\
&\quad + \sqrt{g_2(1-\eta_2)} \hat{v}_1^{\dagger} + \sqrt{G_2(1-\eta_1)} e^{i\varphi} \hat{v}_2
\end{aligned} \tag{5.57}$$

where G_i is the power gain for FWM$_i$ process, and $g_i = G_i - 1\ (i = 1, 2)$, φ is interference phase inside nonlinear interferometer. $\hat{v}_1$ and $\hat{v}_2$ are vacuum fields due to the introduction of the attenuations η_1 and η_2.

Let us evaluate the first parameter, i.e. the degree of intensity-difference squeezing (*DS*) between the two output beams, which is given by the ratio of the variance on the intensity difference $\text{Var}[\hat{N}_1 - \hat{N}_2]$ to the same variance at SQL $\text{Var}[\hat{N}_1 - \hat{N}_2]_{SQL} = \langle \hat{N}_1 + \hat{N}_2 \rangle$, a ratio value smaller than 1 means the presence of quantum correlation between the two output beams. In I-type nonlinear interferometric SPR sensor given by Figure 5.32a, the *DS value* between the two output beams can be calculated as

$$DS_a = \frac{(G\eta_1 - g\eta_2)^2 + Gg(\eta_2 - \eta_1)^2 + g\eta_2(1-\eta_2) + G\eta_1(1-\eta_1)}{(G+g)(G\eta_1 + g\eta_2) + 4Gg\sqrt{\eta_1\eta_2}\cos\varphi} \tag{5.58}$$

where G_1 is assumed to be equal to G_2 in the following discussions; thus, Eq. (5.58) will be reduced to a simplified form of $1/[4Gg(1+\cos\varphi)+1]$ when no losses are present in Figure 5.32a. Following similar procedures, the *DS* between the two output beams in Figure 5.32b can be calculated as

$$DS_b = \frac{DS_{bn}}{DS_{bd}} \tag{5.59}$$

where the numerator DS_{bn} and denominator DS_{bd} of DS_b are

$$\begin{aligned}
DS_{bn} &= [2Gg\eta_2(1+\cos\varphi) - \eta_1(G^2 + g^2 + 2Gg\cos\varphi)]^2 \\
&\quad + (\eta_2 - \eta_1)^2 Gg(G+g)^2(1+\cos\varphi)^2 + (\eta_2 - \eta_1)^2 Gg\sin\varphi^2 \\
&\quad + 2Gg\eta_2(1-\eta_2)(1+\cos\varphi) + \eta_1(1-\eta_1)(G^2 + g^2 + 2Gg\cos\varphi)
\end{aligned} \tag{5.60}$$

and

$$DS_{bd} = 2Gg\eta_2(1+\cos\varphi) + \eta_1(G^2 + g^2 + 2Gg\cos\varphi) \tag{5.61}$$

respectively. Similarly, Eq. (5.59) will also be reduced to the simplified form of $1/[4Gg(1+\cos\varphi)+1]$, when no losses are present in Figure 5.32b. In order to make a direct comparison with a quantum SPR sensor based on a single FWM process in the red dashed box in Figure 5.32a, its *DS* can also be expressed as

$$DS_S = \frac{2G_S^2(\eta_1 - \eta_2)^2 + G_S\eta_1(1 - 2\eta_1) + g_s\eta_2(1 - 2\eta_2) + 2G_S\eta_2(2\eta_1 - \eta_2)}{G_S\eta_1 + g_s\eta_2} \tag{5.62}$$

where G_s is the power gain in a single FWM process and $g_s = G_s - 1$. To make a fair comparison, the intensities of the sensing beams involving the plasmonic interaction should be equal. Specifically, the intensities of the sensing beams from a quantum SPR sensor based on a single FWM process, I-type, and II-type nonlinear interferometric SPR sensors are $G_s\eta_1$, $G^2\eta_1 + g^2\eta_2 + 2Gg\sqrt{\eta_1\eta_2}$ ($\varphi = 0$), and $\eta_1(G^2 + g^2 + 2Gg)$ ($\varphi = 0$), respectively. For the sake of simplicity, firstly, $\eta_1 = \eta_2$ is assumed, and this assumption can also preserve the quantum correlations that existed in the three different sensors; secondly, if the seed beam is amplified by a single FWM process with power gain $G_s = 3$, then the amplification factor $G^2 + g^2 + 2Gg$ from nonlinear interferometer should also be equal to 3; thus, the power gain G of each single FWM process in nonlinear interferometer is equal to $(1+\sqrt{3})/2 \approx 1.366$. Doing this can guarantee that the intensities of the three sensing beams are always equal. In this sense, the values of Eqs. (5.59) and (5.62) are equal, which means that quantum performance of a single FWM process with power gain $G_s = 3$ is equivalent to the one of II-type nonlinear interferometer with $G = 1.366$. Therefore, the quantum performance of a quantum SPR sensor based on a single FWM process with power gain $G_s = 3$ can be fully replaced with the one of II-type nonlinear interferometric SPR sensor with $G = 1.366$ in Figure 5.32b, and thus the sensing performances will be mainly compared between I-type nonlinear interferometric SPR sensor and II-type nonlinear interferometric SPR sensor.

Based on the above discussions, the sensing principle of *DS* values with respect to refractive index will be shown as follows. As is depicted in Figure 5.33a, the *DS* values of I-type and II-type nonlinear interferometric SPR sensors are always below SQL (the cyan dashed line), while the *DS* value of I-type nonlinear interferometric SPR sensor (the red curve) is much smaller. This is because the two beams generated from FWM_1 process still preserve quantum correlation after experiencing the identical attenuation, and their correlation degree will be enhanced by the next FWM_2 process. While for II-type nonlinear interferometric SPR sensor, the quantum correlation shared by the two output beams will only be degraded by the attenuation process. Similarly, the dependence of I-type and II-type nonlinear interferometric SPR sensors on the incident angle for the two different samples: $n_s = 1.301$ (black curve) and $n_s = 1.302$ (red curve) is shown in Figure 5.33b, it can be seen that both of the *DS* values obtain the maximum values near resonance angle due to the higher loss and the minimum values far from resonance angle due to the lower loss. This phenomenon can be explained as follows. As shown in Figure 5.31b,c, based on the operational principle of a conventional SPR sensor, the reflected beam will experience more attenuation near resonance angle than far from resonance angle, and more attenuation means the introduction of more loss to the reflected beam; this will inevitably lead to the disappearance of more quantum correlations. Meanwhile,

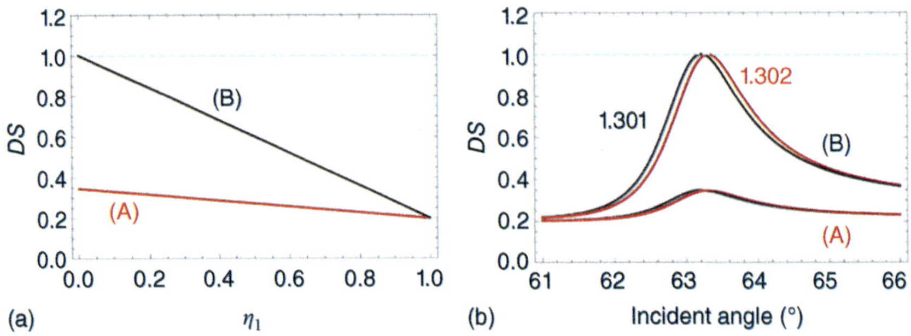

Figure 5.33 (a) The *DS* values of I-type (A) and II-type (B) nonlinear interferometric SPR sensors as a function of attenuation η_1. (b) The *DS* values of I-type (A) and II-type (B) nonlinear interferometric SPR sensors as a function of incident angle for two different samples: $n_s = 1.301$ (black curve) and $n_s = 1.302$ (red curve). The cyan dashed line: SQL. Other details can be seen in Figure 5.31.

quantum correlation is characterized by *DS* value; a lower (higher) quantum correlation corresponds to a larger (smaller) *DS* value. Thus both of the *DS* values are totally different between near resonance angle and far from resonance angle. Note that the maximum *DS* values of I-type and II-type nonlinear interferometric SPR sensors are 0.3462 and 0.9989, respectively; this means that near resonance angle, the correlation degree of I-type nonlinear interferometric SPR sensor is improved by a factor of $10\log(0.9989/0.3462) = 4.6\,\text{dB}$ compared with that obtained with II-type nonlinear interferometric SPR sensor.

The aforementioned sensing principle is based on angle demodulation, intensity demodulation as the other demodulation technique, i.e. the dependence of the *DS* value on refractive index will be shown as follows. As shown in Figure 5.34, the *DS* values decrease with the increase of refractive index. Such behaviors can be explained as follows. As shown in Figure 5.31d, a low (high) refractive index corresponds to a low (high) reflection coefficient and a high (low) loss, and a high (low) loss indicates the decreasing (increasing) quantum correlations. To sum up, the correlation degree of I-type nonlinear interferometric SPR sensor near resonance angle is improved by a factor of $10\log(0.9989/0.3462) = 4.6\,\text{dB}$ compared with that obtained with II-type nonlinear interferometric SPR sensor.

The first sensing parameter, i.e. the *DS* values of the two output beams generated from I-type and II-type nonlinear interferometric SPR sensors as a function of refractive index, has been analyzed. In this section, the estimation precision δn_s as the second sensing parameter can also be used to characterize the sensing performance, which can be obtained through an error propagation analysis, such that

$$\delta n_s = \frac{\langle \Delta \widehat{S} \rangle}{\left| \frac{\partial \langle \widehat{S} \rangle}{\partial n_s} \right|} \tag{5.63}$$

where $\widehat{S}$ consists of some measurement combinations of the two output beams and $\langle \ldots \rangle$ denotes the mean value. In general, this parameter is used to study the estimation ability of refractive index via a sensor. A smaller value indicates a

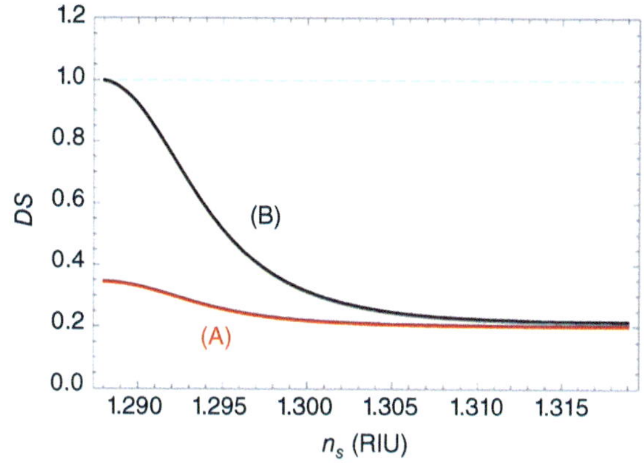

Figure 5.34 The *DS* values of I-type (A) and II-type (B) nonlinear interferometric SPR sensors as a function of refractive index. The cyan dashed line: SQL.

reduction in the uncertainty of the estimation and thus an increase in the sensitivity. For the two-mode squeezed state, the signal $\hat{S}$ can be written as the form of intensity difference between the two output beams. So the denominator of Eq. (5.63) $\left|\frac{\partial\langle\hat{S}\rangle}{\partial n_s}\right|$ can be simplified as $N_S\left|\frac{\partial|r_{spr}|^2}{\partial n_s}\right|$, where N_s is the intensity of the sensing beam. Note that the intensities of all the sensing beams are set to be equal, and thus the sensing performance of the different SPR sensors only depends on the numerator, i.e. nonclassical features existed in the different SPR sensors. Therefore, to quantify the effect of nonclassical features, i.e. quantum correlations, on the estimation precision enhancement, it is necessary to define a ratio of the precision based on SQL to the precision from quantum correlations. The ratio is defined as

$$\mathfrak{R} = \frac{\delta n_{s(SQL)}}{\delta n_s} = \frac{\langle\Delta\hat{S}\rangle_{(SQL)}}{\langle\Delta\hat{S}\rangle} \tag{5.64}$$

where $\mathfrak{R}$ greater than 1 reveals an enhancement in the estimation precision δn_s, or equivalently, a quantum noise reduction compared with the classical scenario. For the cases of I-type and II-type nonlinear interferometric SPR sensors, their estimation precision ratios $\mathfrak{R}_s$ are the reciprocal of the square root of the corresponding DS_s, i.e. $\mathfrak{R} = \frac{1}{\sqrt{DS}}$. As shown in Figure 5.35a, the $\mathfrak{R}$ values of I-type nonlinear interferometric SPR sensor are always greater than that obtained with II-type nonlinear interferometric SPR sensor, resulting from a much stronger quantum noise reduction existed in I-type nonlinear interferometric SPR sensor mentioned before. Similarly, the $\mathfrak{R}$ values of I-type (A) and II-type (B) nonlinear interferometric SPR sensors as a function of incident angle for the two different samples: $n_s = 1.301$ (black curve) and $n_s = 1.302$ (red curve), are also shown in Figure 5.35b, a tiny change in the refractive index from 1.301 (black curve) to 1.302 (red curve) can be identified with the higher $\mathfrak{R}$ values by observing a shift in the resonance angle using I-type nonlinear interferometric SPR sensor compared with II-type nonlinear interferometric SPR sensor. However, I-type nonlinear interferometric SPR sensor show the estimation precision enhancement even in the high loss condition, i.e. near resonance angle, in which I-type nonlinear interferometric SPR sensor shows the distinct differences compared

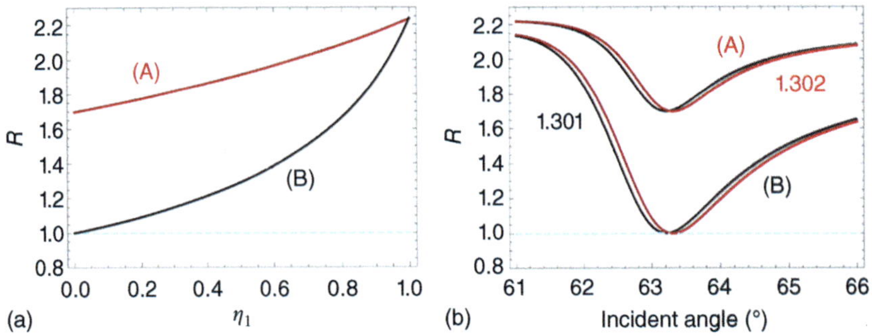

Figure 5.35 (a) The $\Re$ values of I-type (A) and II-type (B) nonlinear interferometric SPR sensors as a function of η_1. (b) The $\Re$ values of I-type (A) and II-type (B) nonlinear interferometric SPR sensors as a function of incident angle for the two different samples: $n_s = 1.301$ (black curve) and $n_s = 1.302$ (red curve). The cyan dashed line: 1.

with II-type nonlinear interferometric SPR sensor. This is mainly due to the fact that two beams involving plasmonic interaction in I-type nonlinear interferometric SPR sensor will still preserve quantum correlation after experiencing identical attenuation and will be further enhanced by the next FWM_2 process. While the quantum correlation shared by the two beams' output from II-type nonlinear interferometric SPR sensor will only be degraded after the plasmonic interaction. This point can also be confirmed in Figure 5.33a; meanwhile, estimation precision ratio is defined as the reciprocal of the square root of the corresponding *DS*, and thus I-type nonlinear interferometric SPR sensor will show estimation precision enhancement compared with II-type nonlinear interferometric SPR sensor in the high loss condition. In other words, the minimum values from I-type and II-type nonlinear interferometric sensors are 1.7 and 1.0, respectively, which means that the ability of the estimation precision enhancement of I-type nonlinear interferometric SPR sensor near resonance angle is enhanced by a factor of $10\log[1.7/1.0] = 2.3$ dB compared with the one of II-type nonlinear interferometric SPR sensor.

If the intensity demodulation instead of angle demodulation is involved in the discussions of the estimation precision enhancement (the results are shown in Figure 5.36), the $\Re$ values from both I-type and II-type nonlinear interferometric SPR sensors increase with the increase of refractive index, while they show the distinct differences in the low refractive index range mentioned before, and their values saturate at the value of 2.24 in the high refractive index limit. In a word, the estimation precision ratio of I-type nonlinear interferometric SPR sensor under the high loss condition can obtain a maximum 2.3 dB enhancement compared with II-type nonlinear interferometric SPR sensor.

After *DS* and estimation precision ratio from both I-type and II-type nonlinear interferometric SPR sensors are discussed, the third sensing parameter, i.e. *SNR*, will be studied in this section. For two-mode squeezed states, its *SNR* can be defined as the ratio of the sensing beam intensity to the intensity-difference noise between the two output beams. This definition has the following advantages: firstly, the sensing beam intensity rather than intensity-difference between the two beams as the signal

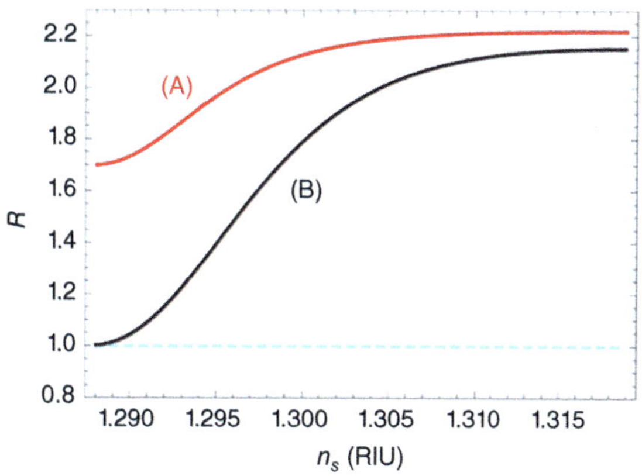

Figure 5.36 The ℜ values of I-type (A) and II-type (B) nonlinear interferometric SPR sensors as a function of refractive index. The cyan dashed line: 1.

can obtain a maximum value; secondly, intensity-difference noise between the two output beams can obtain a minimum value compared with the noise of the single sensing beam because each beam from the twin beams has a thermal noise level above SQL. Thus, such cooperation between high signal and low noise level enables the *SNR* from nonlinear interferometric SPR sensors to be further enhanced than that obtained with classical scenario. Following this basic idea, the *SNR* values of I-type and II-type nonlinear interferometric SPR sensors can be given by

$$SNR_a = \frac{G^2\eta_1 + g^2\eta_2 + 2Gg\sqrt{\eta_1\eta_2}\cos\varphi}{(G\eta_1 - g\eta_2)^2 + Gg(\eta_2 - \eta_1)^2 + g\eta_2(1-\eta_2) + G\eta_1(1-\eta_1)} \tag{5.65}$$

and

$$SNR_b = \frac{\eta_1(G^2 + g^2 + 2Gg\cos\varphi)}{DS_{bn}} \tag{5.66}$$

respectively. Under the condition $\eta_1 = \eta_2$, the *SNR* of the classical scenario is 0.5, which can be realized by splitting a beam in a coherent state with a power equal to the total power of the two correlated beams, using one beam as the sensing beam to interact with the attenuated total reflection setup using the Kretschmann configuration, and the other beam experiencing the identical attenuation as the reference beam, then directing them both to a differential detector. Thus, this balanced detection system makes it possible to cancel all the sources of classical noise and obtain an accurate measure of the *SNR* of classical scenario. As shown in Figure 5.37, the *SNR* values of both I-type and II-type nonlinear interferometric SPR sensors are greater than that obtained with classical scenario, while the *SNR* value of I-type nonlinear interferometric SPR sensor is always greater than that obtained with II-type nonlinear interferometric SPR sensor. The maximum differences in *SNR* values can be seen from the high-loss situation, similar to the discussion of estimation precision enhancement. Specifically, the minimum *SNR* values from I-type and II-type nonlinear interferometric SPR sensors are 1.73 and 0.60, respectively; this corresponds directly to a 10 log10[1.73/0.60] = 4.6 dB *SNR* improvement from I-type nonlinear interferometric SPR sensor under the high-loss condition compared with II-type nonlinear interferometric SPR sensor.

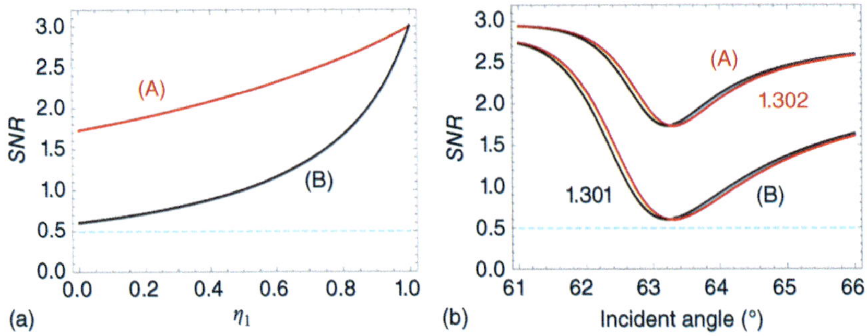

Figure 5.37 (a) The *SNR* values of I-type (A) and II-type (B) nonlinear interferometric SPR sensors as a function of η_1. (b) The *SNR* values of I-type (A) and II-type (B) nonlinear interferometric SPR sensors as a function of incident angle for the two different samples: $n_s = 1.301$ (black curve) and $n_s = 1.302$ (red curve). The cyan dashed line: 0.5.

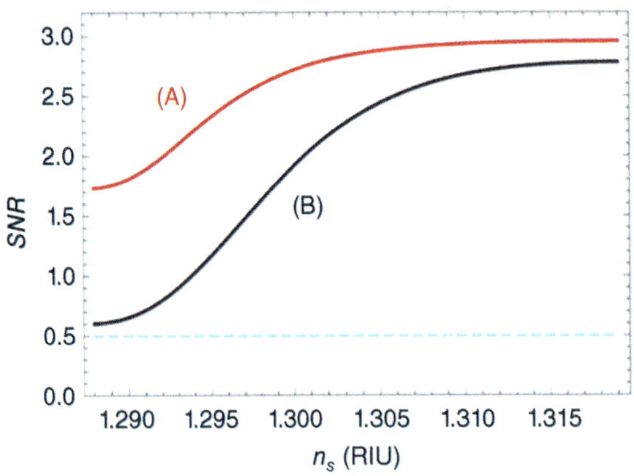

Figure 5.38 The *SNR* values of I-type (A) and II-type (B) nonlinear interferometric SPR sensors as a function of refractive index. The cyan dashed line: 0.5.

The *SNR* values as a function of refractive index are shown in Figure 5.38. Similar to the dependence of estimation precision enhancement on refractive index, here the *SNR* values of both I-type and II-type nonlinear interferometric SPR sensors also increase with the increase of refractive index, but their values saturate at the value of 3.0 in the high refractive index limit. In conclusion, the *SNR* value of I-type nonlinear interferometric SPR sensor under the high-loss situation is enhanced by a factor of 4.6 dB compared with II-type nonlinear interferometric SPR sensor, while this advantage will disappear in the high refractive index limit.

In conclusion, we have theoretically characterized the sensing performance of I-type and II-type nonlinear interferometric SPR sensors and demonstrated the equivalence between II-type nonlinear interferometric SPR sensor with power gain $G = 1.366$ and a quantum SPR sensor based on a single FWM process with power gain $G_s = 3$. The distinct advantages of I-type nonlinear interferometric SPR sensor near resonance angle are presented as follows. Firstly, the *DS* as the first evaluation parameter, the *DS* value of I-type nonlinear interferometric SPR sensor is improved by a factor of 4.6 dB than that obtained with II-type nonlinear interferometric SPR sensor; secondly, estimation precision ratio as the second evaluation parameter, the estimation precision ratio of I-type nonlinear interferometric SPR sensor is enhanced by a factor of 2.3 dB than that obtained with II-type nonlinear interferometric SPR sensor; lastly, *SNR* as the third parameter, the *SNR* value of I-type nonlinear interferometric SPR sensor is improved by a factor of 4.6 dB than that obtained with II-type nonlinear interferometric SPR sensor. The results presented here may find other potential sensing applications based on a nonlinear interferometer, i.e. nonlinear interferometric bending, pressure, and temperature sensors.

5.5 Summary and Outlook

In this chapter, firstly, we give the comprehensive descriptions about SU(1,1) interferometer, including experimental implementation of phase-locking, enhancement of phase sensitivity, experimental realization of entanglement enhancement, and QNC; secondly, we also describe other types of nonlinear interferometers, including a NSI, a hybrid interferometer consisting of a nonlinear FWM process and a linear beam splitter, a nonlinear beam splitter, and interference-induced quantum-squeezing enhancement; lastly, a nonlinear interferometric SPR sensor has been theoretically proposed and demonstrated its sensing advantages by using the sensing parameters such as *DS*, estimation precision, and *SNR*.

The above discussions about the different kinds of nonlinear interferometers enrich the background knowledge about optical interferometric sensing, especially for nonlinear interferometric sensing. Based on these basic configurations, different sorts of nonlinear interferometric sensors (bending, temperature, pressure sensors, etc.) can also be efficiently constructed to improve the sensing parameters, e.g. *DS*, estimation precision, and *SNR*.

References

1 Yurke, B., McCall, S.L., and Klauder, J.R. (1986). SU(2) and SU(1,1) interferometers. *Phys. Rev. A: At. Mol. Opt. Phys.* 33: 4033.
2 Plick, W.N., Dowling, J.P., and Agarwal, G.S. (2010). Coherent-light-boosted, sub-shot noise, quantum interferometry. *New J. Phys.* 12: 083014.
3 Jing, J., Liu, C., Zhou, Z. et al. (2011). Realization of a nonlinear interferometer with parametric amplifiers. *Appl. Phys. Lett.* 99: 011110.
4 Wang, H., Marino, A.M., and Jing, J. (2015). Experimental implementation of phase locking in a nonlinear interferometer. *Appl. Phys. Lett.* 107: 121106.
5 Liu, S., Lou, Y., Xin, J., and Jing, J. (2018). Quantum enhancement of phase sensitivity for the bright-seeded SU(1,1) interferometer with direct intensity detection. *Phys. Rev. Appl.* 10: 064046.
6 Xin, J., Qi, J., and Jing, J. (2017). Enhancement of entanglement using cascaded four-wave mixing processes. *Opt. Lett.* 42: 366–369.
7 Xin, J., Wang, H., and Jing, J. (2016). The effect of losses on the quantum-noise cancellation in the SU(1,1) interferometer. *Appl. Phys. Lett.* 109: 051107.

8 Xin, J., Liu, J., and Jing, J. (2017). Nonlinear Sagnac interferometer based on the four-wave mixing process. *Opt. Express* 25: 1350–1359.

9 Liu, S. and Jing, J. (2017). Hybrid interferometer with nonlinear four-wave mixing process and linear beam splitter. *Opt. Express* 25: 15854–15860.

10 Fang, Y., Feng, J., Cao, L. et al. (2016). Experimental implementation of a non-linear beam splitter based on a phase-sensitive parametric amplifier. *Appl. Phys. Lett.* 108: 131106.

11 Liu, S., Lou, Y., and Jing, J. (2019). Interference-induced quantum squeezing enhancement in a two-beam phase-sensitive amplifier. *Phys. Rev. Lett.* 123: 113602.

12 Wang, H., Fu, Z., Ni, Z. et al. (2021). Nonlinear interferometric surface-plasmon-resonance sensor. *Opt. Express* 29 (7): 11194–11206.

6

Spatial-frequency-shift Super-resolution Imaging Based on Micro/nanomaterials

Mingwei Tang[1,2] *and Qing Yang*[1,2]

[1] *Zhejiang University, College of Optical Science and Engineering, International Research Center for Advanced Photonics, State Key Laboratory of Extreme Photonics and Instrumentation, 38 Zheda Road, Hangzhou, Zhejiang, 310027, China*
[2] *Zhejiang Lab, Intelligent Perception Research Institute, Zhongtai Street, Hangzhou, Zhejiang, 311100, China*

6.1 Introduction

Optical microscopy is probably one of the most significant technical accomplishments in the history of humankind. Since its invention by Hans Lippershey and Zacharias Janssen around 1590, optical microscopy has revolutionized many aspects of science and technology, especially in life sciences, for its superiority in minimal invasiveness and excellent compatibility with live-cell imaging. However, classical linear optical microscopy is still restrained by the optical resolution limitation formulated by Abbe in his revolutionary paper published in 1873 [1]. Von Helmholtz derived a more general expression of this concept, stating that resolution under conventional illumination is described by the illumination wavelength (λ_0) and the numerical aperture (NA) of the system: $R = \frac{\lambda_0}{2NA}$ [2]. The resolution of a traditional optical microscope (i.e. the smallest distance at which two point like objects could be discriminated) under visible light illumination can only reach ~200 nm in the lateral dimensions and ~500 nm in the axial dimension using an objective lens ($NA = 1.4$) at 550 nm wavelength. These values are fundamentally restricted by optical diffraction (or uncertainty principle from the perspective of quantum physics), which leads to the absence of subwavelength information about the object. The increasing demand for improved resolution has been inspiring numerous attempts to break the resolution limit. However, it was not until recently that fruitful developments emerged. In this chapter, we focus on the discussion of lateral super-resolution microscopy.

The confocal laser scanning microscopy (CLSM) invented in the 1960s utilizes focused laser for excitation and a pinhole for detection to shape the point-spread function (PSF) of the system and increases the resolution of a microscope down to around $1/\sqrt{2}$ of the diffraction limit [3–5]. The combination of CLSM and fluorescent microscopy has become a powerful and indispensable tool for life science and material science since its invention [6].

Optical Imaging and Sensing: Materials, Devices, and Applications, First Edition.
Edited by Jiang Wu and Hao Xu.

By further advancing the confocal system, super-resolution fluorescent microscopy circumvents the diffraction limit by distinguishing the fluorescent emission of different areas or fluorophores in an additional dimension, e.g. spectroscopic characteristics or time [7]. Reversible saturable optical fluorescence transitions (RESOLFT) microscopy represents a typical class of this absolute far-field strategy, which reversely silences fluorophores at the predefined diffraction-limited region and shrinks the PSF [8–14]. Several newly invented methods share this principle, e.g. stimulated emission depletion (STED) microscopy [8, 10], ground-state depletion (GSD) microscopy [15, 16], saturated pattern excitation (SPE) microscopy [17], and dynamic saturation optical microscopy (DSOM) [18]. Another category of the method relies on the precise location of the single emitter by activating individual molecules stochastically within the diffraction-limited region at different time intervals. Such single-molecule localization microscopy (SMLM) includes stochastic optical reconstruction microscopy (STORM) [19, 20], photoactivated localization microscopy (PALM) [21], and fluorescence photoactivation localization microscopy (FPALM) [22]. Similar to these methods, super-Resolution Radial Fluctuations (SRRF) [23] extracts the superresolution information from fluorescence fluctuations rather than localization. The methods mentioned above rely heavily on special fluorescent markers [24, 25], thus limiting their employment in label-free and fluorescent-probes-limited scenarios [26–28]. A detailed description of these methods can be found in other reviews [29–36]. Another intriguing attempt worth noting is to employ the carefully designed amplitude or phase zone plate to achieve sub-diffraction limit focusing in the far field, that is, the so-called super-oscillatory lens [37–41]. Through the judiciously designed diffractive unit, the focal spot size in a certain region of the target plane is controllable in lateral (from infinite small to $0.38\lambda/NA$) [42] and longitudinal directions [43]. Combined with the confocal technique, both labeled and label-free super-resolution microscopy could be realized in a purely noninvasive manner [37, 43, 44]. However, the decreased focus spot is usually accompanied by increased sidelobes, which imposes practical difficulties in its applications in super-resolution imaging. For more information on this planer diffractive lens, please refer to the latest reviews [38, 45].

To enhance the lateral resolution in far-field microscopy without the slow scanning process, people innovatively turn to manipulation in the spatial-frequency (SF) domain instead of the spatial domain. This method, termed as spatial-frequency-shift (SFS) method [46], can realize high-speed and wide-field high-resolution imaging. The classical technique is structured illumination microscopy (SIM), which uses the Moiré effect to improve the imaging resolution by illuminating the sample with patterned light [17, 47–50]. Limited by the pattern periodicity realized by the interference of two free-space light, the lateral resolution can only reach down to approximately $\lambda/4$. Another way of increasing the resolution is Fourier ptychographic microscopy (FPM) [51–53], which recovers high-resolution images from multiple low-resolution images taken under oblique illuminations [54–56]. By shifting the high-SF information into the collectible area in the SF space defined by the objective NA, these methods operate in SF space, stitching together several variably illuminated low-resolution images. Although these methods have many advantages, including low illumination density, good biocompatibility, simple sample preparation process, fast imaging speed, and wide field of view (FOV) compared with other super-resolution imaging methods [7, 57–63], deep subwavelength super-resolving ability was not provided due to the limited SFS range restricted by free-space light illumination. Thus, evanescent-wave illumination, instead of free-space light, is necessary to further break the diffraction limit in order to cover larger SF range in the SF space.

Evanescent waves were first employed by the microscopy community in the 1940s [64]. The in-plane wavevector of an evanescent wave is larger than that of a propagating wave and corresponds to finer details of the object. In other words, the sub-diffraction-limited details could be visualized once evanescent waves are collected. Near-field scanning optical microscopy (NSOM) is the first technique invented to capture the evanescent waves in the near-field region, which uses nanometer-sized probes to scan point by point and collect the corresponding evanescent waves on the specimen's surface [65–73]. However, this technique has a relatively low throughput and requires postprocessing of massive scanning data. Super-resolution microscopy benefits from evanescent-wave illumination combined with SFS method. Generally speaking, super-resolution microscopy utilizes mainly two kinds of evanescent waves for illumination: evanescent fields at dielectric waveguide boundaries [74] and surface plasmon polaritons (SPPs) or localized plasmons (LPs) at metal/dielectric interfaces [75, 76]. Several papers have reported evanescent-wave-assisted SFS methods using delicately designed structures [46, 77, 78], including evanescent-fields-illuminated SFS label-free microscopy [55, 78, 79], evanescent-fields-illuminated SFS-labeled microscopy [80, 81], plasmonic structured illumination microscopy (PSIM) [56, 82–84], localized plasmonic structured illumination microscopy (LPSIM) [85, 86], and metamaterial-assisted illumination nanoscopy [87, 88].

This chapter aims to provide a systematic discussion on the evanescent-wave-assisted SFS super-resolution microscopy techniques based on micro/nanomaterials and metamaterials. It should be noted that some techniques, such as nonlinear effects, are beyond the scope of this paper, although they are very prospective methods that can fundamentally break the diffraction limit [47, 49].

6.2 The Principle of SFS Super-resolution Imaging Based on Micro/nanomaterials

For SFS methods, evanescent-wave illumination provides a significant SFS of the object's spatial spectrum, as shown in Figure 6.1 [46]. The detected spatial spectrum of objects can be described as:

$$F_{fs}(\vec{k}) = F_0(\vec{k} - \vec{k}_s) \cdot TF(\vec{k}) \tag{6.1}$$

where $F_0(\vec{k})$, $TF(\vec{k})$ are the Fourier spectrum of the object and the transfer function (TF) of the imaging system, $\vec{k}_s$ is the SFS provided by the evanescent-wave illumination. It should be noted that the TF should be adapted to optical transfer function

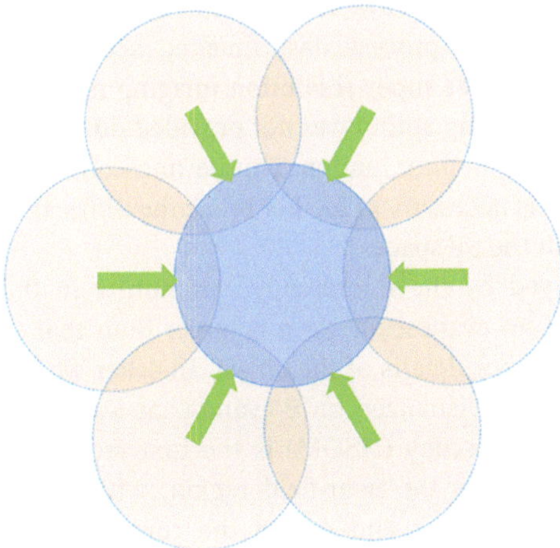

Figure 6.1 Mechanisms of evanescent-waves-assisted SFS method. SFS methods shift the high spatial spectrum of objects step by step to the detectable region of the objective and stitch them together in the SF space with the reconstruction algorithm. The blue circle represents the SF space detected by the conventional objective, while the orange one represents the SF space detected by the SFS method.

(OTF) or coherent transfer function (CTF) in different imaging systems, e.g. we use the CTF in evanescent fields-illuminated SFS label-free microscopy, while in the case of evanescent-fields-illuminated SF-labeled microscopy we use the OTF. As shown in Figure 6.1, the spatial spectrum centered by $\vec{k}_s$ will be shifted to the lower SF region and transferred to propagating modes, thereby presenting super-resolution information to the far field. The final resolution of SFS methods is then $R_{fs} = \frac{2\pi}{k_c+k_s}$.

Evanescent fields occupy the near-field region of dielectric waveguides due to the imaginary transverse wavevector [89]. The magnitude of the in-plane wavevector can be tuned by the waveguide geometry and can thus provide the SF required to break the diffraction limit for SFS-based new microscopy techniques.

Employing evanescent-wave illumination allows the super-resolution imaging based on SFS method. To date, a variety of approaches utilizing specially designed substrates to generate evanescent waves for illumination and shift undetectable SF signal into the passband of conventional microscope have been proposed. These super-resolution imaging substrates include nanowires (NW) [55, 90], polymers [79], photonic waveguides [81, 91–96], wafers [97], SPPs [56, 82, 83, 98], LPs [85, 86, 99], and metamaterials [87, 88]. Evanescent-waves illumination SFS microscopy can be easily constructed using on-chip techniques and has excellent potential for practical applications.

6.3 Super-resolution Imaging Based on Nanowires and Polymers

In 2013, Hao et al. [100] utilized evanescent fields along a micro-optical fiber as a near-field illumination source and achieved super-resolution imaging in the direction perpendicular to the microfiber. The laser-guiding microfiber was precisely positioned on the sample surface using a piezoelectric positioning stage. It provided a SFS magnitude proportional to the effective refractive index of the guided mode. Based on the SFS effect, a slot pair structure of 225-nm central

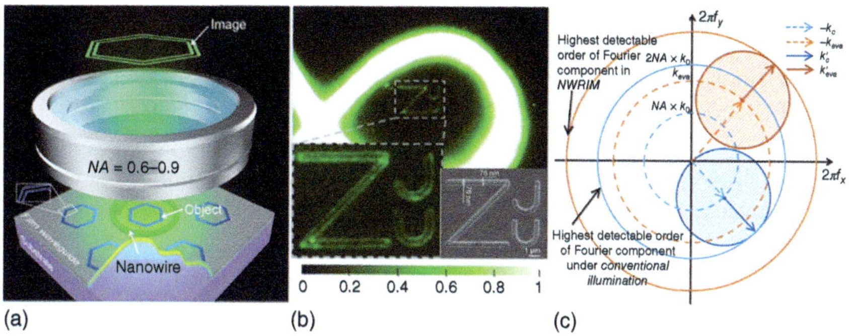

Figure 6.2 Evanescent-wave illumination SFS microscopy. (a) Schematic of the configuration and imaging process. (b) Evanescent-wave illumination SFS images (bottom left) and SEM (bottom right) of an arbitrary ZJU pattern on an Al_2O_3 film. $NA = 0.85$. (c) The underlying mechanism of the evanescent-wave illumination SFS method is represented in the spatial-frequency space. The blue and yellow solid circles in (c) indicate the optical transfer function (OTF) of the conventional microscope system and that of the evanescent-wave illumination SFS microscopy. Source: Reproduced with permission [55], © 2017, American Physical Society.

width was successfully resolved under 600 nm-wavelength light illumination using a 0.8 *NA* objective ($0.61\ \lambda/NA = 457$ nm). However, the 1D geometry limits the practical 2D imaging due to the difficulty in microfiber positioning.

In 2017, Liu et al. [55] proposed an on-chip evanescent illumination design for label-free 2D sub-diffraction-limited imaging with a large FOV. This NW ring illumination microscopy (NWRIM) configuration enables omnidirectional evanescent-wave illumination and is compatible with standard microscopes for efficient super-resolution imaging.

Figure 6.2a shows the schematic of the NWRIM, in which a CdS NW serves as the local light source [101–103]. When the NW is pumped by a 405-nm continuous-wave laser, the excited fluorescent light (center wavelength of ~520 nm) could efficiently couple into the 200-nm-thick TiO_2 film waveguide beneath the NW and illuminate the samples on the waveguide. Thus, scattered light can be collected by a far-field objective, contributing sub-diffraction-limited spatial information to the final image. As shown in Figure 6.2b, the system successively resolved predefined line patterns with 152-nm center-to-center distance slots using a 0.85 *NA* objective ($0.61\lambda/NA = 373$ nm). Figure 6.2c shows the detectable SF components in NWRIM and conventional microscopy, respectively. For conventional microscopy, both the wavevector of illumination and detection are confined by the *NA* of the objective. Consequently, the detectable Fourier components in far field are limited to a circle with a radius of $2NA \times k_0$. While in NWRIM with an illumination wavevector of k_s, the highest detectable Fourier components are expanded to $k_s + NA \times k_0$, where the illumination wavevector is expressed as:

$$k_s = k_0 \times N_{eff} \tag{6.2}$$

here, N_{eff} is the effective refractive index of the evanescent illumination waves, which could be larger than the *NA* of the objective lens if a high-refractive-index waveguide is adopted, so that a higher resolution can be achieved correspondingly. They have also employed an Al_2O_3–SiO_2 double-layer waveguide further to extend the propagation length of the evanescent wave and thus the FOV. Later, by optimizing the fabrication process of both the TiO_2 waveguide and the CdS NW, they realized an NWRIM with ~6000 μm² FOV and 122-nm center-to-center resolving ability [90]. However, due to the lack of control over the illuminating directions and spectrum calculation, frequency-aliasing (the spatial spectrum for different directions and SFSs are mixed) is inevitable in the NW-ring illumination arrangement. Therefore, we can't get the correct super-resolution imaging using this way.

The same group [79] reported an improved set-up in 2019, using a polygonal-geometry waveguide chip to reduce the frequency-aliasing problem. A fluorescent polymer film beneath the TiO_2 waveguide illuminates the sample with evanescent waves traveling in 16 precisely controlled directions. A 2D distortion-less image over ~200 μm² FOVs can be reconstructed due to the combination of oblique illumination and multiwavelength evanescent-wave illumination, which ensures wide coverage of the SF space.

The fabrication process of the polymer film-based substrate is shown in Figure 6.3a. A 90-nm-thick poly(9,9′-dioctylfluorene-alt-benzothiadiazole) (F8BT) film was first spin-coated on the surface of a 0.5-mm-thick, 1.5 cm × 1.5 cm K9 substrate and dried using a heating stage at 80 °C for 5 minutes. Next, laser direct writing technique was used to fabricate an array of polygonal patterns. Wavelength of the pulsed laser was 780 nm, and its duration was less than 120 fs with a repetition rate of 80 MHz. Measured laser power was about 310 mW, and machining precision of the system was smaller than 100 nm. Then a 70-nm-thick titanium dioxide film was deposited by electron-beam (E-beam) evaporation at room temperature to form a planar waveguide on the glass substrate. The refractive index of the deposited titanium dioxide film was 2.05, which supports an evanescent wave with a lateral wavevector of $k_s = 2.05k_0$. An array of chips can be fabricated in a batch, as schematically shown in Figure 6.3c.

To reduce the frequency-aliasing problem, the film is excited to produce illumination evanescent fields with multiple wavevector components (Figure 6.3b). The directions of the wavevector are controlled by exciting the edges of the polygonal patterns of the film. To achieve the tunability of the wavevector magnitude, the SFS raw images of the wide spectrum illumination are filtered using band-pass filter at 500 ± 2 nm ($FWHM$ = 10 ± 2 nm) or 632.8 ± 2 nm ($FWHM$ = 10 ± 2 nm). Vertical and oblique illumination are also applied to illuminate the sample, making sure there is enough coverage of frequency information from low-frequency range to high-frequency range (Figure 6.3d). Compared with microfiber and NW light sources, the fluorescent polymer films have superior luminescence and cost efficiencies; they are facile to fabricate; and scale-up for mass production is feasible. The wideband fluorescence emission of F8BT (from 480 to 700 nm for excitation in the blue) covers a large SF range, which is vital for the reconstruction of distortion-free images from the sample.

The principle of the SFS label-free recovery process is retrieving the phase map from multiple-intensity images, which alternates between the spatial domain (x–y) and the SF domain ($k_x - k_y$), as shown in Figure 6.4.

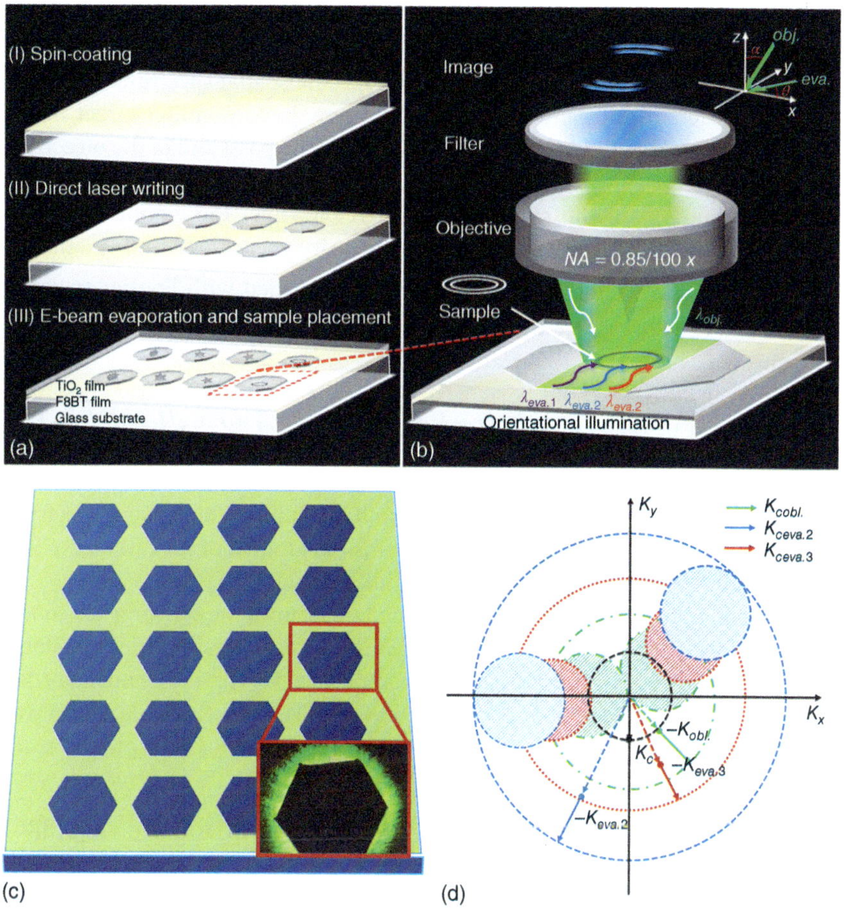

Figure 6.3 (a) Fabrication procedures for SRCs to produce evanescent waves at multiple wavelengths/directions. (b) Schematic diagram of the evanescent field illumination and oblique illumination process. Under a given illumination direction, the object (here a circular ring pair) is partially resolved at different wavelengths using appropriate bandpass filters. In the diagram, θ represents the illumination direction, and α is the incidence angle for oblique illumination. And *obl.* Means oblique illumination, and *eva.* represents evanescent illumination. (c) Schematic and microscopic image of the chip. (d) Schematic of reconstruction method. K_c, $K_{cobl.}$, $K_{ceva.2}$, and $K_{ceva.3}$ are the cut-off wavenumbers of the microscope system for different illumination modes and wavelengths; K_{obl} is the wavevector under oblique illumination; $K_{eva.2}$ and $K_{eva.3}$ are wavevectors for evanescent field illumination at wavelengths of λ_2 and λ_3, respectively. Source: [79], © 2019, John Wiley & Sons, CC BY 4.0.

(Step 1): The recovery process begins with making a guess of the super-resolution image in the spatial domain: $\sqrt{I_h}e^{i\varphi_h}$. The initial guess will not fail the reconstruction, but a good guess will affect the convergence rate and the signal-to-noise ratio (SNR) of the final image. Usually the wide-field, low-resolution image is used to get a fast convergence, but this will bring many noises in the background of the

Fourier space. Here we use the all-one image as the guess to get a clean SF background.

(Step 2): Second, we use the CTF to select the specific spatial spectrum and apply inverse Fourier transformation to generate a new low-resolution target image $\sqrt{I_l}e^{i\varphi_l}$. This process also keeps noise from outside the passband of the objective.

(Step 3): The phase φ_l of the low-resolution image is used as an estimate for the phase map φ_h of the sample $S(x)$. Therefore, only the amplitude of the low-resolution image is updated by the intensity measurement $I_{k_s,\phi}$, that is $\sqrt{I_l}e^{i\varphi_l} \rightarrow \sqrt{I_{k_s,\phi}}e^{i\varphi_l}$. The complex image is then applied with Fourier transformation and used to replace the subregion in the spatial spectrum. This step can be expressed as:

$$F^{m+1}(k) = F^m(k) \times (1 - CTF\,(k - k_s)) + CTF\,(k - k_s) \times \mathcal{F}\left(\sqrt{I_{k_s,\phi}}e^{i\varphi_l}(k - k_s)\right) \tag{6.3}$$

(Step 4): In the next step, we repeat steps 2 and 3 to update the spatial spectrum with all the raw images. The images number is the product of the number of SFS magnitudes $N(k_S)$ and the number of azimuthal angles $N(\phi)$.

(Step 5): Finally, steps 2–4 were repeated several times until the convergence condition was satisfied. To get the intensity of the sample, we inverse Fourier transform the spatial spectrum and get the sample's complex field $\sqrt{I_h}e^{i\varphi_h}$. The intensity distribution of the thin sample is the amplitude $\sqrt{I_h}$. The presented algorithm was tested on simulated data, as displayed in the top of Figure 6.4 for a ground truth input and a successful intensity reconstruction.

We demonstrated the practical capabilities of our super-resolution chip (SRC) to image irregular samples, such as multiwall carbon nanotubes (MWCNTs). We deposited bundles of MWCNTs (diameter: 10–20 nm; length: 200–250 μm) on the waveguide surface. The sample was illuminated with evanescent waves at wavelengths of 532 and 632.8 nm from 16 directions. Figure 6.5a presents a scanning electron microscopy (SEM) image of MWCNTs sample 1. As demonstrated in Figure 6.5b, subwavelength details in Figure 6.5a are not resolved by conventional wide-field microscopy. Figure 6.5c shows an image of sample 1 under evanescent illumination along the direction of the white arrow at 532 nm, in which some details are already visible. As demonstrated in Figure 6.5d, MWCNTs sample 1 is well reconstructed using our method. However, the penetration depth of evanescent waves restricts the imaging depth, as demonstrated in sample 2 (Figure 6.5e–h). Figure 6.5e represents an SEM image of MWCNT sample 2, and clearly, some fibers are crossing over others and therefore not in contact with the substrate. Thus, the evanescent illumination mode does not contribute any information to those parts of the sample (Figure 6.5g), and high SF detail will be omitted in the reconstruction (Figure 6.5h). The intensity profiles are presented in Figure 6.5i, validating good agreement between SEM and reconstruction results.

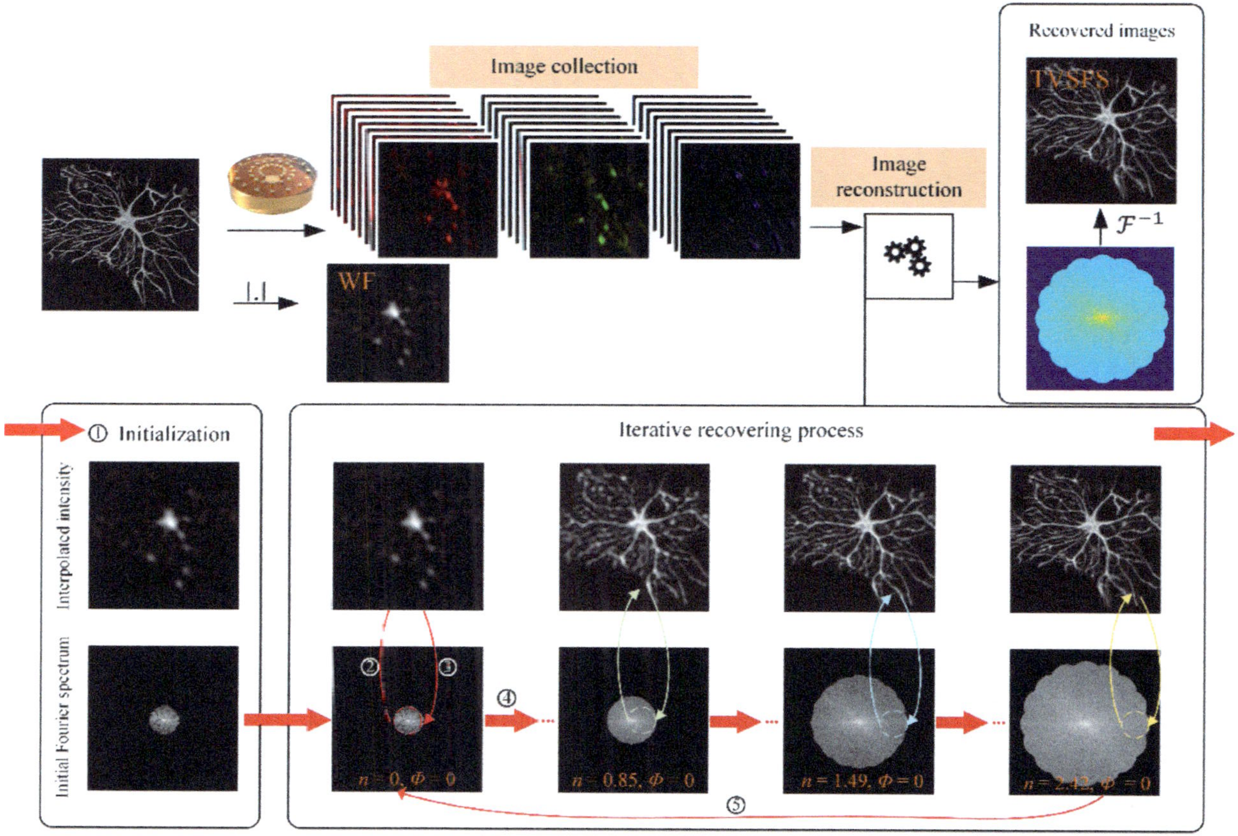

Figure 6.4 Iterative recovery procedure of TVSFS label-free imaging. The reconstruction algorithm is illustrated in five steps. Step 1: initialize the high-resolution image, $\sqrt{I_h}e^{i\varphi_h}$. Step 2: low filter in the Fourier space and inverse Fourier transform to generate the low-resolution image $\sqrt{I_l}e^{i\varphi_l}$. Step 3: replace the intensity $\sqrt{I_l}$ with the intensity measurement $I_{k_s,\phi}$, that is, $\sqrt{I_{k_s,\phi}}e^{i\varphi_l}$, and update in the Fourier space. Step 4: repeat steps 2–3 for other plane-wave incidences (total of $N(k_s) \cdot N(\phi)$ intensity images). Step 5: repeat steps 2–4 until the convergence.

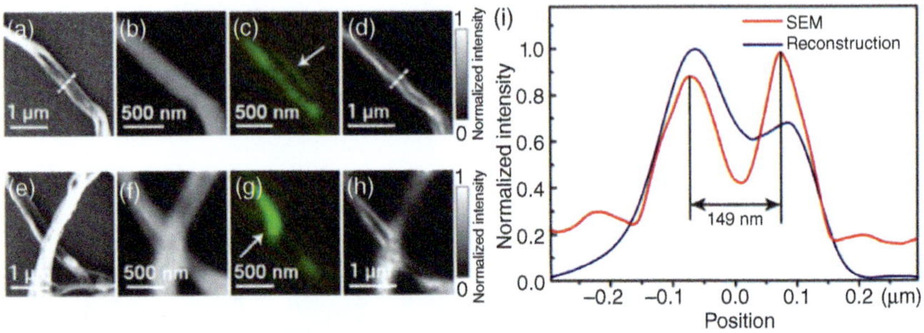

Figure 6.5 (a) SEM image of multiwall carbon nanotubes (MWCNTs). (b) Optical image under vertical illumination. (c) Image for evanescent-wave illumination at 532 nm along directions of white arrow. (d) Reconstructed image using all illumination modes. Panels (e–h) show corresponding images for another sample with two MWCNT strands crossing over one another. Norm. intensity denotes normalized intensity. (i) Normalized intensity profiles along white dashed lines in (a) and (d). Source: [79], © 2019, John Wiley & Sons, CC BY 4.0.

6.4 Super-resolution Imaging Based on Photonic Waveguides

Photonic waveguide is an effective approach toward integration and high-throughput imaging, which is important and very useful for practical applications in materials, biology, and medicine. Recently, photonic waveguide-based microscopy that confines the guided light using a high refractive index material on a silicon (Si) chip has attracted more and more interest due to its low cost, high compatibility, and capability to be integrated with existing circuit design and fabrication processes. The subwavelength evanescent field on the top of the waveguide offers uniform illumination to the sample, which extends over a wide FOV only limited by the attenuation of the light in the waveguide. The waveguide-based evanescent field excitation microscopy decouples the illumination and detection paths, significantly reducing the complexity of the optical imaging system and demonstrating the advantages of high integration, large FOV, and high-throughput for both fluorescent-labeled imaging and label-free super-resolution imaging.

6.4.1 Label-free Super-resolution Imaging Based on Photonic Waveguides

Similar to super-resolution imaging based on polymers described in Section 6.3, the label-free super-resolution imaging can also be implemented using photonic waveguide (such as Si_3N_4) in conjunction with a multispectral illumination. This concept [91] was first proposed by our group and further experimentally demonstrated by another group [92, 104].

As is shown in Figure 6.6a, the working region is designed to be octagonal with eight input waveguide channels connected to each edge of it. The samples are put on

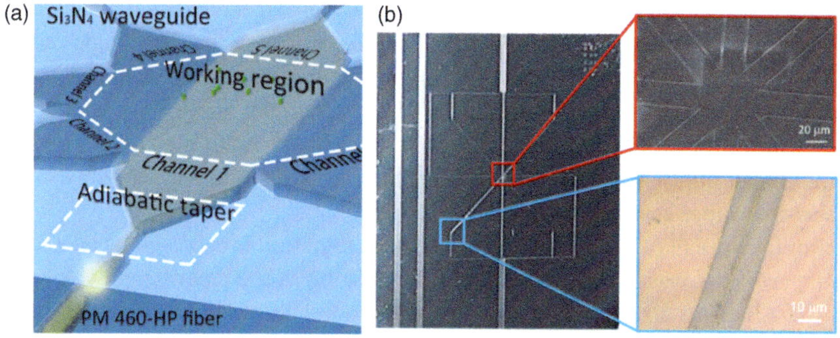

Figure 6.6 (a) The scheme of photonic waveguide for label-free super-resolution imaging. The coupling system and the octagon-viewing area of the multiwavelength-illuminated waveguide. Source: Reproduced with permission [91] Copyright 2019, IOP Publishing. (b) The photonic waveguide chip fabricated by our group.

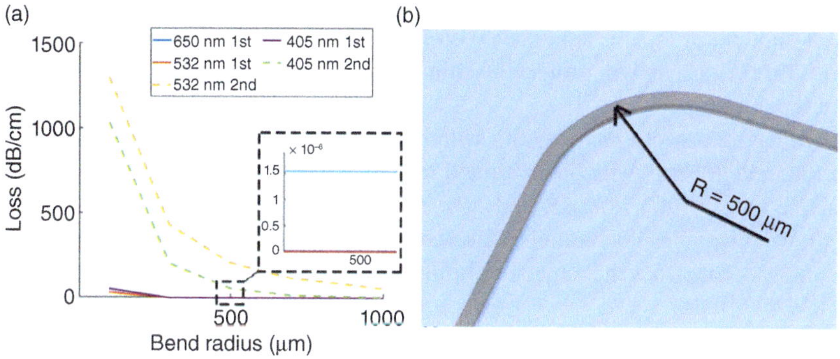

Figure 6.7 (a) Bending loss of the fundamental mode (the solid line) and the second-order mode (the dashed line) for three employed wavelengths in the rib waveguide. The second-order mode is not supported for 780 nm. (b) Schematic of the bend waveguide structure with radius of 500 μm as mode filter. Source: Reproduced with permission [91] Copyright 2019, IOP Publishing.

the working region with the single-mode evanescent-wave illumination from eight input waveguide channels. Figure 6.6b shows images of the fabricated chip.

The waveguide needs to be designed for single-mode operation due to the requirement for accurate frequency shift for image reconstruction. We simulated and found that high-order modes will change the wave front and effective refractive index of the illumination, which results in superposition of the frequency spectrum with different SFS orientation and magnitude. For the visible regime, the single-mode condition is about hundreds of nanometers wide for Si_3N_4 waveguide, which is difficult for achieving with the photolithography. To simplify the fabrication process, rib waveguide structure was utilized to make it possible for single-mode propagation with relatively wider waveguide (up to 1 μm). The mode control is realized using bend structure in rib waveguide. Figure 6.7a shows that the bending loss for fundamental TE mode (the solid line) is much lower than that of the second-order mode (the dashed line) in rib waveguide. Thus, 500-μm-radius bend structure (Figure 6.7b) can work as an excellent mode filter for all three applied wavelengths. The single-mode behavior of the waveguide provides illumination with narrow frequency band and approximately plane wave front with certain incident angle. These features help to improve the accuracy of frequency shift and acquire images with high SNR, making our multiwavelength evanescent-wave illumination waveguide a feasible platform for label-free super-resolution imaging.

To improve the FOV, a wider waveguide for illumination is achieved by adiabatically tapering the waveguide from 1 to 50 μm. As long as the taper length is longer than the adiabatic tapering length, the single-mode behavior will be maintained despite the waveguide width's change. The effective FOV is the overlap area of the eight incident directions, which is an octagonal region in the center of the waveguide with an area of around 2000 μm^2. From the simulation, we conclude that a 10 mm length tapering can bring single-mode behavior (more than 99%). As shorter wavelength leads to stronger diffraction, the 405 and 532 nm wavelength light request for shorter adiabatic tapering length than 780 nm light. After a lithography process, a SU-8 photoresist layer is spin-coated on it, with a window region at the working region. The SU-8 layer will protect the chip from propagation loss caused by scattering of the dust.

The imaging process includes the switching of three wavelengths, for example, 780, 532, and 405 nm, which correspond to fundamental TE modes with effective refractive indexes of 1.78, 1.85, and 1.91 in the Si_3N_4 waveguide. For each wavelength, we turn on each channel of the waveguide and collect the SFS raw images, and finally reconstruct the super-resolution image using the method described in Figure 6.4.

6.4.2 Labeled Super-resolution Imaging Based on Photonic Waveguides

In conventional SIM and total internal reflection fluorescence (TIRF)-SIM, the same objective lens is used for excitation and collection. Therefore, a large FOV and high resolution cannot be simultaneously obtained (see Figure 6.8a). Besides, the commercial liquid immersion objective lens has a limited *NA* of ~1.7, which restricts the resolution of TIRF-SIM to around 80 nm [105]. The chip-based SIM (cSIM) breaks this limitation by using a waveguide to generate standing evanescent-wave interference pattern for SIM. The SFS is no longer limited by the objective lens but rather depends on the effective refractive index of the waveguide (usually larger than 2.0), and a larger FOV can be obtained by choosing a low-*NA* objective lens (see Figure 6.8b).

To avoid the missing band in large-SFS imaging under evanescent illumination of large wavevector, the shift value needs to be tuned broadly. Through wavevector manipulation based on interfering two evanescent waves propagating along different azimuthal directions, the superposed effective wavevector can be tuned broadly, which is determined by the sine of the half-angle between the two propagation directions (Figure 6.9). The SFS of cSIM can be expressed as: $k_s = 2n_f \sin\frac{\theta}{2}/\lambda_{ex}$, where λ_{ex}

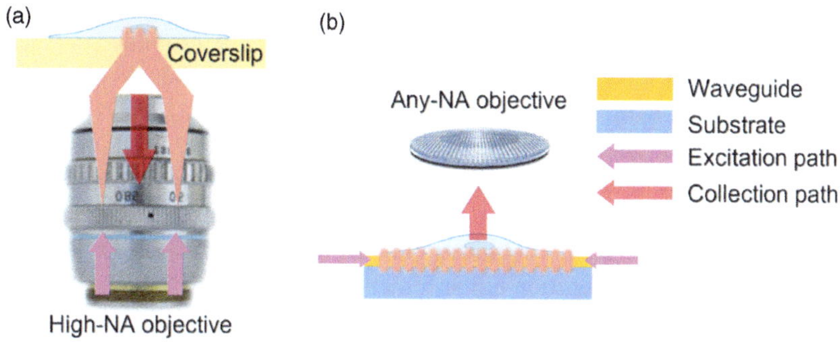

Figure 6.8 Conventional SIM and TIRF-SIM use the same objective lens for excitation and collection (a), while chip-based SIM uses the photonic waveguide for excitation and any NA objective for collection (b).

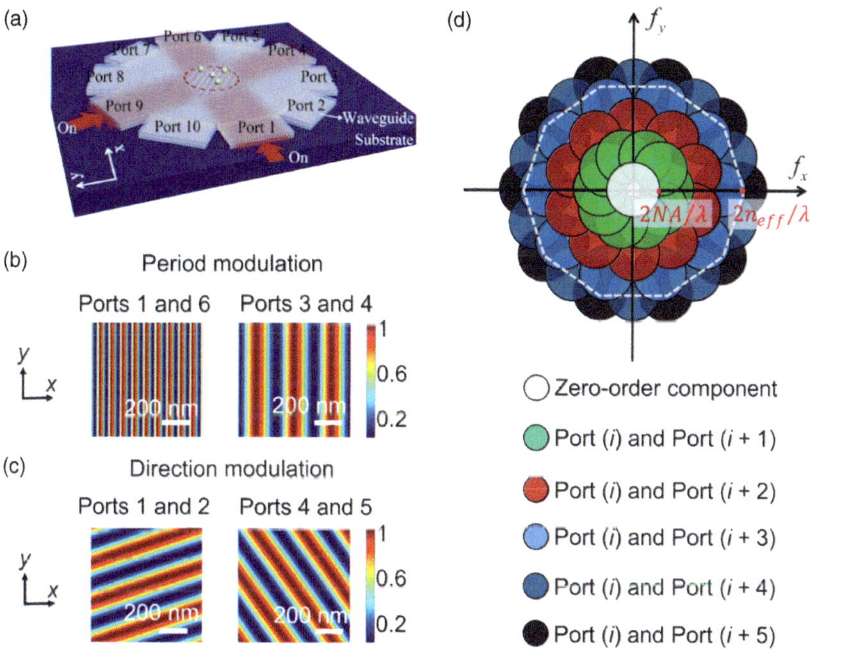

Figure 6.9 Large and tunable SFS of STUN in the lateral dimension. (a) Schematic setup of a STUN chip based on GaP decagon waveguide. The FOV is circled by a red dashed line. (b, c) Examples of the interference patterns with different input ports on, showing the period modulation and direction modulation, respectively. (d) The detected spatial-frequency ranges in STUN. Different colors represent the detectable components with different input ports on $i = 1, 2, 3, 4, 5$; $\lambda_{illu} = 561$ nm; $\lambda_{emis} = 600$ nm; $n_{eff} = 3.38$; $NA = 0.9$. Scale bars in (b) and (c): 200 nm. Source: Reproduced with permission [93] Copy right 2021, Springer Nature.

is the excitation wavelength, n_f is the effective refractive index of the guided mode, and θ is the angle of interference. The effective refractive index of the guided mode is usually a little less than the refractive index of the waveguide material. By using natural dielectric materials with a high refractive index, like Si_3N_4 ($n = 2$) [81, 92], Ta_2O_5 ($n = 2.1$), TiO_2 ($n = 2.6$), SiC ($n = 2.6$), or even GaP ($n = 3.4$) [93, 97, 106], the resolution of cSIM can be potentially reduced to ~50 nm (e.g. TiO_2 waveguide, $NA = 1.7$, $\lambda_{ex} = 405$ nm).

Helle et al. [81] reported a cSIM demonstration using Si_3N_4 waveguide ($n_f = 1.7$). They reported a 1.2 times resolution enhancement compared with TIRF-SIM, corresponding to a lateral resolution of 117 nm ($NA = 1.2$, $\lambda_{ex} = 660$ nm). To obtain uniform illumination over a large FOV, a single-mode waveguide is adiabatically tapered [107, 108] to a wide waveguide (theoretically applies for waveguides with over several mm^2 areas). By integrating phase modulation (such as on-chip thermo-optical modulation) [81], cSIM may have opened an avenue for high-throughput, large-scale producible, and miniaturized super-resolution imaging.

Our group demonstrated the potential to achieve a resolution down to 65 nm using GaP waveguide combined with a 0.9-*NA* objective [93]. Besides, we propose a 3D super-resolution imaging method, called the SFS tunable nanoscopy (STUN), based on photonic waveguide by using the SFS effect in the vertical dimension, as illustrated by Figure 6.10a. The vertical SF of the evanescent illumination covers a range much wider than the vertical SF passband of the objective and is mixed with the spatial frequencies of the object, which provides the basis to achieve the ultrahigh resolution in the vertical dimension. Figure 6.10b shows the intensity distribution of the evanescent wave on the GaP waveguide chip, and Figure 6.10c shows its SF spectrum. The red dot line in Figure 6.10c represents the position of the axial cutoff SF($0.56/\lambda$) of a 0.9 *NA* objective. The ultrawide range of the SF spectrum shown in Figure 6.10c infers that deep subwavelength vertical resolution is possible.

Consequently, the high-SF spectrum information in the vertical dimension of the object has been shifted to the passband of the objective by the illumination. To extract the high-SF information of the object, we introduce a sectional saturation effect. The saturation effect is a nonlinear process, which can change the line shape of the vertical intensity distribution of evanescent illumination. It is possible to saturate the surficial fluorophores by increasing the input laser intensity, and in this saturated region, the SF of the "effective illumination intensity distribution" is down-tuned, as shown by the orange zone in Figure 6.10d. The high-SF region is moved away from the surface, which is like a vertical scanning process to enable sectional imaging. The "effective illumination intensity distribution" can be considered as the coefficients in the equation set formed by tens of images captured at different saturation intensities, and the unknown vertical spatial distribution within the penetration depth of the evanescent wave can be solved [106].

This sectional method has compatibility with chip-based illumination, which is difficult with the commonly used methods that require modulation of the incident angle to the total internal reflection interface [109, 110]. Pulsed excitation with durations shorter than the lifetime of fluorophores can be utilized for the sectional saturation effect, promising fast acquisition and low phototoxicity.

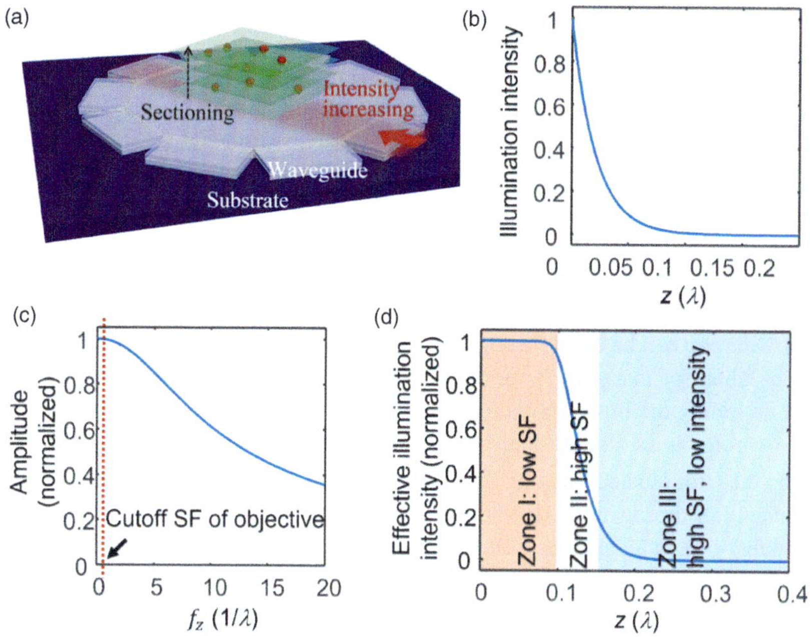

Figure 6.10 Mechanism of the SFS tuning in vertical dimension based on sectional saturation effect for super-resolution sectional imaging. (a) The schematic diagram. One port is chosen for input. (b, c) The exponential intensity distribution of the evanescent wave on a GaP chip and its spatial-frequency spectrum. The red dotted line represents the position of the axial cutoff SF (~$0.56/\lambda$) of the objective. (d) Emission intensities (or effective illumination intensities) along the vertical dimension with the surficial objects saturated. Objective $NA = 0.9$. $\lambda_{illu.} = 561$ nm. $\lambda_{emis.} = 600$ nm. "SF" denotes "spatial frequency." Source: Reproduced with permission [93] Copyright 2021, Springer Nature.

6.5 Super-resolution Imaging Based on Wafers

6.5.1 Principle of Super-resolution Imaging Based on Wafers

Continued research in fields such as materials science and biomedicine requires the development of a super-resolution imaging technique with a large FOV and deep subwavelength resolution that is compatible with both fluorescent and nonfluorescent samples. Existing on-chip super-resolution methods focus exclusively on either fluorescent or nonfluorescent imaging, and, as such, there is an urgent requirement for a more general technique that is capable of both modes of imaging. To realize labeled and label-free super-resolution imaging on a single scalable photonic chip, Tang et al. [97] proposed a universal super-resolution imaging method based on the tunable virtual-wavevector SFS (TVSFS) principle. Using this principle, imaging resolution can be improved more than threefold over the diffraction limit of a linear optical system. Here, diffractive units are fabricated on the chip's surface to provide wavevector-variable evanescent-wave illumination, enabling tunable SFSs in the Fourier space. A large FOV and resolutions of $\lambda/4.7$ and $\lambda/7.1$ were achieved for label-free and fluorescently labeled samples using a gallium phosphide (GaP) chip. With its large FOV, compatibility with different imaging modes, and monolithic integration, the proposed TVSFS chip may advance fields like cell engineering, precision industry inspection, and chemical research.

For deep subwavelength imaging resolution, the magnitude of the illumination wavevector should be large. However, when the SFS magnitude (k_s, which equals the magnitude of illumination wavevector divided by 2π) is more than double the radius of the maximal detection range of the collection objective in the frequency domain (k_c), a gap is observed in the Fourier space (top-right corner of Figure 6.11a), leading to artifacts in the final image, and sometimes even causing image reconstruction to fail [93]. This gap can be filled through wavevector tuning, using either wavelength tuning for label-free imaging [79] or multi-angle tuning [81, 93] for labeled imaging. To date, no tuning methods compatible with both labeled and label-free samples, or capable of producing a tunable SFS vector with magnitude more than $\frac{2}{\lambda}$ for label-free imaging and $\frac{4}{\lambda}$ for labeled imaging, have been reported. In this study, we develop a deep wavevector tuning method (with SFS magnitude tuning range beyond the limit of $\frac{2}{\lambda}$ for label-free imaging and $\frac{4}{\lambda}$ for labeled imaging) that is compatible with both fluorescent and nonfluorescent super-resolution imaging.

The improved resolution obtained with the TVSFS method can be understood by analyzing the Fourier space. The bottom of Figure 6.11b,c depicts the Fourier space for label-free imaging and labeled imaging, respectively. The colored circles represent spectra obtained using different SFS illuminations, while the circle at the

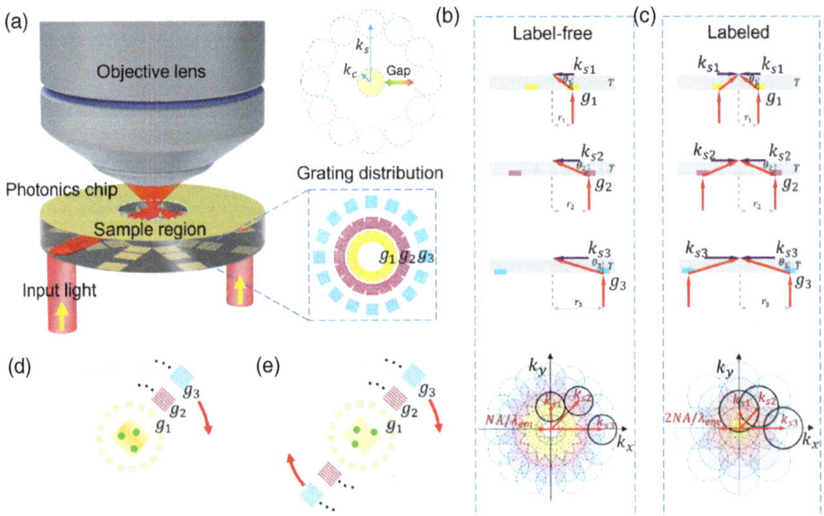

Figure 6.11 Physical scheme of chip-based TVSFS super-resolution imaging. (a) Schematic illustration of chip-based TVSFS super-resolution imaging. Top right: gap in the Fourier space obtained with high-refractive-index materials, explaining the necessity for multilevel tuning for SFS imaging. k_c and k_s are the cutoff and SFS wavevectors. (b, c) Illustration of TVSFS imaging in the Fourier space and the principle of SFS tunability. (d) Single-beam illumination for label-free imaging. (e) Double-beam illumination for labeled imaging.

center (in solid lines) represents the spectrum of an image taken with conventional microscopy. The latter spectrum contains only information from spatial frequencies less than NA/λ_{em} for label-free imaging, and $2NA/\lambda_{em}$ for labeled imaging, where λ_{em} is the wavelength of the emission light. To achieve isotropic and distortion-less deep super-resolution imaging, the Fourier space needs to be enlarged and filled using omnidirectional SFS tuning, performed with enough spectral overlap. From the theoretical calculations, the spectral overlap between two successive illuminations should exceed 25%. At least three successive SFSs are required to obtain a resolution of $\lambda/4.7$ for label-free imaging and $\lambda/7.1$ for labeled imaging. For omnidirectional label-free imaging, each SFS magnitude requires at least 16-directional illumination, while 8-directional illumination is needed for omnidirectional labeled imaging. On our chip, multilevel tuning is obtained using gratings with different periods and orientations. As shown in Figure 6.11a, well-designed and period-tunable subwavelength gratings are fabricated at predefined locations on the surface of the chip. A sample region is defined on the other side of the chip, where illumination from different directions overlaps. The lateral distance between the grating and the sample region, r, should increase with the thickness of the chip substrate (T) and the angle of the light deflected by the gratings (θ), since $r = T \cdot \tan(\theta)$, as shown in the coupling map in Figure 6.11b,c. The gratings on the chip's surface deflect the incident, propagating light, introducing evanescent waves with different wavevectors illuminated on the sample region. The SFS magnitude of an evanescent wave introduced by the gratings is defined as $k_s = \frac{n_{PC} \cdot \sin\theta}{\lambda_{ex}}$ for label-free imaging and $k_s = \frac{2n_{PC} \cdot \sin\theta}{\lambda_{ex}}$ for labeled imaging, where n_{PC} is the refractive index of the photonic chip and λ_{ex} is the wavelength of the excitation light. Hence, small SFS magnitudes are obtained by the large-period gratings close to the center of the chip, which can deflect the input light at a small angle. Conversely, for large SFS magnitudes, the corresponding gratings have a small period and are located near the edge of the chip. Sixteen gratings were designed for multi-angle directional illumination. Hence, label-free and labeled subwavelength resolution imaging can be realized using the same chip by switching between single-beam illumination (Figure 6.11d) and double-beam illumination (Figure 6.11e). The process for reconstructing a complete TVSFS image consists of combining spectra from different orientations, from low SFS to high SFS. For label-free TVSFS imaging, plane wave evanescent illumination with 3 SFS magnitudes and 16 rotational angles is made incident on the sample. The coherent interaction between this illumination and the sample generates a scattering pattern correlated to a specific part of the sample's spectrum. Label-free TVSFS imaging of a single frame requires acquisition of 49 images (one acquisition for every SFS illumination and one for vertical illumination) of scattered patterns. With labeled TVSFS imaging, two evanescent illumination beams meet at the center of the top surface of the chip and interfere with each other to form periodic structured light for illumination (Figure 6.11e). Illumination from eight angles is enabled through fabrication of an array of gratings with differing orientations, while the phases of the interference arms are controlled such that three raw images (three phases) are required for one SFS magnitude per rotational angle. The number of SFS directions required for labeled imaging is halved compared to the number needed for label-free imaging because the radius of the transfer function for incoherent imaging is doubled to make overlapping easier.

For every frame acquired, the detected spatial spectrum can be described as:

$$F_d = F_o(\vec{k} - \vec{k}_s) \cdot TF(\vec{k}) \tag{6.4}$$

where $\vec{k} = (k_x, k_y)$ represents the SF in the lateral plane, $F_o(\vec{k})$ is the Fourier spectrum of the object, and $TF(\vec{k})$ is the transfer function of the imaging system, which is bound by the NA of the objective lens. $\vec{k}_s$ represents the SFS vector used in acquisition, and the magnitude can be expressed as:

$$k_s = \frac{1}{p} = \frac{1}{\lambda_{ex}} n_{PC} \sin\theta = \frac{1}{\lambda_{ex}} n_{PC} \frac{r}{\sqrt{r^2 + T^2}} \tag{6.5}$$

for label-free imaging, and

$$k_s = \frac{2}{p} = \frac{2}{\lambda_{ex}} n_{PC} \sin\theta = \frac{2}{\lambda_{ex}} n_{PC} \frac{r}{\sqrt{r^2 + T^2}} \tag{6.6}$$

for labeled imaging, where T is the thickness of the substrate, r is the lateral displacement between the grating and the center of the chip, λ_{ex} is the wavelength of the illumination source, and p is the period of the gratings. Here, the refractive index of the substrate n_{PC} changes with the wavelength of the incident light. The illumination wavevectors can be scaled by modifying the design parameters.

Imaging resolution is determined by both the maximal aperture of the system, k_c, and the SFS magnitude that is provided by the evanescent illumination module, k_s, as,

$$\Delta_{xy} = \frac{1}{k_c + k_s} \tag{6.7}$$

where k_c is NA/λ_{em} and $2NA/\lambda_{em}$ for label-free and labeled imaging, respectively. For label-free imaging, the illumination wavelength equals the imaging wavelength (i.e. $\lambda_{em} = \lambda_{ex}$).

Therefore, the theoretical resolution limit with TVSFS can be formulated as:

$$\Delta_{xy} = \frac{\lambda_{em}}{(NA + n_{PC} \cdot \sin\theta)} \tag{6.8}$$

for label-free imaging, and

$$\Delta_{xy} = \frac{\lambda_{em}}{2\left(NA + \frac{n_{PC} \cdot \sin\theta \cdot \lambda_{em}}{\lambda_{ex}}\right)} \tag{6.9}$$

for labeled imaging. If the SFS magnitude is larger than k_c, the resolution exceeds that of conventional wide-field and SIM microscopy. Compared with conventional FPM, which uses real wavevector for multi-SFSs, TVSFS exploits tunable virtual wavevector for deep SFSs and can achieve a resolution three times better than the Abbe diffraction limit.

The FOV of TVSFS is determined by the area of the evanescent wave, which is regulated by the area of the fabricated gratings. In the present work, we obtained 40 μm × 40 μm gratings by etching with a focused ion beam (FIB), a size that can

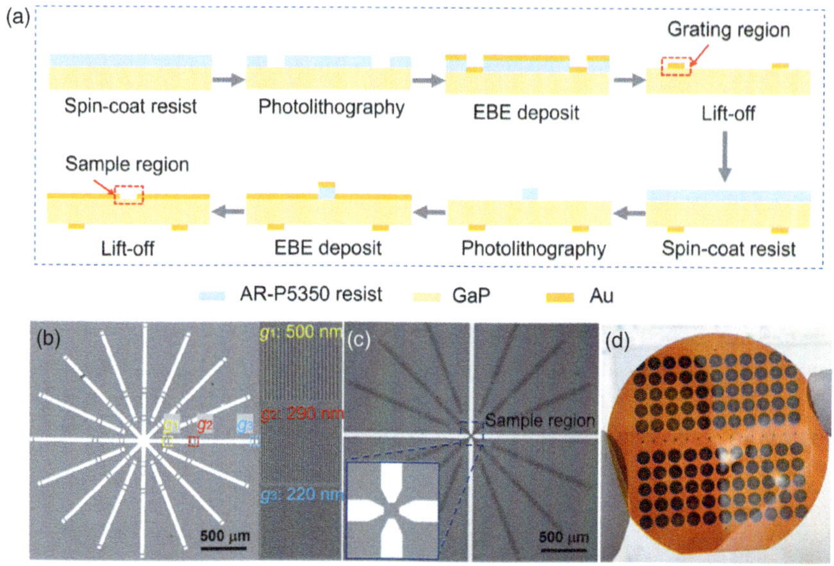

Figure 6.12 Wafer-scale fabrication of a TVSFS super-resolution imaging chip. (a) Schematic explanation of the wafer-scale chip-fabrication process. (b) Optical microscopy image of the bottom surface of the chip. Insets: gratings with periods of 500 nm (g_1), 290 nm (g_2), and 220 nm (g_3), corresponding to the labels in the main figure. (c) Optical microscopy image of the top surface of the chip. Inset: enlarged view of the sample region. (d) Array of super-resolution imaging chips fabricated on a 2 in. GaP wafer.

be increased to greater than 100 μm × 100 μm in future attempts. As the illumination module and the collection objective are decoupled in the TVSFS method, a low-*NA* objective lens can be combined with a large-wavevector illumination module to simultaneously achieve a high resolution and a large FOV.

In this work, we created arrays of TVSFS chips on a 2-in. wafer using standard photolithography and lift-off processes. Background light suppression is the crucial step in successful TVSFS imaging, as the coupling of light with the gratings introduces stray reflections. To achieve this suppression, we developed a light-blocking chip, fabricated using dual-surface lithography followed by metal deposition, as summarized in Figure 6.12a. The bottom surface of the chip is for SF shifting, achieved with the multilevel gratings, while the top surface is for imaging. All areas on the chip surfaces other than the sample or grating regions are covered with metal to block the background light. Besides blocking the stray light, the metal film also absorbs the evanescent wave that incidents onto it. Therefore, the light that bounces back at boundaries outside the FOV can be neglected, and we can get raw images with high contrast. In addition to providing an unlimited FOV, the transparency of the GaP substrate for visible wavelengths makes a dual-surface lithography process using a standard lithography machine possible. We used FIB to fabricate the different gratings at the predesigned positions. Optical images of the top and bottom surfaces of a chip following fabrication are shown in Figure 6.12b,c. With this process, more than 100 chips can be fabricated in a batch and separated for use.

6.5.2 Label-free Super-resolution Imaging Based on Wafers

To investigate the label-free imaging capability of our TVSFS chip, we simulated multilevel tuning of the illumination wavevector. Here, a pattern of Zhejiang University's eagle logo, defined with a double-line profile with line-to-line distance set at 215 nm, was selected as the imaging sample. The spatial distribution of this pattern under multilevel wavevector illumination provided by a 660 nm source is presented in Figure 6.13a. Here, k_{s1}–k_{s3} represents SFS vectors with increasing magnitudes ($1.3k_0$, $2.3k_0$, and $3.1k_0$), and $k_0 = \frac{1}{\lambda_{ex}}$ is the magnitude of SFS vector in free space. These SFS vectors were selected such that there was enough overlap between them in the Fourier space to ensure good convergence during image reconstruction. A conventionally acquired image of the eagle logo, obtained with vertical illumination through a 0.85-*NA* objective, is shown in Figure 6.13b, while the image in Figure 6.13g depicts a reconstruction covering the low-SFS, middle-SFS, and

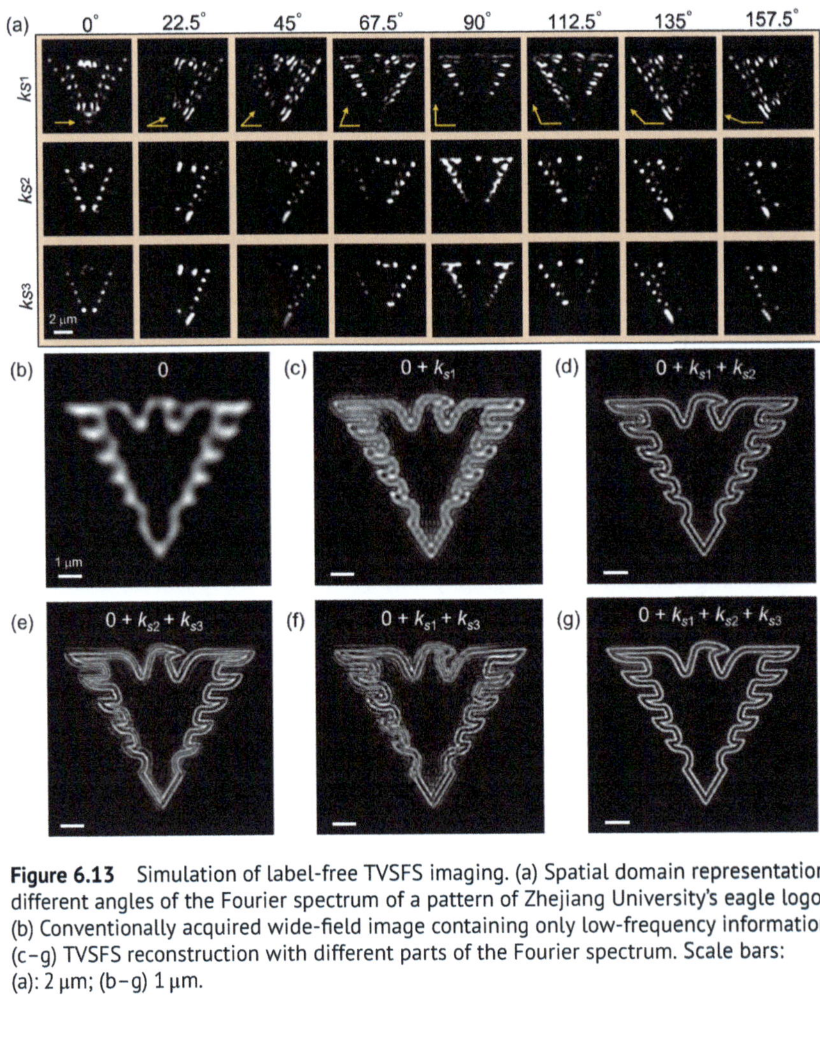

Figure 6.13 Simulation of label-free TVSFS imaging. (a) Spatial domain representations of different angles of the Fourier spectrum of a pattern of Zhejiang University's eagle logo. (b) Conventionally acquired wide-field image containing only low-frequency information. (c–g) TVSFS reconstruction with different parts of the Fourier spectrum. Scale bars: (a): 2 μm; (b–g) 1 μm.

high-SFS spectrum bands. Only the latter image reproduces the double-line details, highlighting the improved resolution obtained with multilevel tuning. In contrast, omitting parts of the spectrum, i.e. without multilevel tuning, results in images with poorer resolution (Figure 6.13c,d), or false reconstruction of the actual sample (Figure 6.13e,f).

To confirm this behavior, we conducted practical experiments recreating the simulations, using the same parameters for comparison. The eagle logo was etched on the surface of a GaP wafer using FIB, as shown in an SEM image in Figure 6.14b. We used a 660-nm-wavelength laser diode for illumination. Diode excitation was modulated using a sawtooth wave, and the resulting laser beam was spatially filtered to suppress speckle noise and improve the SNR of the raw images. Experimentally acquired images (Figure 6.14a) matched well with those obtained in simulations (Figure 6.13a). The minor mismatches between experimental and simulated images can be explained by fabrication defects, since the uniformity of grating fabrication and the refractive index of the photonic substrate are crucial for obtaining a perfect raw image. The final high-resolution image of the sample (Figure 6.14d) was obtained using our spatial spectrum-splicing algorithm (see Figure 6.4). A 4.5× resolution enhancement relative to the resolution of images acquired directly with wide-field illumination can be observed with the label-free TVSFS images (see the Fourier spectra at the bottom-right corner of Figure 6.14b–d). There is an excellent match between the line profiles of the SEM and TVSFS images (Figure 6.14f), validating our method. The image of missing spectrum band SFS (MB-SFS) presents false reconstruction and inferior match with the SEM image (Figure 6.14e,f). Image resolution can be increased further using a shorter wavelength. This is demonstrated in Figure 6.15, which depicts a sample etched with four lines imaged with 561 nm illumination, such that resolution of lengths as small as 120 nm were achieved.

6.5.3 Labeled Super-resolution Imaging Based on Wafers

Unlike label-free imaging where uniform evanescent waves provide coherent illumination, labeled fluorescent imaging requires periodic patterns formed by the interference between two counter-propagating evanescent waves [111]. The counter-propagating evanescent waves used in our experiments were obtained by stimulating grating couples. Both beams were TE-polarized, corresponding to the orientation of the gratings, to enhance the contrast of the interference pattern, which determines the resolution of the final reconstructed image. SFS magnitude is adjusted by changing the grating period. The photonic chips were fabricated with an empty sample region and three periods of gratings. Each period has eight gratings with equally distributed azimuthal angles. The gratings were distributed at predefined positions to ensure that light with SFS magnitudes of $1.8k_0$, $3.2k_0$, and $4.8k_0$ overlapped in the sample region. The SFS magnitudes were designed to ensure enough spectrum overlap between the adjacent shift values. A 1.1-*NA* water-immersion objective lens was used to collect the raw TVSFS images. Light-splitting and phase-shifting (three phases per orientation per SFS magnitude) between the two light paths were achieved by controlling the laser beam using

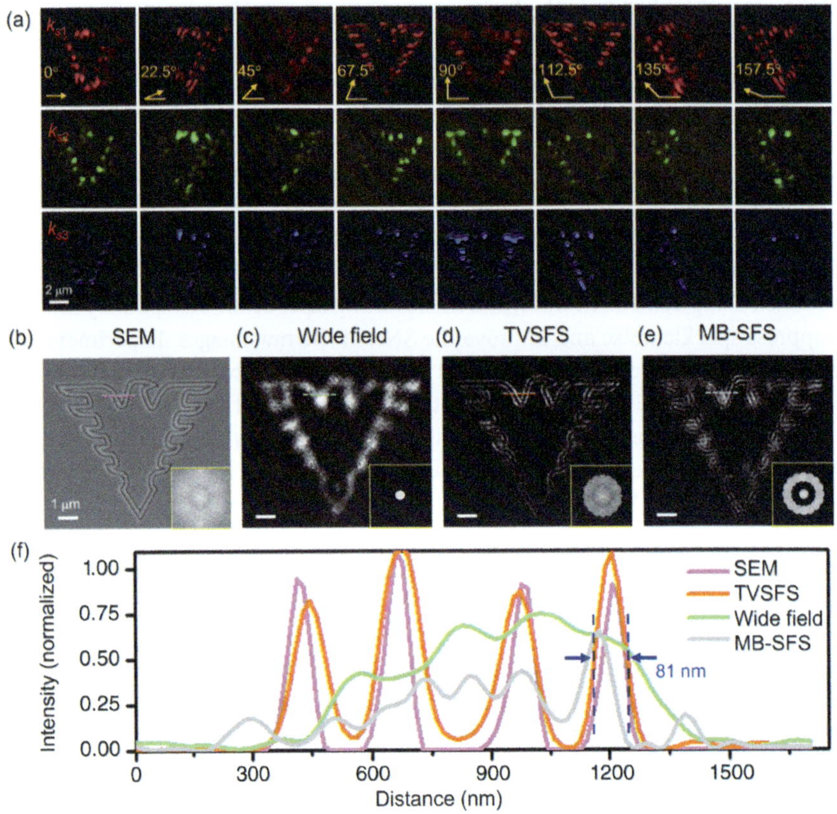

Figure 6.14 On-chip label-free TVSFS imaging of etched "ZJU" eagle logo. (a) Raw, label-free TVSFS images acquired from eight different directions using three wavevectors of varying magnitude. Colors indicate the magnitude of the wavevector used for image acquisition (k_{s1}, k_{s2}, and k_{s3} indicate SFS magnitudes of $1.3k_0$, $2.3k_0$, $3.1k_0$, respectively). (b–e) Images of the eagle logo taken using (b) SEM, (c) conventional wide-field microscopy under vertical illumination, (d) label-free TVSFS imaging, and (e) label-free MB-SFS. Insets in the bottom-right corner of these figures correspond to the Fourier spectra of the acquired images. (f) Line profiles of the region indicated by the dashed line in (b–e). The intensity profiles of the SEM and wide-field images were inverted for better correspondence to the dark-field TVSFS image. Details of the etched lines that are not resolved in the wide-field image (green line) and MB-SFS image (gray line) are clearly resolved in the TVSFS image (red line). The line profile of the latter image matches well with that of the SEM image (magenta line). Scale bars: (a): 2 μm; (b–e) 1 μm.

a spatial light modulator (SLM). For isotropic resolution, the SLM pattern has to be rotated to have four equally spaced orientations. For higher resolutions, the distance between the two beams should be adjusted to fit the gratings with smaller periods (distributed at the margin of the photonic chip).

To demonstrate labeled TVSFS imaging, we conducted experiments using fluorescent 40-nm beads. Samples were illuminated by periodic patterns with varying fringe spacings and reconstructed sequentially to incorporate higher SFS magnitudes. Conventionally acquired images of the fluorescent beads are shown

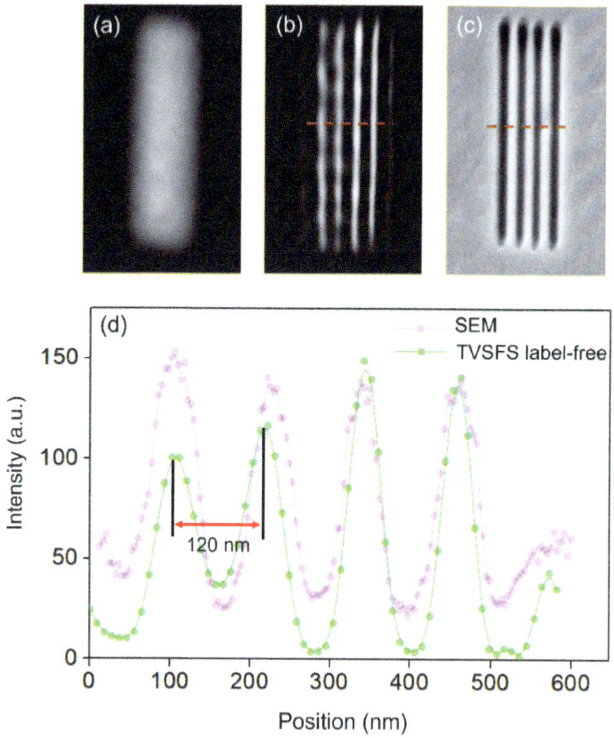

Figure 6.15 Resolution calibration of TVSFS label-free imaging using the 561 nm laser and 1.49-*NA* objective. (a–c) The images of a four-line slits taken under wide field, TVSFS label-free, and SEM, respectively. The four-line slits were fabricated using FIB and have a 120-nm center-to-center distance. (d) The line comparison between (b) and (c) shows a good correspondence between TVSFS label-free and SEM, which could not be achieved by the wide field.

in Figure 6.16a, whereas Figure 6.16b d depicts reconstructed images. The line profiles (Figure 6.16e) from the regions marked with the white dashed lines in Figure 6.16a–d illustrate how resolution improves as larger wavevectors are considered. Finer details can be observed clearly in the TVSFS reconstruction in a region that appears as a single feature in the diffraction-limited image (Figure 6.16a). The intensity profiles indicate that image contrast is improved when the SFS magnitude increases from $1.8k_0$ to $3.2k_0$, and individual beads can be resolved when the SFS magnitude reaches $4.8k_0$. Based on the line profile in Figure 6.16e, the two beads in Figure 6.16e are separated by 93 nm.

6.6 Super-resolution Imaging Based on SPPs and Metamaterials

To further increase the resolution, illumination with higher SF is needed. High-refractive-index dielectric materials have demonstrated a successful choice to

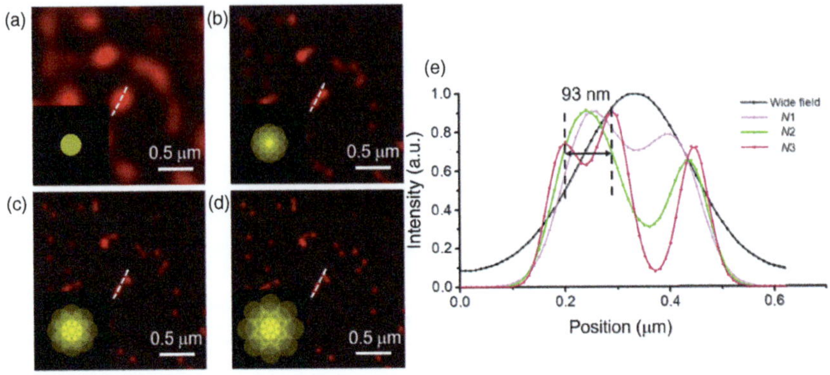

Figure 6.16 Resolving fluorescent beads with labeled TVSFS imaging. (a) Diffraction-limited image of a sample of fluorescent 40 nm beads. (b–d) TVSFS reconstruction of the images using the TVSFS method with maximum SFS magnitudes of $1.8k_0$, $3.2k_0$, and $4.8k_0$. Insets in the bottom left corners of these figures show the corresponding Fourier spectra. (e) Line profiles of the region indicated by the white line in (a–d). Two beads located 93 nm apart are resolved individually in the labeled TVSFS image with a maximum SFS magnitude of $4.8k_0$ but not in the diffraction-limited image. All images were obtained with excitation/emission wavelengths of 639 nm/661 nm.

enhance the resolution by generating evanescent-waves illumination in SFS microscopy, as aforementioned. Plasmonic material is another excellent candidate to form excitation illumination with a large SF, i.e. the SPPs and LPs. SPPs and LPs are demonstrated electromagnetic oscillations generated by collective oscillations of electrons in resonance with a light wave at a conductor/dielectric interface [112, 113]. Once excited, SPPs and LPs can be confined to a deep subwavelength scale, leading to a remarkable enhancement of the local field and allowing the manipulation of light far below the diffraction limit [114]. They are attractive for wide-ranging applications, including subwavelength imaging [115–117], sensing [118–128], subwavelength waveguides [129–133], plasmonic lithography [134–136], photovoltaics [137–140], optical tweezers [141–146], and optical analog computing [147–150].

SPPs were employed in superlens experiments for SFS-based super-resolution microscopy as early as 2007 [151, 152]. In the near-field superlens, the large-wavevector evanescent waves of the sample are resonantly enhanced by the plasmonic superlens slab, thereby generating a super-resolution image on the other side [153]. To bring the near-field information into the propagating far-field regime, the far-field superlens was designed with a sub-wavelength grating on top to downshift the spatial spectrum of the sample [151, 152]. Plasmonic waves could also be applied in FPM to provide a better resolution [52]. In this work, SPPs are excited by Kretschmann setup using a 1.4 NA objective, and 2D super-resolution imaging is realized (sample: grating structure with 240-nm period width) under 640-nm-wavelength light illumination ($0.61\lambda/NA$ = 279 nm). Recently, methods using patterned surface plasmons to illuminate the samples have been theoretically proposed and experimentally demonstrated, like PSIM [56, 82] and LPSIM

[85, 154]. The principle, as well as detailed experimental implements of PSIM and LPSIM, has been discussed in a book edited by Zhaowei Liu [155]. These methods, combined with fluorescent imaging, demonstrate the biocompatibility and will find applications in future research in biomedical study.

6.6.1 SPP-assisted Illumination Nanoscopy

Sharing a similar mechanism with SIM, PSIM uses surface plasmon interference (SPI) [156] instead of conventional laser interference to illuminate the sample. Generally, PSIM has a better resolution compared with SIM due to the much smaller period of SPI, which can bring a higher SFS in the SF space. The SFS k_s of PSIM is determined by k_{spp}, which can be tuned by the permittivity of metal and the surrounding dielectric at a specific illumination wavelength. For a metal/dielectric interface, the dispersion relationship of SPPs can be described as:

$$|k_{spp}| = |k_0|\sqrt{\frac{\varepsilon_m \varepsilon_d}{\varepsilon_m + \varepsilon_d}} \tag{6.10}$$

where $|k_{spp}|$ and $|k_0|$ represent wavevectors of SPPs and free-space light propagating in vacuum, and ε_m and ε_d are the permittivity of metal and dielectric. A comparison of the typical dispersion curves of the SPPs and photons in air and dielectric materials, as plotted in Figure 6.17, shows that k_{spp} is always greater than k_0 and can be extremely large at the surface plasmon resonant frequency ω_{sp}. Thus, this large wavevector of SPPs can provide a more substantial SFS and a higher resolution compared with conventional SIM and evanescent field- illuminated SIM. For insulator–metal–insulator (IMI), the coupling of propagating modes on different interfaces will split the dispersion relationship of SPPs into odd modes and even modes as the metal thickness decreases to tens of nanometers [157, 158]. It is worth noting that the $|k_{spp}|$ of even modes will increase inversely with the metal thickness, which is advantageous for super-resolution imaging applications. Different modes can be excited individually or simultaneously to form large-wavevector interference patterns that can be used in PSIM.

To launch SPPs, the momentum mismatch between excitation photons and plasmons should be compensated. Existing SPP excitation methods include prism coupling, grating coupling, edge/slit coupling, objective coupling, and near-field excitation [112]. Among these methods, grating coupling [159], edge/slit coupling [82], and objective coupling excitation [98, 160, 161] are mostly used in PSIM configuration for their convenience in implementation and possibility of generating regular SPP interference patterns, which are superior to irregular ones in terms of resolution and signal-to-noise improvement in the imaging [111].

The Implementation of PSIM requires illumination patterns to be tuned with multiple directions and multiple phase shifts. Although the coupling structure geometry is fixed, the SPI direction can be tuned by the illumination polarization since the SPP-coupling efficiency for polarization perpendicular to the coupling structure is more significant than that parallel to the coupling structure [162]. To realize the multiple phase shifts in the SPI, controlling the optical delay in individual excitation

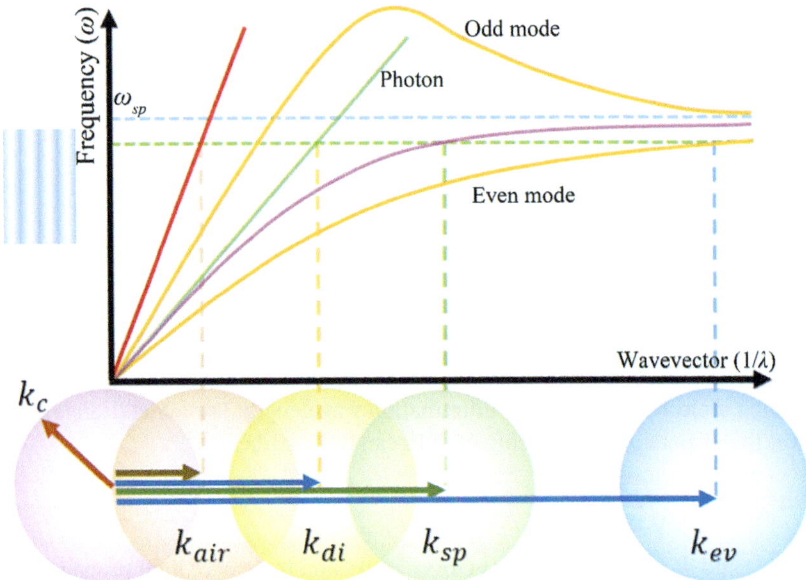

Figure 6.17 Concept of resolution enhancement by structured illumination. Resolution improvement representation in SF space of SIM, evanescent-fields-illuminated SIM, and PSIM. Typical dispersion curves of SPPs on a thick-metal-film/dielectric interface (pink curve) and a dielectric/thin-metal-film/dielectric interface (orange curve). The SPP of dielectric/thin-metal-film/dielectric interface splits into the odd mode and even mode when the thickness of the metal decreases to around 50 nm. The pink circle corresponds to the spatial information within the passband of conventional microscopy with k_c representing its cutoff wavevector. The maximum resolution in conventional SIM, as defined by the sum of the cutoff frequency k_c and the spatial-frequency shift k_{air}, is approximately twice that of wide-field imaging (the red line represents the dispersion curve of photons in the air). The evanescent-fields -illuminated SIM, either TIRF-SIM or cSIM, has a spatial-frequency shift defined by the k_{di}, exceeding that of conventional SIM (the green line represents the dispersion curve of photons in dielectric materials). The wavevector of SPPs is nonlinear with photon energy. Thus, it is possible to recover information from an area more than twice the size of the normally observable region using PSIM. k_{air}, k_{di}, k_{sp}, and k_{ev} represent the spatial-frequency shift obtained in air, dielectric material, single-interface SPPs, and even-mode SPPs. ω_{sp} is the surface plasmon frequency.

beams [159] and controlling optical vortices with topological charges for objective coupling [160] are possible solutions. For edge/slit excitation with multiple periods on the order of micrometers, it is more practical to adjust the angle of incident light to introduce a phase difference $\phi(\theta)$ between two adjacent edges/slits. This angle variance could translate to a phase difference in the excited SPP waves and thus lateral shift of the interference patterns.

6.6.1.1 Metal–Dielectric Multilayer Metasubstrate PSIM

The substrate of PSIM is mainly composed of metal/dielectric multilayer films. In 2012, Wang et al. [159] first experimentally demonstrated the super-resolution imaging of nanoparticles using plasmonic standing waves. A 2D PSF full width at

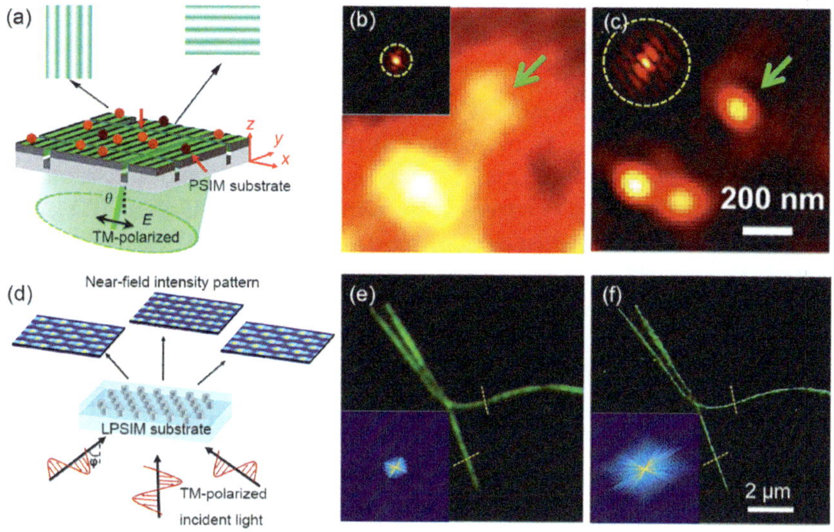

Figure 6.18 Demonstration of resolution improvement in PSIM and LPSIM. (a) Schematic of the PSIM. Green patterns on the top of the substrate are near-field intensity patterns on an object plane generated by p-polarized laser beams in two orthogonal directions. (b) Conventional fluorescence image. (c) Reconstructed PSIM image. Fourier spectra are shown in the top-left corner of (b, c), respectively. The yellow dashed circles in (b) and (c) indicate the optical transfer function (OTF) of the conventional microscope system and that of the PSIM system. Source: Reproduced with permission [82], © 2014, American Chemical Society. (d) Schematic of the LPSIM. Generated near-field intensity patterns on an object plane created by TM-polarized laser beam incident on the 60 nm hexagonal silver disc array on the sapphire substrate at angles of −60°, 0°, and 60°, along one of the three symmetry axes. (e) Diffraction-limited image of green microtubules. (f) Corresponding LPSIM image of (e) with significantly improved resolution. Fourier spectra are shown in the bottom-left corner of (e, f), respectively. Source: Reproduced with permission [86], © 2018, American Chemical Society.

half maximum (FWHM) of 172 nm was demonstrated with fluorescence centered at 645 nm under a 1.42 *NA* objective lens. In 2014, Wei et al. [82] completed another demonstration of PSIM using a silver film with a thickness of 250 nm and periodic slit array spaced by 7.6 μm to launch counter-propagating SPPs. The propagating SPPs form interference patterns and excite the fluorescence beads with diameters of 100 nm distributing randomly on the silver film surface. Six images with different illumination polarization and angles were recorded and spliced in the SF space to form a large Fourier domain. The FWHM of a single 100-nm-diameter fluorescent bead decreases from 327 to 123 nm after SFS processing, yielding a ~2.6-fold resolution improvement (Figure 6.18a–c).

Further efforts have been devoted to improving the resolution of PSIM with more complex substrates, e.g. the metal–dielectric multilayer. Modes in a metal–dielectric multilayer can be represented by the linear combination of individual surface modes. Thus, as the number of metal films increases, the maximum of the modal index will increase accordingly [163]. The usage of metal–dielectric multilayer to launch deep-subwavelength patterns can be applied in photolithography [164]. Similarly, the metal–dielectric multilayers can also be adopted in PSIM for the generation of large wavevector $|k_{spp}|$.

However, the energy transmission rate is limited due to the significant loss of multilayered metal films. To overcome this problem, Yu and coworkers [165] proposed a six-layer structure (with 1 μm^2 cross-section area) consisting of Ag–Al_2O_3–Ag–SiO_2–Ag–H_2O to decrease the number of metal films. FDTD simulation and theoretical calculation demonstrated an interference SPP pattern with a period of 84 nm could be generated at the surface of the substrate. The postprocessing includes illumination with different incident angles and polarization controls. Finally, the image of a 10-nm-diameter quantum dot was reconstructed as a spot with FWHM of 41 nm, which translates to a 5.3-fold improvement compared with conventional epi-fluorescence microscopy. Different from this work, in ref. [166], the symmetrical coupling of two short-range SPP modes in the dielectric film is used, therefore generating SPPs with ultralarge wavevector. The simulation shows the proposed Ag/Al_2O_3/Ag/H_2O multilayer films could have a high resolution of 16 nm when applied in PSIM, which is 13.6-fold compared with conventional epi-fluorescent microscopy. The multilayer structure also has excellent tunability for SPPs by varying the thickness and permittivity of dielectrics.

6.6.1.2 Graphene-assisted PSIM

Graphene has been widely investigated for its several unique properties, such as ultrahigh carrier mobility and excellent tunability in conductivity [167–174]. Similar to noble metals supporting collective free electron oscillations, the electrons in doped graphene can also respond to the electromagnetic field resonantly, leading to graphene plasmons (GPs). Compared with the metal case, the energy loss of the tightly confined GPs is lower, and the propagation length is about a few dozens of wavelengths in the excited mode. Most importantly, the propagation length and the effective refractive index can be dynamically tuned by adjustment of the chemical potential of graphene through applying temperature field [175], magnetic field [176], or electrical field [172]. Thanks to these inherent properties, research on GPs has made remarkable progress [177, 178] and found comprehensive applications in transformation optics [179, 180], nanoimaging [84, 181–183], and tunable metamaterials [175, 184, 185]. Besides, the developments on up-conversion fluorescent nanoparticles that converse near-infrared (NIR) light to visible wavelengths [186, 187] have made it easier to facilitate visible super-resolution imaging using GPs in the NIR range. Graphene can be combined with designed substrates to realize ultrahigh-resolution PSIM.

For the graphene layer, the surface conductivity σ_g can be calculated:

$$\sigma_g = \frac{ie^2 k_B T}{\pi \hbar^2(\omega + i/\tau)} \left(\frac{\mu_c}{k_B T} + 2\ln\left(e^{-\frac{\mu_c}{k_B T}} + 1\right)\right) + \frac{ie^2}{4\pi\hbar} \ln \left| \frac{2\mu_c - \hbar(\omega + i/\tau)}{2\mu_c + \hbar(\omega + i/\tau)} \right| \quad (6.11)$$

where e represents electron charge, k_B is the Boltzmann constant, T is the Kelvin temperature, $\hbar$ is the reduced Planck constant, ω is the radian frequency, μ_c is the chemical potential, and τ is the electron–phonon relaxation time, respectively. The relative permittivity ε_g can be represented as:

$$\varepsilon_g = 1 + i\frac{\sigma_g}{\varepsilon_0 \omega \Delta} \tag{6.12}$$

where ε_0 is the permittivity of vacuum and Δ denotes the graphene thickness. From Eqs. (6.11) and (6.12), the permittivity of graphene depends on the chemical potential, which can be externally tuned by the bias voltage, thus leading to wavevector-tunable GPs.

The graphene-assisted PSIM was first proposed by Zubairy and coworkers [84]. In their scheme, GPs were launched by the grating on a dielectric-monolayer graphene-dielectric substrate, and a wavevector of $45.7k_0$ was obtained. By optimizing parameters like the Fermi energy, wavelength of the incident light, and permittivity of the dielectrics, they obtained a 10-nm resolution compared with conventional microscopy.

To further increase the resolution and make it suitable for practical application, another scheme termed as hybrid graphene on metasurface structure (GMS) was proposed [182]. The physical mechanism of GMS is based on localized surface plasmon enhancement and GPs. The GMS structure consists of a single layer of graphene deposited on a $SiO_2/Ag/SiO_2$ multilayer. The 10-nm-thick silver film enables a highly localized SPP as the excitation source for GPs lying above. Finite-difference time-domain (FDTD) simulation finds that standing-wave GPs with a period of 11 nm can be achieved on graphene for 980-nm-wavelength excitation light, and the resolution could theoretically reach 6 nm.

6.6.2 Localized Plasmon-assisted Illumination Nanoscopy

Different from PSIM, which uses propagating SPPs, the LPSIM uses patterned LPs to illuminate the sample (Figure 6.18d) [85, 86, 99, 154, 188]. The illumination patterns are generated by an array of localized plasmon antennae fabricated by electron-beam lithography. Different from SIM or PSIM, whose SFS is limited by the free-space or propagating SPP dispersion relations, the period of the structured illumination patterns in LPSIM is only limited by the antenna-size geometry, thus an arbitrary SFS in the Fourier space could be obtained. Suppose the pitch of the antenna array is p, then the SFS k_s can be described as: $k_s = 2\pi/p$.

In 2017, Ponsetto et al. [85] first experimentally confirmed the super-resolution ability of LPSIM. The chip is made of silver nanodiscs hexagonally distributed on silica or sapphire substrates. A thin protective layer covered on the silver nanodiscs was used to protect the silver from oxidation and separate the silver from the biological samples, while the thickness was small enough to keep the sample in the range of the evanescent plasmonic field. The size and pitch of nanodiscs were optimized through simulation. In the experiment, the antenna has a diameter of 60 nm and a pitch of 150 nm. They used this chip to image fluorescent polystyrene beads (emission wavelength of 500 nm) and achieved a resolution of 74 nm ($\lambda/(5.6NA)$). The collected images are far-field fluorescent emission modulated by the localized plasmon-generated near-field patterns. To reconstruct the final image, the near-field illuminating patterns should be varied by changing the incident angle. Also, to excite the LPs in the desired direction, the polarization of the incident light should be controlled correspondingly. In 2018, Bezryadina et al. [86] demonstrated a wide-field LPSIM with 50 nm ($\lambda/(6NA)$) spatial resolution at video-rate speed (30–40 Hz) with comparatively low light intensity (100–150 W cm^{-2}). The 60-nm-diameter silver discs with 125 nm pitches were used to achieve the best resolution. The microtubules stained with green fluorescence were imaged in transmission mode with a $100 \times 1.65\,NA$ high-index oil immersion objective, demonstrating the compatibility with live-cell imaging (Figure 6.18e,f). To further improve the resolution, the LPSIM could be combined with the SF-compression method. Bezryadina et al. [188] employed optical tweezers to control the microsphere lens on top of the LPSIM substrate with fluorescent objects and obtained a resolution of 57 nm ($\lambda/10$) with a low NA objective lens.

The LPSIM can provide a large shift in the SF space by using nanoantenna arrays and further decreasing the pitch. However, when the pitch is smaller than $\frac{\lambda}{2NA}$, a missing gap will appear in the SF space, which leads to distortion in the final reconstructed image. Recently, it has been demonstrated that the blind-SIM algorithm is superior to the standard SIM algorithm in the case of information missing in SF space [99]; however, an oversampling should be sacrificed. The blind-SIM algorithm reconstructs the super-resolution image in the real space using a cost-minimization approach. With more information provided, the resolution of the reconstructed image will be better. Nevertheless, it is still hard to achieve the theoretical resolution beyond $\lambda/(6NA)$. In the experiment, they used 27 sub-images in six directions to obtain an imaging resolution of 81 nm ($\lambda/(5.8NA)$), while the theoretical resolution is 52 nm using an array pitch of 135 nm with a collecting objective lens of 1.2 NA. Further, the missing SF could also be filled by designing quasi-periodic structures, each excited by a different illumination wavelength.

Compared with PSIM, the SFS of LPSIM is independent of the wavelength and permittivity character. However, to improve the SNR and the imaging speed, the wavelength and permittivity should be carefully chosen to ensure resonant plasmonic enhancement. The illumination of LPSIM is a paralleled point scanning using nanoantenna fields, separating the illumination and detection lens; thus, a large SFS imaging with a large FOV could be obtained at high speeds. The limit of the LPSIM is large-scale fabrication, the imperfection of which will bring wrong sampling and generate a distorted image. Similar to other near-field illumination super-resolution imaging methods, LPSIM can only detect sample structures with a thickness within the depth of the evanescent waves. From another point of view, the sectioning ability will bring a better imaging contrast by cutting off the out-of-focus noise.

6.6.3 Metamaterial-assisted Illumination Nanoscopy

To further improve the resolution, researchers turn to multilayer metamaterials, which support ultrahigh-SF waves [189]. Hyperbolic metamaterials (HMMs) are the most widely used examples and, benefiting from their anisotropic hyperbolic isofrequency curve, are capable of carrying much higher SF contents than most

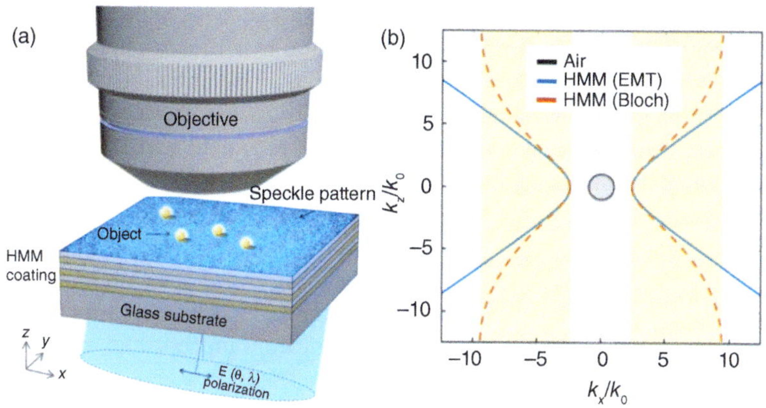

Figure 6.19 (a) A HMM-coated substrate projects ultrafine structured speckles onto objects lying on its top surface. (b) An isofrequency curve of air, an ideal HMM by effective medium theory (EMT), and a practical HMM consist of periodical layered structures of Ag and SiO_2 (Bloch) at wavelength of 488 nm. Both the wavevectors *kx* and *kz* are normalized to the wavevector in air k_0. The allowed *k*-bandwidth is highlighted in gray (air) and in orange (practical HMM).

materials that can be found in nature. The ideal HMM has an unlimited *k*-space, while a practical HMM has a *k*-space limitation from its layer periodicity, as shown in Figure 6.19b. For example, a practically achievable Ag/SiO_2 multilayer with a period of 20 nm can support a highest *k* mode of around $10k_0$ working at 488 nm wavelength.

Zhaowei Liu's group reported the speckle metamaterial-assisted illumination nanoscopy (speckle-MAIN) [88], which uses the Ag–SiO_2 multilayer HMM to generate speckle-like sub-diffraction-limit illumination patterns in the near-field with improved SF. Such patterns, similar to traditional SIM, are then used to excite objects on top of the surface. The HMM, fabricated by sputtering deposition method, consists of three pairs of Ag and SiO_2 layers. To convert the incident plane wave to high-*k* vector waves and generate ultrahigh-resolution speckle on the top surface, they took advantage of intrinsic surface scattering as well as volumetric scattering caused by nonperfect multilayers. To have enough distinguishable speckles, they illuminated the HMM with a random, diffraction-limited optical field generated by either a diffuser or a multimode fiber. The complex fields (diffraction-limited speckles), equal to a composite of plane waves at different angles and phases, will be converted into sub-diffraction-limited speckles after passing through the HMM and excite the fluorophores in a specimen. The fluorescence images excited by different speckle patterns are then directly collected by a standard inverted microscope system (Figure 6.19a). Finally, blind-SIM algorithm was adopted, with the consideration of a statistical prior that the averaged speckle illuminations are homogeneous, to reconstruct the final super-resolution image, similar to but with better resolution than speckle-SIM [190]. They have demonstrated that speckle-MAIN can bring the resolution down to 40 nm by imaging different samples, such as fluorescent beads, quantum dots, and Cos-7 cells.

To achieve a higher resolution, larger wavevector should be provided, which is restricted by the limited unit cell size of HMM. Natural hyperbolic materials such as Bi_2Se_3, Bi_2Te_3, and GaTe could circumvent some of the limitations but were also limited by the large loss. Zhaowei Liu's group proposed a low-loss organic hyperbolic material (OHM) to support remarkably high-*k* optical modes in the visible frequency [87]. The fabricated OHM has a periodicity of 1.6 nm, the hyperbolic dispersion of which supports a frequency $k_{cutoff} = 54k_0$ at $\lambda = 465$ nm, and they demonstrated the OHM-based SIM method with 30 nm resolution capability. Compared to their inorganic counterparts, the OHMs not only provide better deep-SFS illuminating field in visible spectral range but also offer many advantages, such as enhanced photostability of fluorophores near the OHM, easy fabrication process, and excellent biocompatibility for bioimaging applications.

6.7 Summary and Outlook

Just a few years after the first implementation, super-resolution techniques are transforming our understanding of optical physics along with the microscopic biology world. Ongoing developments suggest that an impressive diversity of biological and medical questions could be answered using optical nanoscopy. As complementary techniques to existing far-field nanoscopy like STED and STORM, evanescent wave-assisted super-resolution methods have shown us their power in resolution, imaging speed, and FOV. Compared with conventional imaging methods used for the detection of subwavelength details that rely on relatively sophisticated, cumbersome, and expensive microscopy systems, chip-based super-resolution methods provide new avenues toward cost-effective and easy-to-use super-resolution imaging. The single photonic chip can be integrated with a small optical collection system (such as a smartphone with a camera), to function as a light-weighted super-resolution imaging system. Besides, super-resolution methods based on large-wavevector electromagnetic evanescent waves put less requirement on special fluorescent dyes and will find wide applications in biology, medicine, and semiconductor industry.

Resolution and imaging quality are the priorities of a super-resolution system. For surface wave-based super-resolution methods, the key to improving the resolution and image quality is finding new materials with high refractive index and low optical loss. Metamaterials fill in the gaps by providing the ability to generate large-wavevector evanescent waves or collect evanescent waves efficiently. The realization of a theoretically perfect super-resolution device is possible with judicious design processes. However, the limits in fabrication accuracy and throughput of current technologies, e.g. electron beam lithography [191] and FIBs [192], should be further enhanced to meet the ever-increasing demand for precision and large-scale fabrication [193].

The resolution of the SFS method is closely related to two parameters: the maximum lateral wavevector magnitude of the illumination and the SNR of collected raw images. Similar to other super-resolution methods like STORM, the ultimate resolution of the SFS method is only limited to the SNR of the images we can acquire.

Practically, the high-SFS signal for normal samples is much weaker than that of low-SFS, and the quality of collected raw images will be affected by the photon noise. A lower photon number will cause an unsmooth connection in the Fourier space and for both labeled and label-free imaging. The input power of the illumination could be increased to achieve a better SNR, which may bring the problem of camera saturation to the low-SFS raw images. Therefore, for low-SFS and high-SFS, the illumination intensity should be adjusted and the spectrum signal rescaled accordingly in the final spectrum-splicing process. Another concern regarding the method is that, although resolution can be improved by increasing the lateral illumination wavevector, the resultant shallower field penetration depth would limit the capability for volumetric imaging. Thus, the resolution and imaging depth need to be optimized together, according to the specific application. From another perspective, the shallow illumination depth also brings the advantage of reduced defocus noise, making the SFS method suitable for imaging structures located near the chip surface, for example, the cell membrane.

The SFS super-resolution technique can be improved further in a number of ways. Firstly, the number of SFS magnitudes required for good image reconstruction can be reduced with the adoption of better reconstruction methods, such as deep-learning techniques [194–196], thus decreasing the number of raw images to be captured. This consequently reduces the cost of chip fabrication and increases the speed of imaging. Secondly, the errors in the necessarily imperfect geometries of the waveguide (in different situations, we all use waveguide to transmit the evanescent wave for illumination) may cause two effects: the nonuniformity of the illumination intensity and the introduction of impure wavevectors, which will reduce the resolution and introduce artifacts using the conventional reconstruction algorithm. However, these effects on the image can be relieved by using pattern-estimation reconstruction methods, such as blind-SIM [190]. Thirdly, advanced fabrication techniques may also improve the performance of the SFS super-resolution chips. With minor design changes, an on-chip evanescent light source can be implemented on the device to replace the off-chip light source. Commercial LEDs or vertical-cavity surface-emitting laser (VCSEL) chips [197] can be bonded to the photonic chip, thereby providing cheap, mass-producible, microscope-compatible super-resolution imaging. Finally, the chip-based super-resolution imaging method will give full play to its superiority if combined with integrated technologies. For example, a field-portable super-resolution imaging device can be obtained if we can develop a lightweight and compact optomechanical adaptor to make the super-resolution chip adapt to the existing camera module of a cell phone [198]. In the future, these SFS super-resolution chips may serve as a multifunctional platform on which many functions (e.g. electrical stimulation, microfluidics, and sensing) can be integrated for application in fields such as the medical point of care (POC) [199, 200], biology, materials science, and chemical research.

References

1 Abbe, E. (1873). *Arch. Mikrosk. Anat.* 9: 413.
2 von Helmholtz, H. (1874). *Ann. Phys.* 557.
3 Minsky, M. (1961). Microscopy apparatus. US Patent 3,013,467, filed 7 November 1957 and issued 19 December 1961.
4 Brakenhoff, G.J., Vandervoort, H.T.M., Vanspronsen, E.A. et al. (1985). *Nature* 317: 748.
5 Sheppard, C.J.R., Mehta, S.B., and Heintzmann, R. (2013). *Opt. Lett.* 38: 2889.
6 Pawley, J. (2010). *Handbook of Biological Confocal Microscopy*. New York: Springer Science & Business Media.
7 Sahl, S.J., Hell, S.W., and Jakobs, S. (2017). *Nat. Rev. Mol. Cell Biol.* 18: 685.
8 Hell, S.W. and Wichmann, J. (1994). *Opt. Lett.* 19: 780.
9 Willig, K.I., Harke, B., Medda, R., and Hell, S.W. (2007). *Nat. Methods* 4: 915.
10 Klar, T.A. and Hell, S.W. (1999). *Opt. Lett.* 24: 954.
11 Willig, K.I., Rizzoli, S.O., Westphal, V. et al. (2006). *Nature* 440: 935.
12 Hofmann, M., Eggeling, C., Jakobs, S., and Hell, S.W. (2005). *Proc. Natl. Acad. Sci. U.S.A.* 102: 17565.
13 Waeldchen, S., Lehmann, J., Klein, T. et al. (2015). *Sci. Rep.* 5: 15348.
14 Laissue, P.P., Alghamdi, R.A., Tomancak, P. et al. (2017). *Nat. Methods* 14: 657.
15 Hell, S.W. and Kroug, M. (1995). *Appl. Phys. B: Lasers Opt.* 60: 495.
16 Donnert, G., Keller, J., Medda, R. et al. (2006). *Proc. Natl. Acad. Sci. U.S.A.* 103: 11440.
17 Heintzmann, R., Jovin, T.M., and Cremer, C. (2002). *J. Opt. Soc. Am. A* 19: 1599.
18 Enderlein, J. (2005). *Appl. Phys. Lett.* 87: 097105.
19 Rust, M.J., Bates, M., and Zhuang, X. (2006). *Nat. Methods* 3: 793.
20 Jones, S.A., Shim, S.-H., He, J., and Zhuang, X. (2011). *Nat. Methods* 8: 499.
21 Betzig, E., Patterson, G.H., Sougrat, R. et al. (2006). *Science* 313: 1642.
22 Hess, S.T., Girirajan, T.P., and Mason, M.D. (2006). *Biophys. J.* 91: 4258.
23 Han, Y., Lu, X., Zhang, Z. et al. (2019). *Opt. Exp.* 27: 38337.
24 Han, Y., Zhang, Z., Liu, W. et al. (2021). *Front. Chem.* 8: 601436.
25 Han, Y., Li, M., Zhang, M. et al. (2019). *J. Chem. Technol. Biotechnol.* 94: 2040.
26 Masters, B.R. (2015). *Super-Resolution Microscopy Techniques in the Neurosciences*. Totowa, NJ: Humana Press.
27 Kim, E., Baaske, M.D., Schuldes, I. et al. (2017). *Sci. Adv.* 3: 8.
28 Subramanian, S., Wu, H.Y., Constant, T. et al. (2018). *Adv. Mater.* 30: 21.
29 Hell, S.W. (2007). *Science* 316: 1153.
30 Blom, H. and Widengren, J. (2017). *Chem. Rev.* 117: 7377.
31 Baddeley, D. and Bewersdorf, J. (2018). *Annu. Rev. Biochem.* 87: 965.
32 Sigal, Y.M., Zhou, R.B., and Zhuang, X.W. (2018). *Science* 361: 880.
33 von Diezmann, A., Shechtman, Y., and Moerner, W.E. (2017). *Chem. Rev.* 117: 7244.
34 Schermelleh, L., Heintzmann, R., and Leonhardt, H. (2010). *J. Cell Biol.* 190: 165.

35 Galbraith, C.G. and Galbraith, J.A. (2011). *J. Cell Sci.* 124: 1607.
36 Vicidomini, G., Bianchini, P., and Diaspro, A. (2018). *Nat. Methods* 15: 173.
37 Rogers, E.T.F. and Zheludev, N.I. (2013). *J. Opt.* 15: 094008.
38 Chen, G., Wen, Z.-Q., and Qiu, C.-W. (2019). *Light Sci. Appl.* 8: 56.
39 Berry, M.V. and Popescu, S. (2006). *J. Phys. A: Math. Gen.* 39: 6965.
40 Huang, F.M. and Zheludev, N.I. (2009). *Nano Lett.* 9: 1249.
41 Huang, F.M., Zheludev, N., Chen, Y., and de Garcia, Abajo, F.J. (2007). *Appl. Phys. Lett.* 90: 091119.
42 Huang, K., Ye, H., Teng, J. et al. (2014). *Laser Photonics Rev.* 8: 152.
43 Qin, F., Huang, K., Wu, J. et al. (2017). *Adv. Mater.* 29: 1602721.
44 Rogers, E.T.F., Lindberg, J., Roy, T. et al. (2012). *Nat. Mater.* 11: 432.
45 Huang, K., Qin, F., Liu, H. et al. (2018). *Adv. Mater.* 30: e1704556.
46 Tang, M., Liu, X., Wen, Z. et al. (2020). *Laser Photonics Rev.* 14: 1900011.
47 Gustafsson, M.G. (2005). *Proc. Natl. Acad. Sci. U.S.A.* 102: 13081.
48 Gustafsson, M.G.L., Shao, L., Carlton, P.M. et al. (2008). *Biophys. J.* 94: 4957.
49 Rego, E.H., Shao, L., Macklin, J.J. et al. (2012). *Proc. Natl. Acad. Sci. U.S.A.* 109: E135.
50 Han, Y., Li, M., Qiu, F. et al. (2017). *Nat. Commun.* 8: 1307.
51 Zheng, G., Horstmeyer, R., and Yang, C. (2013). *Nat. Photonics* 7: 739.
52 Liu, Q.L., Fang, Y., Zhou, R.J. et al. (2016). *Opt. Lett.* 41: 5373.
53 Tian, L., Li, X., Ramchandran, K., and Waller, L. (2014). *Biomed. Opt. Express* 5: 2376.
54 Lu, D. and Liu, Z. (2012). *Nat. Commun.* 3: 1205.
55 Liu, X.W., Kuang, C.F., Hao, X. et al. (2017). *Phys. Rev. Lett.* 118: 076101.
56 Wei, F. and Liu, Z. (2010). *Nano Lett.* 10: 2531.
57 Saxena, M., Eluru, G., and Gorthi, S.S. (2015). *Adv. Opt. Photonics* 7: 241.
58 Wu, Y. and Shroff, H. (2018). *Nat. Methods* 15: 1011.
59 Dan, D., Lei, M., Yao, B. et al. (2013). *Sci. Rep.* 3: 1116.
60 York, A.G., Chandris, P., Nogare, D.D. et al. (2013). *Nat. Methods* 10: 1122.
61 Hirano, Y., Matsuda, A., and Hiraoka, Y. (2015). *Microscopy* 64: 237.
62 Komis, G., Mistrik, M., Šamajová, O. et al. (2015). *Nat. Protoc.* 10: 1248.
63 Wegel, E., Gohler, A., Lagerholm, B.C. et al. (2016). *Sci. Rep.* 6: 27290.
64 Di Francia, G.T. (1952). *Il Nuovo Cimento (1943–1954)* 9: 426.
65 Rotenberg, N. and Kuipers, L. (2014). *Nat. Photonics* 8: 919.
66 Kim, J. and Song, K.B. (2007). *Micron* 38: 409.
67 Bek, A., Vogelgesang, R., and Kern, K. (2006). *Rev. Sci. Instrum.* 77: 043703.
68 Hecht, B., Sick, B., Wild, U.P. et al. (2000). *J. Chem. Phys.* 112: 7761.
69 Dunn, R.C. (1999). *Chem. Rev.* 99: 2891.
70 Zenhausern, F., Martin, Y., and Wickramasinghe, H.K. (1995). *Science* 269: 1083.
71 Xie, X.S. and Dunn, R.C. (1994). *Science* 265: 361.
72 Betzig, E. and Trautman, J.K. (1992). *Science* 257: 189.
73 Pohl, D.W., Denk, W., and Lanz, M. (1984). *Appl. Phys. Lett.* 44: 651.
74 Hecht, E. (2016). *Optics*. London, UK: Pearson Education.
75 Gay, G., Alloschery, O., De Lesegno, B.V. et al. (2006). *Nat. Phys.* 2: 262.

76 Hutter, E. and Fendler, J.H. (2004). *Adv. Mater.* 16: 1685.

77 Pang, C., Liu, X., Tang, M. et al. (2022). *Sci. Bull.* 67: 1317.

78 Liu, X., Meng, C., Xu, X. et al. (2018). *Chin. Phys. B* 27: 118704.

79 Pang, C., Li, J., Tang, M. et al. (2019). *Adv. Funct. Mater.* 29: 1900126.

80 Cragg, G.E. and So, P.T. (2000). *Opt. Lett.* 25: 46.

81 Helle, Ø.I., Dullo, F.T., Lahrberg, M. et al. (2020). *Nat. Photonics* 14: 431.

82 Wei, F., Lu, D., Shen, H. et al. (2014). *Nano Lett.* 14: 4634.

83 Zhang, C., Du, L., Xin, Z. et al. (2018). *Laser Photonics Rev.* 12: 1800148.

84 Zeng, X., Al-Amri, M., and Zubairy, M.S. (2014). *Phys. Rev. B: Condens. Matter* 90: 235418.

85 Ponsetto, J.L., Bezryadina, A., Wei, F. et al. (2017). *ACS Nano* 11: 5344.

86 Bezryadina, A., Zhao, J., Xia, Y. et al. (2018). *ACS Nano* 12: 8248.

87 Lee, Y.U., Posner, C., Nie, Z.Y. et al. (2021). *Adv. Sci.* 8: 2102230.

88 Lee, Y.U., Zhao, J., Ma, Q. et al. (2021). *Nat. Commun.* 12: 1559.

89 Jiang, X., Song, Q., Xu, L. et al. (2007). *Appl. Phys. Lett.* 90: 233501.

90 Pang, C., Liu, X., Zhuge, M. et al. (2017). *Opt. Lett.* 42: 4569.

91 Xu, X., Liu, X., Pang, C. et al. (2019). *J. Phys. D: Appl. Phys.* 52: 284002.

92 Ye, D., Tang, M., Liu, X. et al. (2021). *J. Phys. D: Appl. Phys.* 54: 315101.

93 Liu, X., Tang, M., Meng, C. et al. (2021). *Sci. China Phys., Mech. Astron.* 64: 294211.

94 Diekmann, R., Helle, Ø.I., Øie, C.I. et al. (2017). *Nat. Photonics* 11: 322.

95 Archetti, A., Glushkov, E., Sieben, C. et al. (2019). *Nat. Commun.* 10: 1267.

96 Priyadarshi, A., Dullo, F.T., Wolfson, D.L. et al. (2021). *Commun. Mater.* 2: 85.

97 Tang, M., Han, Y., Ye, D. et al. (2022). *Adv. Sci.* 9: 2103835.

98 Wei, S., Lei, T., Du, L. et al. (2015). *Opt. Express* 23: 30143.

99 Bezryadina, A., Zhao, J., Xia, Y. et al. (2019). *Opt. Lett.* 44: 2915.

100 Hao, X., Liu, X., Kuang, C. et al. (2013). *Appl. Phys. Lett.* 102: 013104.

101 Xu, P., Liu, S., Tang, M. et al. (2017). *Appl. Phys. Lett.* 110: 201112.

102 Tang, M., Xu, P., Wen, Z. et al. (2018). *Sci. Bull.* 63: 1118.

103 Duan, X.F., Huang, Y., Agarwal, R., and Lieber, C.M. (2003). *Nature* 421: 241.

104 Ströhl, F., Opstad, I.S., Tinguely, J.-C. et al. (2019). *Opt. Express* 27: 25280.

105 Li, D., Shao, L., Chen, B.-C. et al. (2015). *Science* 349: aab3500.

106 Tang, M., Liu, X., Yang, Q., and Liu, X. (2020). *Chip-Based Wide-Field 3D Nanoscopy Through Tunable Spatial-Frequency-Shift Effect*, vol. 11549. SPIE/COS Photonics Asia. https://doi.org/10.1117/12.2575355.

107 Tinguely, J.C., Helle, O.I., and Ahluwalia, B.S. (2017). *Opt. Express* 25: 27678.

108 Fu, Y., Ye, T., Tang, W., and Chu, T. (2014). *Photonics Res.* 2: A41.

109 Boulanger, J., Gueudry, C., Muench, D. et al. (2014). *Proc. Natl. Acad. Sci. U.S.A.* 111: 17164.

110 Chen, Y., Liu, W., Zhang, Z. et al. (2018). *Nat. Commun.* 9: 4818.

111 Chakrova, N., Heintzmann, R., Rieger, B., and Stallinga, S. (2015). *Opt. Express* 23: 31367.

112 Maier, S.A. (2007). *Plasmonics: Fundamentals and Applications*. New York: Springer.

113 Barnes, W.L., Dereux, A., and Ebbesen, T.W. (2003). *Nature* 424: 824.

114 Zayats, A.V., Smolyaninov, I.I., and Maradudin, A.A. (2005). *Phys. Rep.* 408: 131.

115 Willets, K.A., Wilson, A.J., Sundaresan, V., and Joshi, P.B. (2017). *Chem. Rev.* 117: 7538.

116 Simkhovich, B. and Bartal, G. (2014). *Phys. Rev. Lett.* 112: 056802.

117 Kawata, S., Inouye, Y., and Verma, P. (2009). *Nat. Photonics* 3: 388.

118 Li, M., Cushing, S.K., and Wu, N. (2015). *Analyst* 140: 386.

119 Krupin, O., Wong, W.R., Mahamd Adikan, F.R., and Berini, P. (2017). *IEEE J. Sel. Top. Quantum Electron.* 23: 103.

120 Singh, P. (2016). *Sens. Actuators, B* 229: 110.

121 Spackova, B., Wrobel, P., Bockova, M., and Homola, J. (2016). *Proc. IEEE* 104: 2380.

122 Soler, M., Estevez, M.C., Villar-Vazquez, R. et al. (2016). *Anal. Chim. Acta* 930: 31.

123 Yesilkoy, F., Terborg, R.A., Pello, J. et al. (2018). *Light Sci. Appl.* 7: 17152.

124 Rifat, A.A., Ahmed, R., Mahdiraji, G.A., and Adikan, F.R.M. (2017). *IEEE Sens. J.* 17: 2776.

125 Zijlstra, P., Paulo, P.M., and Orrit, M. (2012). *Nat. Nanotechnol.* 7: 379.

126 Anker, J.N., Hall, W.P., Lyandres, O. et al. (2008). *Nat. Mater.* 7: 442.

127 Schlucker, S. (2014). *Angew. Chem. Int. Ed.* 53: 4756.

128 Saha, K., Agasti, S.S., Kim, C. et al. (2012). *Chem. Rev.* 112: 2739.

129 Lu, H., Gan, X., Mao, D., and Zhao, J. (2017). *Photonics Res.* 5: 162.

130 Gramotnev, D.K. and Bozhevolnyi, S.I. (2010). *Nat. Photonics* 4: 83.

131 Zenin, V.A., Choudhury, S., Saha, S. et al. (2017). *Opt. Express* 25: 12295.

132 Vetrone, F. and Rosei, F. (2017). *Science* 357: 452.

133 Wang, Y., Ma, Y., Guo, X., and Tong, L. (2012). *Opt. Express* 20: 19006.

134 Luo, J., Zeng, B., Wang, C. et al. (2015). *Nanoscale* 7: 18805.

135 Gao, P., Yao, N., Wang, C. et al. (2015). *Appl. Phys. Lett.* 106: 093110.

136 Liang, G., Chen, X., Zhao, Q., and Guo, L.J. (2018). *Nanophotonics* 7: 277.

137 Cheng, K., Wu, Y.Q., Meng, J. et al. (2015). *Sci. Bull.* 60: 541.

138 Atwater, H.A. and Polman, A. (2010). *Nat. Mater.* 9: 205.

139 Linic, S., Christopher, P., and Ingram, D.B. (2011). *Nat. Mater.* 10: 911.

140 Schuller, J.A., Barnard, E.S., Cai, W. et al. (2010). *Nat. Mater.* 9: 193.

141 Bradac, C. (2018). *Adv. Opt. Mater.* 6: 26.

142 Gao, D., Ding, W., Nieto-Vesperinas, M. et al. (2017). *Light Sci. Appl.* 6: e17039.

143 Garces-Chavez, V., Quidant, R., Reece, P.J. et al. (2006). *Phys. Rev. B: Condens. Matter* 73: 5.

144 Volpe, G., Quidant, R., Badenes, G., and Petrov, D. (2006). *Phys. Rev. Lett.* 96: 238101.

145 Pang, Y. and Gordon, R. (2012). *Nano Lett.* 12: 402.

146 Grigorenko, A.N., Roberts, N.W., Dickinson, M.R., and Zhang, Y. (2008). *Nat. Photonics* 2: 365.

147 Zhu, T., Zhou, Y., Lou, Y. et al. (2017). *Nat. Commun.* 8: 15391.

148 Ruan, Z.C. (2015). *Opt. Lett.* 40: 601.

149 Pors, A., Nielsen, M.G., and Bozhevolnyi, S.I. (2015). *Nano Lett.* 15: 791.

150 Silva, A., Monticone, F., Castaldi, G. et al. (2014). *Science* 343: 160.
151 Liu, Z., Durant, S., Lee, H. et al. (2007). *Nano Lett.* 7: 403.
152 Xiong, Y., Liu, Z., Sun, C., and Zhang, X. (2007). *Nano Lett.* 7: 3360.
153 Fang, N., Lee, H., Sun, C., and Zhang, X. (2005). *Science* 308: 534.
154 Ponsetto, J.L., Wei, F., and Liu, Z. (2014). *Nanoscale* 6: 5807.
155 Liu, Z. (2017). *Plasmonics and Super-Resolution Imaging*. Jenny Stanford Publishing.
156 Luo, X.G., Tsai, D.P., Gu, M., and Hong, M.H. (2018). *Adv. Opt. Photonics* 10: 757.
157 Economou, E.N. (1969). *Phys. Rev.* 182: 539.
158 Luo, X. (2015). *Sci. China Phys., Mech. Astron.* 58: 594201.
159 Wang, Q., Bu, J., Tan, P. et al. (2012). *Plasmonics* 7: 427.
160 Tan, P., Yuan, X.-C., Yuan, G., and Wang, Q. (2010). *Appl. Phys. Lett.* 97: 241109.
161 Zhang, C., Min, C., Du, L., and Yuan, X.C. (2016). *Appl. Phys. Lett.* 108: 201601.
162 Liu, Z., Wang, Y., Yao, J. et al. (2009). *Nano Lett.* 9: 462.
163 Avrutsky, I., Salakhutdinov, I., Elser, J., and Podolskiy, V. (2007). *Phys. Rev. B: Condens. Matter* 75: 241402.
164 Xiong, Y., Liu, Z., and Zhang, X. (2008). *Appl. Phys. Lett.* 93: 111116.
165 Cao, S., Wang, T., Xu, W. et al. (2016). *Sci. Rep.* 6: 23460.
166 Cao, S., Wang, T., Yang, J. et al. (2017). *Sci. Rep.* 7: 1328.
167 Novoselov, K.S., Geim, A.K., Morozov, S.V. et al. (2004). *Science* 306: 666.
168 Novoselov, K.S., Geim, A.K., Morozov, S.V. et al. (2005). *Nature* 438: 197.
169 Geim, A.K. and Novoselov, K.S. (2007). *Nat. Mater.* 6: 183.
170 Liu, X.Z., Galfsky, T., Sun, Z. et al. (2015). *Nat. Photonics* 9: 30.
171 Dong, Y.F., Wu, Z.S., Ren, W.C. et al. (2017). *Sci. Bull.* 62: 724.
172 Pang, C., Lu, H., Xu, P. et al. (2016). *Opt. Express* 24: 17069.
173 Xu, P., Zhang, H., Qian, H. et al. (2017). *Opt. Commun.* 395: 76.
174 Qian, H., Ma, Y., Yang, Q. et al. (2014). *ACS Nano* 8: 2584.
175 Low, T. and Avouris, P. (2014). *ACS Nano* 8: 1086.
176 Tymchenko, M., Nikitin, A.Y., and Martin-Moreno, L. (2013). *ACS Nano* 7: 9780.
177 Hu, H., Yang, X., Zhai, F. et al. (2016). *Nat. Commun.* 7: 12334.
178 Huang, L., Liu, J., Deng, H., and Wu, S. (2018). *Adv. Opt. Mater.* 6: 1701378.
179 Vakil, A. and Engheta, N. (2011). *Science* 332: 1291.
180 Zhang, R.R., Lin, X., Shen, L. et al. (2014). *Opt. Lett.* 39: 6739.
181 Yang, J., Wang, T., Chen, Z. et al. (2016). *Sci. Rep.* 6: 37898.
182 Cao, S., Wang, T., Sun, Q. et al. (2017). *Opt. Express* 25: 14494.
183 Cheng, B.H., Chang, K.J., Lan, Y.C., and Tsai, D.P. (2014). *Opt. Express* 22: 28635.
184 Ju, L., Geng, B.S., Horng, J. et al. (2011). *Nat. Nanotechnol.* 6: 630.
185 Iorsh, I.V., Mukhin, I.S., Shadrivov, I.V. et al. (2013). *Phys. Rev. B: Condens. Matter* 87: 075416.
186 Idris, N.M., Gnanasammandhan, M.K., Zhang, J. et al. (2012). *Nat. Med.* 18: 1580.

187 He, M., Pang, X., Liu, X. et al. (2016). *Angew. Chem. Int. Ed.* 55: 4280.
188 Bezryadina, A., Li, J.X., Zhao, J.X. et al. (2017). *Nanoscale* 9: 14907.
189 Narimanov, E. (2016). *ACS Photonics* 3: 1090.
190 Mudry, E., Belkebir, K., Girard, J. et al. (2012). *Nat. Photonics* 6: 312.
191 Chen, Y. (2015). *Microelectron. Eng.* 135: 57.
192 Kim, C.-S., Ahn, S.-H., and Jang, D.-Y. (2012). *Vacuum* 86: 1014.
193 Moser, H.O. and Rockstuhl, C. (2012). *Laser Photonics Rev.* 6: 219.
194 Jin, L., Liu, B., Zhao, F. et al. (2020). *Nat. Commun.* 11: 1934.
195 Nguyen, T., Xue, Y., Li, Y. et al. (2018). *Opt. Express* 26: 26470.
196 Zhang, Q., Liang, C., Tang, M. et al. (2023). *Opt. Exp.* 31: 3719.
197 Mei, Y., Weng, G.E., Zhang, B.P. et al. (2017). *Light Sci. Appl.* 6: e16199.
198 Wei, Q., Qi, H., Luo, W. et al. (2013). *ACS Nano* 7: 9147.
199 Zhu, H., Isikman, S.O., Mudanyali, O. et al. (2013). *Lab Chip* 13: 51.
200 Cetin, A.E., Coskun, A.F., Galarreta, B.C. et al. (2014). *Light Sci. Appl.* 3: e122.

7

Monolithically Integrated Multi-section Semiconductor Lasers: Toward the Future of Integrated Microwave Photonics

Jin Li and Tao Pu

College Of Communications Engineering, Army Engineering University of PLA, Nanjing, 210007, China

7.1 Introduction

Microwave photonics (MWP) is a cross-discipline combining both the advantages of microwave technology and photonics technology, in which microwave signals are generated, processed, transmitted, and controlled by optical methods [1–3]. During last decades, with deeply researching on microwave signals in optical domain, it has attracted a lot of attention because of the unique characteristics of the photons such as wide bandwidth, low-loss transmission, and anti-electromagnetic interference, resulting in an extensive application in telecommunication, radar, and sonar, as well as electronic warfare [4]. However, at the same time of sharing its advantages, what impresses the researchers and industries most are high cost and poor stability, heavily limiting its practical usage in the abovementioned domains. To be specific, basically, all the demos of the MWP systems are designed and accomplished by utilizing relatively expensive, large-volume, and power-consuming discrete optoelectronic components, which are critically sensitive to complex external environmental disturbances, such as vibration and temperature gradients. Therefore, they can't fully take the place of mature electronic instruments, i.e. electronic integrated circuits (EICs), in many designed applications like the generation of microwave signals. However, limited by the nature of electrical circuits, all the electrical methods of generating, processing, as well as transmitting high-frequency microwave signals are rather challenging. The operation of microwave signals in optical domain compared with electrical domain is still recognized as the better candidate for the scenarios of high-frequency and high-quality microwave signals requirement, though MWP systems composed of discrete devices bring about some unavoidable problems mentioned before. In consequence, for future high-performance and large-bandwidth microwave applications, it is the urging requirement to resolve the imperfections in the reliability, convenience, as well as the cost of MWP systems. Fortunately, with the rapid development of photonic integration circuits (PICs) technology, making discrete optoelectronic components toward the direction of integration on one chip is a promising approach to overcoming these problems.

Optical Imaging and Sensing: Materials, Devices, and Applications, First Edition.
Edited by Jiang Wu and Hao Xu.

Integration of MWP systems is an inevitable trend for a better use of MWP [5–7]. As a novel field, integrated microwave photonics (IMWP) converging the advantages of both MWP and the PICs, has been appearing with great charm since it was proposed. It aims at the incorporation of MWP subsystems in PICs, such as laser sources, external modulators, optical amplifiers, and photodetectors, which is crucial for the implementation of low-cost, high-stability, and advanced analog optical front-ends and, therefore, be conductive to achieve the aforementioned evolution goals [5]. Actually, the benefits it brings about are not merely the reduction of the footprint, power consumption, and the physical complexity. Compared to the two schemes above, i.e. EICs and the discrete MWP systems, IMWP combines the strengths of both schemes, which can realize a great performance improvement and simultaneously keep the entire system within a highly integrated configuration [4]. For example, by confining light to a very small space, the interaction between the photons and the matter can be enhanced greatly, and this is usually realized through nonlinear optical processes, and thus could be beneficial to bring about new technologies applied for the IMWP [6]. Until now, based on the IMWP, a large number of multifunctional and reconfigurable PICs applied for microwave signal generation, processing, transmission, and control have been developed, which are significant to push MWP systems from laboratory demonstrations toward practical applications and finally promote further development of MWP [7].

Since IMWP has so many potential advantages, the use of PICs will focus on technologies that can be used with integrated waveguides, and in the IMWP research, the first component supposed to be considered would be the laser source. Laser sources, as active devices, are of great importance in all optoelectronic components because they are not only portable suppliers of optical waves but also equipped with distinct electro-optic conversion properties. The nature of the MWP is the interconversion between the light wave and microwave based on the optoelectrical devices, and the capability of the interconversion usually determines the properties of the MWP systems. Specifically, the semiconductor laser is considered as the most universal and attractive light source among various lasers, which is attributed to the unique characteristics of small SWAP (size, weight, and power), a high degree of stability, and especially easy integration [8, 9].

Owing to the inside mechanism of optical gain dependent on the interaction of photon and carrier, the semiconductor lasers belong to inherently nonlinear optoelectronic devices and thus, rich varieties of nonlinear dynamics, including stable locking [10, 11], period oscillation [12, 13], self-pulsation [14], quasi-periodic pulsation [15, 16], frequency locking [17], and chaotic oscillation [18, 19], can be observed once the semiconductor lasers suffer from the external disturbance, such as optical injection, optical feedback, as well as current modulation. During the last two decades, by adjusting the semiconductor lasers to operate under appropriate working conditions, people have investigated extensively and thoroughly in realizing high efficient electro-optic conversion (modulation response enhancement, chirp suppression, nonlinear reduction, and analog signal transmission) [20–67], tunable and wideband microwave and millimeter-wave signal generation (single-tone or frequency-modulated microwave waveforms) [56, 68–118], as well as microwave signal processing (filtering, true time delay, frequency conversion, signal format conversion, and so on) [116, 119–124].

Integration of semiconductor lasers is an unavoidable trend for the development of conventional configurations using discrete lasers. Despite the unique characteristic improvement in optical injection or optical feedback systems, they are mostly researched and applied based on at least two discrete semiconductor lasers or one semiconductor laser accompanied by a reflective mirror. Thus, a large bulk, high power consumption, and complex structures are unavoidable. These defects naturally exist in discrete configurations and are hard to overcome. Furthermore, the polarization state and environmental disturbance are also inevitable obstacles, which are destabilizing elements highly affecting the performances of optical systems. These questions mentioned above largely limit the practical usage of semiconductor lasers, leading to them only being studied deeply in the laboratory. The integration is a great solution in terms of these shortcomings.

Integration of semiconductor lasers is also beneficial for the development of IMWP [20]. As an important light source module, the advancement of the monolithially integrated multi-section semiconductor lasers (MI-MSSLs) can greatly promote the development of IMWP. Firstly, attributing to the integration on one chip, they are equipped with some natural advantages including light weight, compact size, low power consumption, high reliability, and efficient reconfigurability. Secondly, in theory, basically, all MWP functions can be attained with the help of MI-MSSLs and are even better compared with conventional single-section lasers. So far, many researches have convinced that optimizing the laser structural design and replacing the conventional single-section lasers with MI-MSSLs can allocate more prominent improvement in the whole optoelectrical performances. For example, large-bandwidth analog link transmissions with high linearity can be easily realized with the enhanced direct modulation response by making full use of the photon–photon resonance (PPR) effect in the MI-MSSLs [45, 56, 63]. What's more, multiple MWP functions can be realized simultaneously by using one monolithically integrated laser chip, indicating its great flexibility and widespread applicability. Last but not least, due to the laser structure variation induced more irritative interaction between photons and photons, as well as photons and electrons, MSSLs own more abundant nonlinear dynamics than traditional single-segment semiconductor lasers, and hence, they have more enormous potential in the practical applications, prompting itself a promising candidate applied into the IMWP. In view of these innate advantages, they will be the necessary trend of technology development and would be excellent candidates for many significant application prospects ranging from broadband wireless access networks to emerging MWP-based radars.

Usually, different kinds of MI-MSSLs are composed of diverse sections, including active feedback section, phase tuning section, and distributed feedback (DFB) laser section. They share the same substrate and can be realized on one chip passively and actively based on the maturity of semiconductor integration process. It is worth mentioning that based on the reconstruction equivalent chirp (REC) technique, monolithically integrated mutually injected semiconductor lasers (MI-MISLs) can be fabricated just by following the similar procedures of standard semiconductor lasers, thus making it possible to be a more efficient and cost-effective choice for realizing various MWP functions. Until now, there have been many successful researches reporting the novel MI-MSSLs at length and simultaneously considering their potential applications in MWP, including electro-optical conversion effect enhancement, photonic microwave generation, and MWP processing. Moreover, the monolithically integrated distributed feedback multi-section multi-wavelength semiconductor laser arrays (MI-DFB-MS-MW-SLAs) are also discussed intensely for the fact that they can parallel integrate several discrete multichannel semiconductor lasers with complex structures in an array chip, which are well-suited to realize multiple MWP functions at the same time, especially in the transmission application of wavelength division multiplexing radio-over-fiber (WDM-ROF) networks nicely. However, to the best of our knowledge, a comprehensive overview of the current state-of-the-art in this field is still vacant. Hence, in this chapter, our aim is to review the basic laser structures of the MI-MSSLs, describe their main photonic measures applied in the field of microwaves, and report on some of the most salient advances obtained during the last years in this emergent field.

This chapter is organized as follows. In Section 7.2, three major types of MI-MSSLs, including passive feedback lasers (PFLs), active/amplified feedback lasers (AFLs), as well as MI-MISLs, will be analyzed and discussed. Their distinctive advantages and limitations are compared separately. In Section 7.3, the electro-optic conversion capability enhancement utilizing the PPR effect instead of the carrier–photon resonance (CPR) effect is discussed, especially in aspects of increasing modulation bandwidth, reducing nonlinearity, and suppressing chirp. Additionally, some analog photonic microwave transmission experiments are enumerated and examined, verifying the feasibility of the MI-MSSLs in the application of ROF systems. Within Section 7.4, a considerable amount of review of previous work on the topic of main optical techniques for generating single-tone microwave signals based on the nonlinear dynamics of MI-MSSLs is given, including free-running state, mode-beating self-pulsations (MB-SPs) state, period-one (P1) oscillation, and sideband injection locking. In addition, the generation of specific frequency-modulated microwave waveforms, such as linearly chirped microwave waveforms (LCMWs), frequency-hopping (F-H) microwave waveforms, and triangular wave signals, are presented as well, which have been widely applied in the radar and solar scenarios. Furthermore, in view of the MHz-level linewidth, the generated microwave signals are worthy of being further optimized. The desirable method of adding extra feedback loops (covering optical–electrical feedback or all-optical feedback loops) is also explored to boost the microwave signal quality. Section 7.5 outlines the existing methods available for the generation of tunable microwave photonic filters (MPFs) as a significant photonic microwave processing technique. Ulteriorly, Section 7.6 pays special attention to the MI-DFB-MS-MW-SLAs, which are considered as one of the most predominant light sources in future PICs and WDM-ROF networks. Finally, Section 7.7 concludes this chapter with several perspectives on the future prospects and challenges of the MI-MSSLs.

7.2 Monolithically Integrated Multi-section Semiconductor Laser (MI-MSSL) Device

During the last 20 years, a great number of MI-MSSLs have been proposed to realize better performances in comparison with the conventional single-section semiconductor laser over MWP applications. Various MI-MSSLs with different laser structures, distinctive lengths of each section, as well as varying materials have been fabricated, presenting diverse dynamic properties. They are not only analyzed theoretically but also verified by experiments. According to the compositions of the laser structures, we divide them into three categories in general, i.e. PFLs as in [47, 53], AFLs as in [56, 88, 97], and MI-MISLs as considered in [58, 61, 82]. In this section, some of the essential or seminal developments on the three kinds of MI-MSSLs are reviewed, and their pros and cons would be deeply demonstrated as well.

7.2.1 Monolithically Integrated Optical Feedback Lasers (MI-OFLs)

Figure 7.1 shows a typical schematic diagram of a delayed optical feedback system, which is well known for the fact that the structure is the prototype of the monolithically integrated optical feedback lasers (MI-OFLs). In such an optically delayed system, a semiconductor laser source is subject to optical feedback from an external cavity through a mirror and the performances of the semiconductor laser will be accordingly changed with the optical feedback signal.

The feedback optical signal can be characterized by the following relation [90]:

$$E_{in}(t) = Ke^{i\phi}E_{out}(t-\tau) \tag{7.1}$$

The real number K and ϕ are the separate representations of the strength and phase of the feedback optical field when the light goes back to the initial laser source. The delay time τ is another characteristic parameter, which is related to round-trip time of the photons dependent on the length of the external cavity, i.e. the twice distance between the mirror and the semiconductor laser. Once the three parameters change, the optical feedback signal will change correspondingly, leading to a drastic variation inside the semiconductor laser. Herein, numerous nonlinear behaviors such as chaos [125] and pulsations [126] can be observed in such an OFL system. Due to its vivid dynamics as well as wide application, this configuration has been studied extensively in many literatures [125–128]. However, as a result of discrete device composition, the optical feedback system is significantly sophisticated and heavily

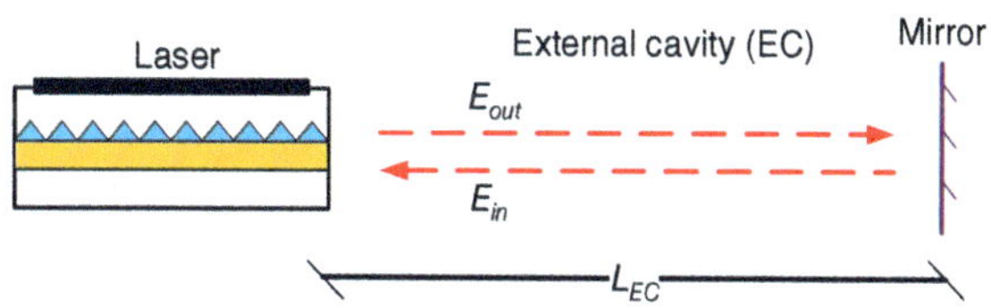

Figure 7.1 Scheme of one delayed optical feedback system.

unstable, largely limiting the system's practical usage. MI-OFLs are proposed on the basis of this model by integrating external cavity with DFB laser source on one chip to settle the shortcomings of discrete components and by depending on whether there exists an additional amplifier section, we can divide the MI-OFLs into two categories, including PFLs and the AFLs.

To get a further understanding of the dynamic mechanisms in the MI-OFLs, we can describe the internal dynamic variation in the DFB laser section by the following set of delay differential equations.

$$\frac{dE}{dt} = \frac{1}{2}(1 + i\alpha)g(N - N_{th})E + \frac{K}{\tau_{in}}E(t - \tau)e^{-i\phi}$$
$$\frac{dN}{dt} = \frac{I}{eV} - \tau_c^{-1}N - \Gamma g(N - N_0)|E|^2 \tag{7.2}$$

E is the amplitude of the complex optical field $Ee^{i\omega t}$. I and N denote the injection current and carrier density in DFB laser section, respectively. N_{th}, N_0, α, τ_{in}, τ_c, g, V, and Γ represent the carrier density at threshold point and at transparent point, linewidth enhancement factor, round-trip time in laser cavity, carrier lifetime, gain coefficient, volume of active region, and the confinement factor for the laser section, separately. Based on the above equations, abundant nonlinear dynamic behavior can be achieved by numerical simulation, and the corresponding change relationships of the parameters are all available [91]. Also, some theoretical analysis and research have been undergone in [55, 97], but the simulation tool they used is the mature commercial package *VPI-transmission Maker*.

7.2.1.1 Passive Feedback Lasers (PFLs)

The PFL is the simplest configuration of multi-segment semiconductor lasers. It only needs to integrate a passive reflection section after the active section to obtain additional optical reflection in laser's physical structure, and thus the short cavity length of the active luminescent segment can be allowed, which is recognized as the easiest way to advance the maximum of the relaxation resonance frequency.

Figure 7.2 shows the original structure of the PFL, which is composed of a DFB section and one integrated feedback (IFB) section without grating structures. Here, the IFB section is only made up of one phase tuning section. In practical work, the DFB section is driven by direct current (DC) to bring about the light oscillation. The cleaved facet of the phase tuning section acts as an end mirror in external cavity regime to provide passive optical feedback. The counter-propagating optical field is not coupled to the carriers or to each other in the phase tuning section, but only experiences some losses and a phase shift. Consequently, by changing the current

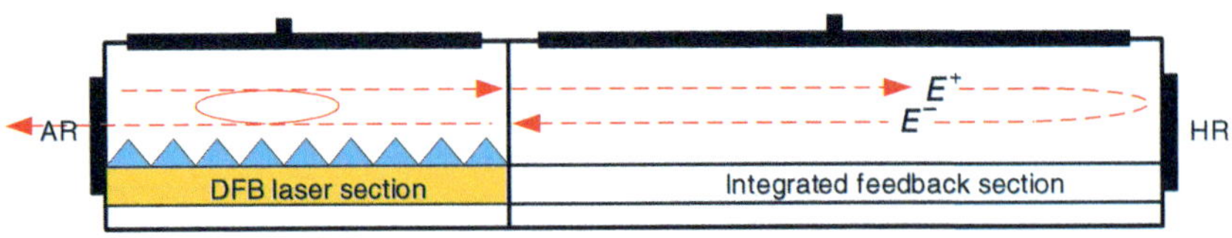

Figure 7.2 Scheme of one PFL, consisting of a DFB section and an IFB section.

injection into the phase tuning section, the strength and the phase of the feedback optical signal can be controlled through the indirect influence of the effective refractive index n_{IFB}^{eff}, leading the PFLs to enter new oscillation states, from hysteresis effects, regular self-pulsations to chaotic output [129]. Besides, similar to the discrete OFL system, the length of the IFB section l_{IFB} is also an affecting factor, which is supposed to be selected properly to guarantee an effective tunability of K as well as ϕ. To be specific, the equations of the K and ϕ in PFLs are given as follows [45]. R represents the reflectivity at the facet of the IFB section. From the formulas, we can find that the K and ϕ cannot be tuned independently, which largely restricts the maximum of the feedback strength and the operating regions.

$$K = R \cdot e^{-2l_{IFB}\alpha_{IFB}} \tag{7.3}$$

$$\phi = 2\pi \frac{2l_{IFB}n_{IFB}^{eff}}{\lambda_0} \tag{7.4}$$

To achieve a better understanding of the mechanisms governing the laser dynamics in a typical PFL, the theoretical investigations and the corresponding experimental identifications of the behavior of the PFL have been carried out in [45–48, 126, 129–131]. The optical feedback signal generated from the IFB section could significantly improve the modulation properties of the laser under the influence of PPR effect [45]. A high 3-dB modulation bandwidth up to 30 GHz was achieved, underlining the potential of the PFL concept for analog direct modulation in ROF links [49]. Although the PFL presents nice performance improvements [125, 126, 129], they also reveal the following defect in terms of low feedback strength, which corresponds to a low-frequency pulsation generation. Generally, in PFLs, the feedback strength is usually very small in the realized devices because of the huge optical loss arising in the IFB cavity. It is original from the imperfect coupling at the active–passive interface, scattering losses in the passive feedback section, as well as the reflectivity induced by the refractive index discontinuity between the semiconductor waveguide and air (≈0.3) at the rear facet [45]. To overcome this limitation, an appropriate high-reflection (HR) coating at the facet of the IFB section edge is usually adopted to allocate a high feedback strength [132]. Furthermore, an additional advantage brought about is that more light extraction output from the side of the DFB section can be achieved due to breaking out the symmetrical light output structure, largely advancing the efficiency of the PFL device. Unfortunately, the single-longitudinal-mode (SLM) yield will be degenerated and the wavelength accuracy will be deteriorated because the reflection phase at the HR-coated facet is unavoidably randomized and cannot be accurately controlled [53, 54]. It is reported that the DFB lasers integrated with passive distributed-reflector (DBR) mirrors functioned as IFB section can achieve accurately controlled reflection phase and thus, they have 100% SLM yield in theory [41]. However, in the fabrication process, the upper waveguide and the active layer are usually removed by dry etching in the fabrication of the passive section. Hence, the abovementioned DFB lasers integrated with passive waveguides all need butt-joint regrowth process to realize the fine connection between the active waveguide and passive waveguide, which heavily increases

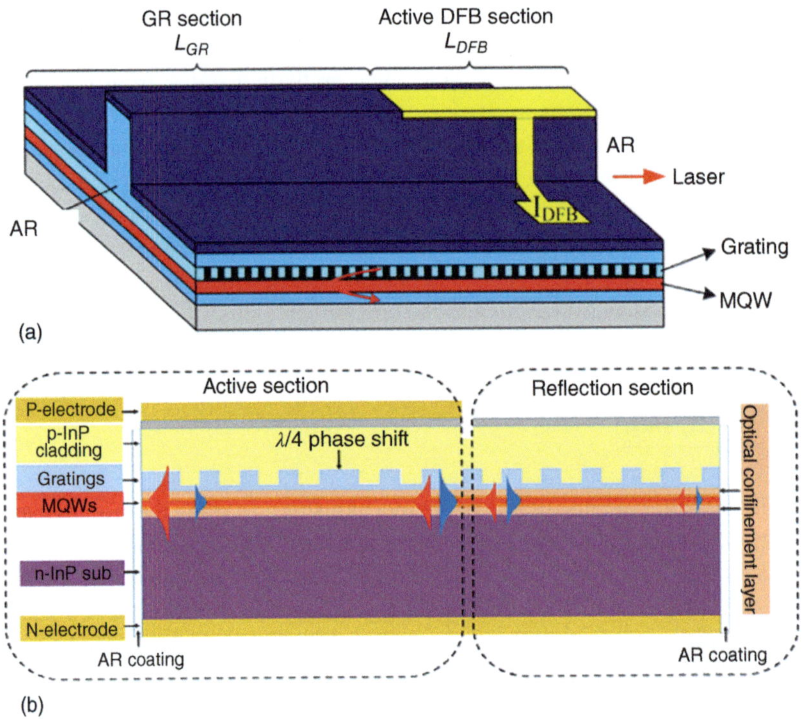

Figure 7.3 Multiple schemes of the PFL structure: (a) Source: Zhang et al. [52], (b) Source: Liu et al. [54].

the fabrication complexity, especially in the condition that aluminum-containing quantum wells are utilized for high differential gain [54].

To solve this imperfection, recent work by [52–54] extends techniques to supply the sufficiently strong delayed optical feedback, great single-mode characteristic, as well as simple fabrication process, whose structure is depicted in Figure 7.3. The work is built conceptually on the approach of [45] but differs in the way by comprising an active single-mode complex-coupled DFB laser used as light source and one phase tuning section with grating reflection (GR). The active section length is still shortened in order to achieve high relaxation oscillation frequency as well as modulation bandwidth. Both sections share the same wafer structure, such as the multi-quantum wells (MQWs) and the same upper and lower optical confinement layers. Consequently, the fabrication process is expected to be much easier due to the neglection of the butt-joint regrowth technology. It should be noted that the GR section acting as a distributed reflector is unbiased without any current injection when the novel GR laser is normally working. In spite of no extra current being provided into the GR section, the GR section will be optically pumped to near transparency with the help of the optical emission from the DFB section in practical use, and thus, the loss derived from absorption can be eliminated basically, promoting a higher reflection strength. Additionally, both gratings in the two sections are designed with the same grating period, and the phase is continuous across the

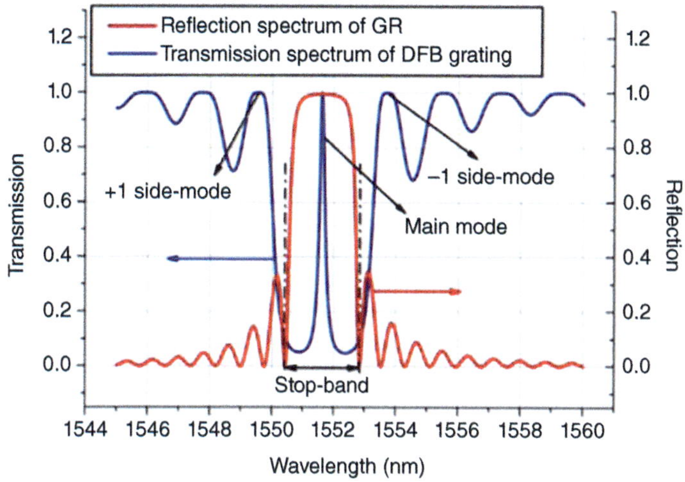

Figure 7.4 Reflection spectrum of GR section and transmission spectrum of the DFB grating.

interface between the two adjacent sections. Therefore, it can be found that the stop-band width of GR section is narrower than that of the grating in the active DFB section owing to its longer grating length, and, the main optical mode is exactly right located at the center of the stop-band of the grating reflection spectrum, while the side modes lie outside of the gain zone, as Figure 7.4 shows [52]. Hence, the reflectivity for the main optical mode is higher than that of other optical side modes. In other words, we can consider the GR section as a built-in optical bandpass filter and much higher side mode suppression ratio (SMSR) would be achieved under this influence, leading to a great SLM yield. An experiment was designed to demonstrate the functionality of the GR section and above 55-dB SMSR was achieved. They also found that the GR section could bring along other advantages, such as increasing the slope efficiency by about 34% and decreasing the threshold current by about 34% compared with the standard DFB laser besides enhancing the SMSR [54]. Concurrent work [57] has also considered a similar approach based on the same multi-section laser structure in which further results of increasing slope efficiency from 0.15 to 0.23 mW mA^{-1} and the decreasing threshold current from 33 to 24 mA, as well as $SMSR > 60$ dB, were observed. These data can prove that in a well-designed PFL, both the output slope efficiency and the threshold current can be significantly improved. Furthermore, the notable bandwidth enhancement is also recorded due to the light re-injection into the DFB section from the side of GR section.

Further, a more efficient implementation has been proposed by [51] to increase the output efficiency of GR semiconductor lasers by injecting the dc currents into both the DFB section and passive GR section. In this scheme, the bias current of the passive grating reflector is always set below the transparent current in order to avoid any possible laser oscillation and meanwhile guarantee the DFB section with high reflectivity. Experimental results showed that the output optical power had a significant improvement with a fixed current of DFB section when the GR section was equipped with current injection. In addition, as the current of GR section goes up, the slope efficiency of the curve of the output power relative to the current of DFB section can be further advanced, which greatly improves the practicability of GR lasers. In [51], further improved results of 39% increase of slope efficiency and >10 mA decrease of threshold current were successfully recorded.

7.2.1.2 Amplified/Active Feedback Lasers (AFLs)

As emphasized above, the PFL is the simplest MI-MSSLs to advance the whole performance of single-section conventional laser. However, an important but inevitable problem within all PFLs is the absence of separate control of the feedback strength and the feedback phase, which usually leads to limited nonlinear dynamic regions, relatively small relaxation resonance frequency, as well as low-frequency microwave signal generation [90]. Consequently, a new laser structure is described for addressing this issue based on the PFLs, which is named as AFLs.

Generally, an AFL chip is designed with reference to the Lang–Kobayashi (L–K) model [89]. The new laser structure is similar in principle to PFLs. It is composed of one DFB laser section with a compound optical feedback cavity but uses an additional amplifier section combined with the phase tuning section to form the IFB cavity, as Figure 7.5 shows. Both of them are independently biased just like the DFB laser section. The free-carrier transitions-induced refractive index modification can

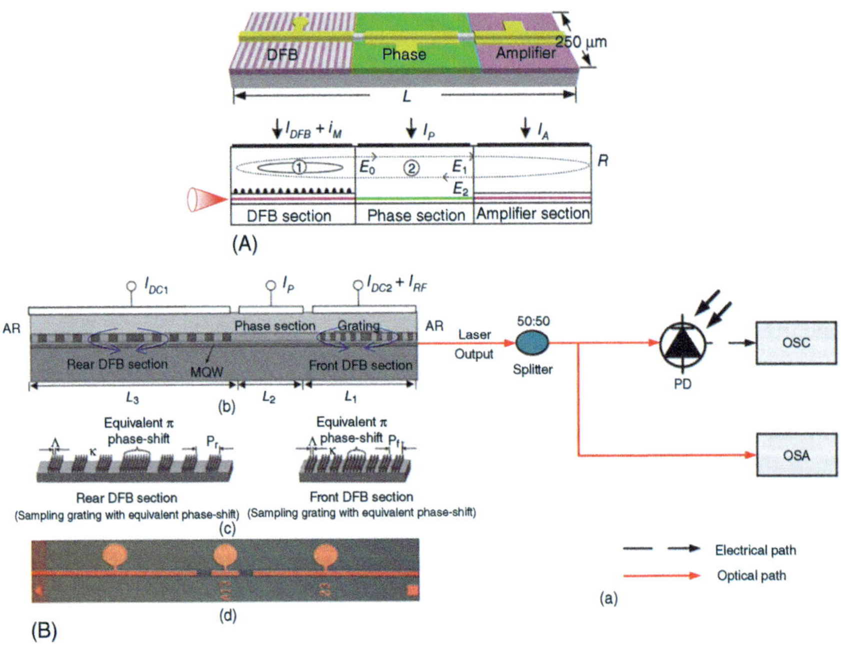

Figure 7.5 (A) Schemes of the AFL structure. (B) (a) Experimental setup using the laser module, (b) schematic of laser structure, (c) REC grating structure, and (d) photograph of the laser chip: (a) Source: Yu et al. [55], (b) Source: Li et al. [56]/with permission of Optica Publishing Group.

be achieved by tuning the injection current in the phase tuning section, while the optical feedback signal in AFLs is amplified by utilizing an active medium in the amplifier section, whose design is overall the same as the DFB laser section except for the grating structure [91]. The increasement of the amplifier section can not only compensate for the optical losses but also allow for the control of feedback strength via injecting current into the amplifier section. Therefore, the most outstanding advantage of AFLs compared to PFLs is the separate control of adjusting the phase and strength of the feedback to the emitted light oscillated from DFB section, owing to the introduction of an amplifier section as a special feature. For a better understanding of the mechanisms of AFLs, a series of reasonable functional expressions of the feedback strength and the feedback phase in the AFLs is provided [55]. R is the reflectivity at the facet of the amplifier section and g_A is the mean amplifier gain, which can be adjusted through the current of the amplifier feedback section.

$$K = \sqrt{R}e^{[-\alpha_p L_p + (g_A - \alpha_A)L_A]} \tag{7.5}$$

$$\phi = \arg(\sqrt{R}) + \phi_p + \alpha_H g_A L_A \tag{7.6}$$

Obviously, for a well-designed AFL with fixed length of each section, the core parameters K and ϕ can be adjusted easily by varying the currents in the amplifier section and phase tuning section. More importantly, as a result of the addition of the amplifier section, the feedback strength owns a wider tuning range than that in PFLs. By tuning the two parameters to different combinations, distinct operation states in AFLs, including single-mode lasing, multi-mode beating, and chaos, can be recorded. Especially, MB-SPs oscillation interests researchers most for its potential application in generating high-frequency microwave signals, which will be stressed in detail in Section 7.4.

An in-depth and systematic theoretical understanding, as well as experimental analysis of various nonlinear dynamics of the AFLs, were concluded, and it has been verified that there exists a self-adjusting effect based on the additional changes of carrier density when the carrier density in the active feedback is high [89]. Attributing to the self-adjustment in the feedback strength and the feedback phase, the AFLs are convinced to be more stable in the generation of high-frequency pulsations than the PFLs. Despite the success of this laser structure in certain aspects, AFLs also have their own shortcomings. At first, the active layer in the passive phase tuning section is supposed to be removed by means of dry etching for fabricating the passive section between two adjacent active sections during the fabricating process of the AFLs, which increases the complexity of production. Subsequently, some of them also need the butt-joint growth technique between the adjacent active and passive waveguides, which has the same limitation just like the PFLs [95].

7.2.2 Monolithically Integrated Mutually Injected Semiconductor Lasers (MI-MISLs)

Optical injection systems have long been widely researched for the advanced characteristics of improving the direct modulation bandwidth, stability, and linearity of semiconductor lasers [38]. Besides, semiconductor lasers, which are subject to optical injection, are deeply analyzed for numerous dynamic characteristics, experiencing from stability (S), P1, period-two (P2), four-wave-mixing (FWM), quasi-period (QP) to chaos (C). In the light of the various natural performances, people have carried out lots of experiments and made huge progress in three key domains of MWP, including microwave signal generation, processing, and transmission. Unfortunately, they also expose many problems due to the discrete system structures. First, external modulators are often applied in optical injection systems to achieve the injection locking state, leading to more complex system structures, larger volume as well as higher power consumption. Second, the operating wavelength and the polarization of the oscillating light in the master laser (ML) and the slave laser (SL) must be carefully controlled, or the optical injection system can only be employed in a limited usage scenario. Third, in a typical optical injection system, the light from the ML is unidirectionally injected to the SL by optical isolators or optical circulators, whose isolation is on the order of 30–40 dB with an insertion of 1–3 dB. Thus, the maximum injection ratio, which is correlated to the working state of the optical injection systems, is confined due to the huge isolator loss and low optical coupling efficiency. Often, an optical amplifier is used to boost the optical power from the ML to achieve a high injection ratio, but the system complexity also increases. To overcome these issues, another class of MI-MSSLs with built-in mutual injection system is proposed.

Originally, the proposed structure of MI-MISLs only consisted of two back-to-back DFB semiconductor laser sections, which were first invented in Defense Advanced Research Projects Agency RF Lightwave Integrated Circuits (DARPA RFLICs) program, as Figure 7.6a shows [58, 59]. Six years later, W.W. Chow et al. proved the feasibility of implementing injection locking systems without optical isolator through the strong-coupling theory, as Figure 7.6b shows [39]. That is to say,

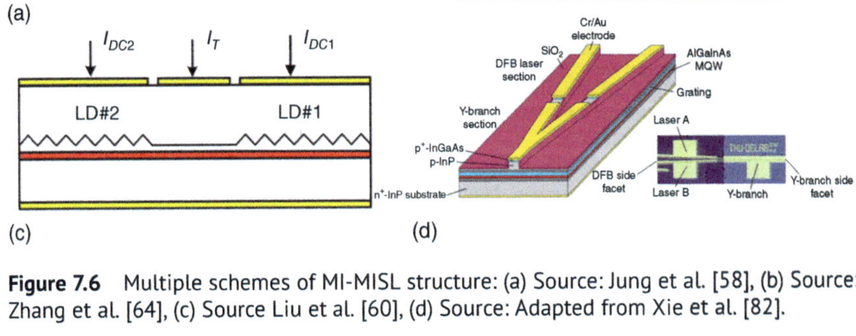

Figure 7.6 Multiple schemes of MI-MISL structure: (a) Source: Jung et al. [58], (b) Source: Zhang et al. [64], (c) Source Liu et al. [60], (d) Source: Adapted from Xie et al. [82].

MI-MISLs can eliminate the optical isolation and the optical circulator. In addition, due to the rather short distance between the two DFB lasers, their polarizations are automatically matched, making the polarization controller not necessary in MI-MISLs. Hence, both the simplification and reliability improvement of the systems are guaranteed. Due to the shortage of the optical isolation/circulator, the light derived from two individual DFB laser sections will drastically interact with each other, and thus, the MI-MISLs can enter a new state of mutual injection. Under this condition, both lasers will try to force the other to operate at their own wavelength, and there is no obvious distinction between the ML and SL. In most cases, for the sake of differentiating each twin from the other for convenience, we recognize the laser modulated by the radio frequency (RF) signal as the front (slave) laser section and the laser with only dc bias as the rear (master) laser section. The oscillating light is output from the facet of the front (slave) laser section in the MI-MISLs besides the Y-branch structures in practical use. Conventionally, the working wavelength of the oscillating light is controlled by adjusting the corresponding control current or working temperature of the semiconductor laser through the effect of heat and carrier density variation. Unluckily, this method does not work in MI-MISLs for the reason that the two DFB laser sections are fabricated together on the same Peltier refrigerating tablet and packaged in one chip, and thus both temperatures of two DFB laser sections are fixed strictly by an external temperature control source, sharing the same value. Though the laser wavelength in this device can only be tuned independently by the respective injection current, a better environmental stability of the MI-MISLs is ensured when compared with the conventional optical injection system constituted by separated DFB lasers. In addition, it should be noted that due to thermal crosstalk, the other laser section would also be influenced, and the relevant oscillation wavelength will be shifted subsequently besides changeable output power once one of the laser section currents is adjusted.

Unlike the case of unidirectional optical injection, the operating behavior of the MI-MISLs is mainly determined by the combined effects of the coupling strength κ, coupling time delay τ as well as the detuning frequency Δf between the two lasers. The time delay τ corresponds to the necessary time that the emitting light from one laser section travels to the other one. Ulteriorly, different from the discrete optical injection systems subject to the long coupling regimes, the optical transmission path between the two DFB laser sections is quite small on a scale of hundreds of micrometers, leading to an ultra-short coupling time delay, and thus the MI-MISLs are classified into ultra-short coupling regimes, where the coupling time delay is comparable to the resonance frequency period ($1/\Delta f$) and the various nonlinear dynamics are generally decided by the detuning frequency and the relaxation oscillation frequency under mutual injection. Actually, in the MI-MISLs, not only the detuning frequency between the two DFB sections but also the relaxation oscillation frequency is different from that in the free-running state and constantly varies in terms of different working conditions as a result of the mutual injection-induced frequency shift, leading to the complicated dynamic characteristics. However, it is difficult for researchers to extract these parameters to judge the working condition of the MI-MISLs. In most studies, the injection currents are given and considered because they can provide more perceptual intuition.

In 2013, a novel development of one three-section MI-MISL is presented in [60, 61], whose structure is depicted in Figure 7.6c. Compared to the previous structure just composed of only two DFB laser sections, one more 400-μm-long grating-less phase-tuning section is added to tune the coupling strength and the coupling time delay between the two 400-μm-long DFB lasers, which are found to have significant influences on determining the locking or unlocking behavior of the MI-MISLs sensitively. By tuning the current of the phase-tuning section to make sure the MI-MISLs are under appropriate coupling conditions, the selectivity of the dynamic lasing modes can get an enhancement, arising from the constructive and destructive interferences of the optical fields in the two coupled laser cavities. It is worthwhile to point out that, just like the variation when regulating the currents of two DFB laser sections, as the control current of the phase-tuning section increases, the lasing modes of both DFB laser sections will shift to longer wavelength simultaneously, owing to the heating effect. In addition, due to the thermal-induced refractive index change, the coupling time delay will change and the coupling phase will also alter accordingly, showing up different dynamical scenarios.

To get a further understanding of the dynamic mechanisms in the MI-MISLs, the following set of delay differential equations is established to describe the internal particle variations within two DFB laser sections under mutual optical injection [60, 61].

$$\begin{aligned}
\frac{dE_1}{dt} &= \frac{1}{2}(1+i\alpha)g(N_1-N_{th})E_1 + \frac{\kappa}{\tau_{in}}E_2(t-\tau)e^{i(\omega_2\tau+\Delta\omega t)} \\
\frac{dN_1}{dt} &= \frac{I_1}{eV} - \tau_c^{-1}N_1 - \Gamma g(N_1-N_0)|E_1|^2 \\
\frac{dE_2}{dt} &= \frac{1}{2}(1+i\alpha)g(N_2-N_{th})E_2 + \frac{\kappa}{\tau_{in}}E_1(t-\tau)e^{i(\omega_1\tau-\Delta\omega t)} \\
\frac{dN_2}{dt} &= \frac{I_2}{eV} - \tau_c^{-1}N_2 - \Gamma g(N_2-N_0)|E_2|^2
\end{aligned} \tag{7.7}$$

$\Delta\omega = \omega_1 - \omega_2$ means the frequency separation of the two lasers at the free-running condition, and the other parameters defined here are same as mentioned before in MI-OFLs. Generally, in the numerical analysis, the two mutually injected laser sections are viewed to have the same structural parameters as well as a symmetrical coupling, except for the initial frequencies at the free-running state. Based on the above equations, one numerical simulation model is demonstrated in [60] and rich varieties of dynamics are reproduced from the stable mutual locking state, FWM, P1 oscillation, P2 oscillation, QP to C. Just like the MI-OFLs, all these so-called optical instabilities stem from the delayed optical injection from a laser to its counter-laser.

Y-branch integrated dual-wavelength laser diode is another specific laser structure belonging to MI-MISLs, in which the two angled DFB lasers are associated with a Y-branch coupler as Figure 7.6d presents [82–86]. As a result of the symmetric geometry, the output of each laser merges into the Y-junction coupler section, and thus, light injection can be realized from one DFB laser into another DFB laser by reflection at the end facet of the Y-branch section. Light is exported through the Y-branch section of the MI-MISLs. The injection level is controlled by adjusting the current of the Y-branch section, which has an important influence on the tuning range of the generated microwave signal frequencies. The wavelength difference between the DFB sections can be tuned by changing the heat sink temperature or by varying the injection currents of two DFB laser sections as well as the Y-branch section.

Based on these MI-MISLs, researchers have made great progress in the generation of high-quality microwave signals and tunable MPFs with huge bandwidth. Additionally, when the two DFB sections are in the state of optical injection locking, the modulation response of the MI-MISLs can show a broad 3-dB modulated bandwidth with a good flatness, which is a significant figure-of-merit in the analog fiber link transmission. It is also seen that for a fixed initial frequency detuning, the resonance peak shifts to a higher frequency with an increased coupling strength when the absorption in the phase tuning section is reduced [65]. Moreover, great performances, i.e. better SLM operation, frequency-chirp reduction under direct modulation, as well as improved nonlinear distortion can be allocated easily, making this type of structure more suitable for analog ROF applications. Furthermore, light injection means fewer carriers are required to reach threshold gain, leading to a lower relative intensity noise (RIN).

Table 7.1 summarizes some comparisons within the three representative laser structures in terms of some of the most important characteristics. It can be seen that though MI-MISLs are equipped with numerous advantages, the inevitable fact impossible to ignore is that these devices subject to mutual injection require not only two DFB sections with precise wavelength detuning but also an additional mode coupler, which greatly increases the fabrication process and finally leads to a low device yield [95].

7.3 Electro-optic Conversion Characteristics

7.3.1 Modulation Response Enhancement

Compared with external modulation, directly modulated semiconductor lasers are of great interest in fiber communication systems, especially in very-short-reach (VSR) ROF links due to their compactness and low cost [20]. However, in a typical single-section semiconductor laser, once the modulation frequency applied to lasers is beyond 10 GHz, the direct modulation characteristics are heavily limited because of the low relaxation oscillation frequency. Furthermore, it is well-known that analog link transmission with central subcarrier frequency up to 20 GHz has been already common as the frequency resources become more and more nervous in recent years. To meet the urgent need for greater speed, much progress has been carried out to improve the modulation performance of the directly modulated semiconductor lasers. Primarily, the modulation bandwidth in directly modulated lasers is supposed to be enhanced, and to realize this goal, it is essential to make the relaxation oscillation frequency as high as possible.

The following formula demonstrates the calculation expression of relaxation resonance frequency [53]. D, W, and L are the thickness, width, and length of the

Table 7.1 Comparison of different MISLs including PFLs, AFLs, and MI-MISLs.

Laser structures	Composition	Advantages	Shortcomings	Nonlinear dynamics
PFLs	Passive feedback section + DFB laser section	The simplest laser structure	K: limited, Not separate control of K and ϕ, Butt-joint growth process	Single-mode steady state, DQS-SPs, MB-SPs, Chaos
AFLs	Active feedback section + phase tuning section + DFB laser section	$0 < K < 1$, Separate control of K and ϕ	Butt-joint growth process (some)	Single-mode steady state, DQS-SPs, MB-SPs, Period doubling, Chaos
MI-MISLs	At least two DFB laser sections	Natural dual-wavelength mechanism, No optical isolator/circulator, Automatic polarization match and optical alignment, Environmentally robust	Precise wavelength detuning, An additional mode coupler between two DFB lasers	Stable locking, P1 oscillation, Frequency locking, Quasi-periodicity, Chaos

Function. DQS-SPs, dispersive Q-switching self-pulsations; MB-SPs, mode-beating self-pulsations; P1, period-one; G, generation; T, transmission; P, processing.

active DFB semiconductor laser, respectively. dg/dn is the material differential gain. I_{th} is the threshold current of the laser. By adjusting the relevant parameters, the relaxation oscillation frequency can be improved effectively.

$$f_r \propto \sqrt{\frac{\Gamma\, {}^{dg}/_{dn}}{DWL}(I - I_{th})} \tag{7.8}$$

Conventionally, one favored approach to broaden the relaxation oscillation frequency is utilizing the enhancement effect of CPR by optimizing the standard semiconductor laser parameters, such as curtailing the laser cavity length and increasing the differential gain of the MQW active layer as well as RC time constant [43]. Especially, it is widely believed that reducing the laser cavity length is the simplest way to increase the relaxation resonance frequency, thus leading to a high 3-dB modulation bandwidth. However, it is also relatively challenging to make DFB semiconductor lasers with short cavity lengths for that it means small equivalent reflection, resulting in high threshold gain and finally degrading the direct modulation bandwidth [54]. Thus, it should be compromised between the reasonable threshold gain and a high modulation bandwidth when related to design a desired laser. Moreover, it is relatively rigorous and hard to fabricate lasers with short cavity lengths as far as conventional cleavage technology is concerned. In addition, it is a pity that further investigations have shown the advancement of modulation bandwidth based on the principle of CPR frequency does not have a qualitative improvement, which is typically in the range of 8–12 GHz.

In order to overcome the CPR frequency limitation of the direct modulation lasers and further strengthen laser modulation bandwidth far beyond the CPR frequency, another alternative method is proposed, which involves enhancing the relaxation resonance frequency by taking advantage of the interaction of the neighboring optical longitudinal cavity modes. This technique with much potential is recognized as PPR effect [44]. Until now, PPR effect has been exploited extensively through optical feedback with the help of an integrated optical waveguide [40, 42, 45–48, 53–56], or by means of optical-injection locking with ML and SL sources [58, 59, 62–65]. As to the optical feedback mechanisms, there exist two kinds: passive feedback and active feedback. Actually, no matter in MI-MISLs, or MI-OFLs, a drastic interaction between the photon and photon can be ensured, and a high PPR frequency can be realized simultaneously, which potentially exceeds the usual CPR frequency a few times and largely relaxes the CPR effect induced high-speed frequency limitation to a great extent. Therefore, the approach of utilizing the MI-MSSLs instead of one-section semiconductor lasers is recognized as a promising candidate to significantly enhance the direct modulation bandwidth with much flexibility nowadays.

Until now, there has been a large body of literature surrounding the modulation characteristics enhancement. In 2011, one PFL with a flat modulation response and high modulation bandwidth of more than 30 GHz was reported well for moderate injection currents of DFB laser sections ranging from 50 to 70 mA [47]. The highest −3 dB modulation bandwidth was observed at 37 GHz, which is largely enhanced by a factor of 3 compared with the initial CPR-induced modulation bandwidth of 12 GHz as Figure 7.7 shows. Similar result has also been obtained in [40] for a two section buried-heterostructure passive feedback laser (BH-PFL) device with a flat modulation response and high modulation bandwidth up to 34 GHz at lower DFB driving currents in 2012. Additionally, an in-depth theoretical analysis as well as the experimental verification were carried out in the PFL with the focus of improving the modulation bandwidth in semiconductor lasers by passive feedback [45, 50]. State-of-the-art results described that the PPR frequency can be tuned efficiently with the feedback phase and the feedback strength of the reflected optical signal by changing the injection current imposed on the IFB section. The feedback phase plays a role in the generation of the PPR frequency. The feedback strength, which depends on the field loss within the IFB section, at the facet of the IFB section, and at the interface of both sections, is responsible for getting a dominance of the PPR frequency. Within each 2π phase region, the increase of the phase tuning current is accompanied by an enhancement of the feedback strength, corresponding to an increased modulation bandwidth. An operation of the PFL with the PPR frequency of 28–40 GHz was found with the phase current of 13–18 mA [45]. However, it doesn't work for the position of PPR in different 2π

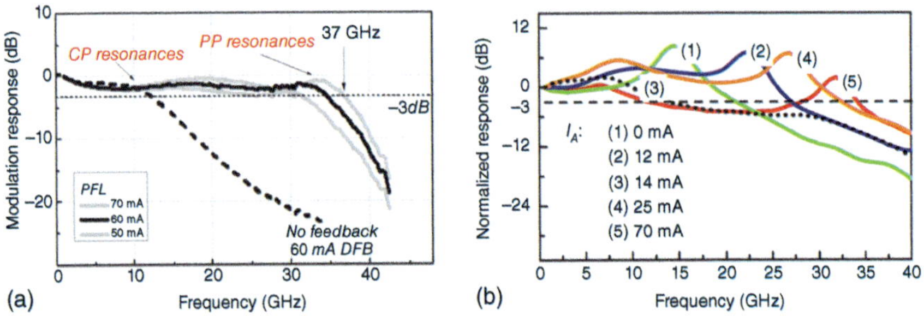

Figure 7.7 (a) The comparison diagram of the enhanced modulation response of a PFL based on the CPR effect (dotted line) and the PPR effect (solid lines) separately, and (b) the enhanced modulation response in an AFL with the variation of control current of the amplified section.

phase periods, which can be explained by a reduced feedback strength resulting from the high carrier and thermal effects induced higher optical losses. Attributing to the natural limitation of the tunable range of the feedback strength in PFLs, the tuning range as well as the maximum of the PPR frequency are greatly narrowed. Thus, it is a wise choice for PFLs to control the cavity length properly in order to obtain a higher feedback strength. Moreover, an additional coating with HR at the facet of the feedback section is a beneficial choice to obtain a high PPR frequency by strengthening the optical feedback.

Different from the limited feedback strength in PFLs, AFLs are famous for their higher modulation bandwidth arising from the improved feedback strength due to the existence of an additional amplified feedback section in the constitution. Attributing to the PPR effect, one AFL with an enhanced −3-dB modulation bandwidth ranging from 12.8 to 19.1 GHz was achieved successfully with the aid of a normal low-frequency wafer [56]. Some similarities can be found between this work and the work of [55], which thoroughly reported both theoretical simulations and experimental verifications to explore the effect of PPR on the enhancement of the modulation bandwidth in an AFL. An enhanced −3 dB bandwidth of 27 GHz with an in-band flatness of ±3 dB was successfully demonstrated at 13 °C, which is nearly two times its original bandwidth determined by CPR frequency. If the PPR peak is selected properly by means of tuning the injection currents, the modulation bandwidth can be further enhanced to greater than 40 GHz in theory. It is found that when the phase tuning section and DFB section currents are fixed, the increasing feedback strength can result in the increasement of the optical mode spacing, corresponding to the enhanced PPR frequency peaks as the control current of the amplified feedback section increases. Figure 7.7b depicts the corresponding simulation results that by varying the current of amplified feedback section from 0 to 70 mA, the PPR peak will increase from 15 to 32 GHz and the modulated bandwidth can be enhanced from 20 to 35 GHz in the meantime. In comparison, the PFL with the same section lengths except an amplified feedback section only has a limited frequency tuning range of less than 4 GHz by theoretical simulation. Hence, it can be concluded that the tuning range of the PPR frequency is largely improved with an optimized feedback strength.

Another study conducted by [55] also found that while the PPR frequency is relatively high and far away from the CPR frequency, it will inevitably exist a damped gap between the CPR frequency and PPR frequency in the frequency response. Seriously, for the application in analog communication systems, the modulation response should not only have a high 3-dB modulation bandwidth but also have a flat S_{21} frequency property. Therefore, further optimization of the device structure will be indispensable to ensure flat behavior in a wide tuning range and get a balanced damping between the CPR frequency and the PPR frequency. In other words, a compromise should be considered between the in-band flatness and the high PPR frequency. By selecting a proper current combination, the gap between the CPR and PPR peaks can be filled and a flat response curve will be obtained.

Besides the optical feedback through an integrated optical waveguide, optical injection locking technique is another potential candidate to improve the modulation bandwidth [62]. In this case, the relaxation resonance frequency can be tuned by changing the detuning frequency of the ML relative to the SL. Thus, it has been recognized that with the help of the MI-MISLs under proper conditions, the modulation bandwidth can be significantly increased in the same way. Actually, in MI-MISLs, owning to the heating effect, the wavelengths of the front DFB laser section and the rear DFB laser section drift to the longer wavelength simultaneously when one of the DFB section currents is increased. They travel different distances, and thus different detuning frequencies are generated. The relaxation resonance frequency can be changed subsequently, resulting in a variable 3-dB modulation bandwidth. This issue was explored by [63] using one monolithic injection-locked DFB laser with two gain sections, and the resonance frequency ranging from 11 to 23 GHz under injection-locked state was successfully demonstrated. Here, the frequency of enhanced resonance peak is defined as the frequency difference between the main lasing mode of the SL section under mutual injection locking and a weak sidelobe resulting from the red-shifted cavity mode of SL section. More recent work by [64] with relaxation resonance frequency changing from 26.8 to 20.9 GHz, was also reported by using the same two-section MI-MISL structure when the current of the rear laser section rises from 110 to 170 mA. The particular phenomenon of the relaxation resonance frequency not being correlated positively to the injection current results in the frequency detuning here, which is defined as the wavelength separation between the ML mode and the side-mode of SL. Among them, the maximum modulation bandwidth of the MI-MISLs increases to 31.6 GHz, which is about 20 GHz higher than that under the free-running state. Altering the phase tuning section current can also have an indirect influence on the modulation response by changing the coupling strength in the three-section MI-MISLs. Since the coupling strength between the two DFB lasers is positively correlated with the phase tuning section current on account of the reduced absorption, the modulation response of the MI-MISLs is significantly enhanced with the phase tuning section current increasing, indicating an enhanced modulation bandwidth. However, something needs to be clarified that, just like the PFLs and AFLs, the coupling phase also has an effect on the working state of the MI-MISLs, which is changeable with the current injected into the phase tuning section. Conceptually similar work has also been carried out by [65] in which a tunable resonance frequency of about 20–34.3 GHz was recorded with the phase tuning section current altered from 5 to 23 mA under mutual injection locking state. Special emphasis should be placed on the fact that the highest resonance frequency of 34.3 GHz observed is more than fivefold of that under free-running state. However, it is a pity that a flat modulation response with enhanced 3-dB modulation bandwidth has not been reported until now.

Photonic microwave transmission named as ROF links is one of the important hotspots in the application of MWP, which can also benefit from MISLs for extremely dependent on the enhanced modulation bandwidth characteristic [27, 48, 49, 66–69]. Historically, microwave signals are directly generated, transmitted, and distributed in electrical circuits through electrical transmission lines, such as the coaxial cable for customers. However, due to the high propagation loss with a typical value of 360 dB km^{-1} @ 2 GHz, microwave signals are only transmitted over a finite distance, which is rather impractical when applied in the long-haul link. More importantly, the propagation losses increase with the signal frequency. The contradiction between the increasing urgent demand for large bandwidth and high capacity and the limited reach of transmission is apparently exposed with the arrival of 5G/6G era. Based on the mature photonic microwave signal generation techniques as well as the low loss fiber with a typical value of 0.2 dB km^{-1}, the novel method of transmitting microwave signals by utilizing high-frequency optical carriers called as ROF link has been proposed to circumvent the dilemma in distribution distance limitations of microwave signals in the electric domain. However, cost and size of the needed optical sources as well as external modulators are important issues when the transmission system refers to the external modulation, which restricts the practical application of transmission system. Utilizing direct modulation mode to load RF signals onto the optical carrier provides an optional choice capable of reducing the insert loss, power consumption, and complexity. Yet, direct modulation single-section semiconductor lasers are heavily limited in their practical use as a result of the narrow modulation bandwidth. Luckily, the question can be settled extraordinarily by using the PPR effect in multistage integrated lasers due to the enhanced electro-optic conversion characteristic. Thus, next, we will focus our attention on the photonic microwave transmission experiments on the condition of MI-MSSLs and verify the feasibility in the application of link transmission by discussing and analyzing some experimental results. A considerable amount of review of previous work in the subject area is given below as listed in Table 7.2.

A 5.3545 Gbaud differential 16-quadrature amplitude modulation (QAM) half-cycle Nyquist single-cycle subcarrier modulation (SCM) signal with a 3.5-GHz subcarrier is applied to the PFL to test specific transmission performance in [66]. The error vector magnitude (EVM) measurement results of back-to-back and after 61-km optical fiber transmission are 5% and 13%, respectively, manifesting the feasibility of the passive feedback module. Similar transmission work has also been pursued in [56] based on one directly modulated AFL, whose schematic of ROF link and analysis of the received signal are displayed in Figure 7.8. One 20 M Symbol s^{-1} 32-QAM signal with 19 GHz carrier is delivered via 25 km fiber transmission and the ultima result of the EVM of the whole link is 3.91%, indicating an excellent received

Table 7.2 Overview of MISLs for modulation bandwidth enhancement and ROF link transmission.

Integrated structure	Achievements: Modulation bandwidth enhancement	Achievements: ROF link transmission (EVM/BER)	Year and reference
PFL	25–40 GHz (simulation) 30–37 GHz (experiment)	×	2011 [47]
PFL	28–40 GHz (experiment)	×	2007 [45]
PFL	29 GHz (experiment)	13% (5.3545 Gbaud 16-QAM, 3.5 GHz, 61 km)	2013 [66]
AFL	27 GHz (experiment) 20–35 GHz (simulation)	×	2015 [55]
AFL	12.8–19.1 GHz (experiment)	3.91% (20 M symbol s^{-1} 32-QAM, 19 GHz, 25 km)	2017 [56]
MI-MISL	23 GHz (experiment)	×	2003 [63]
MI-MISL	20–34.3 GHz (experiment)	×	2015 [65]
MI-MISL	31.6 GHz (experiment)	2.94% (40 M symbol s^{-1} 32-QAM, 6 GHz, 50 km)	2017 [64]
MI-MISL	4.72 GHz (experiment)	3.76×10^{-6} (10 Gbit s^{-1} OFDM, 25 km)	2011 [67]

constellation diagram. Performance improvement of transmitting orthogonal frequency division multiplexing (OFDM) signal by one MI-MISL over differing transmission lengths is also experimentally verified by C. Browning et al. [67]. It has been shown that after 25 km of standard single mode fiber (SSMF), the bit error ratio (BER) decreases from 1.3×10^{-3} to 3.76×10^{-6} when mutual injection is used, which is improved significantly by 3 orders of magnitude than that before optical injection in the MI-MISL. The observed result corresponds to a reduction of 3.83% in average EVM, indicating a viable and reliable transmission. Another study conducted by [64] also achieved a similar finding. In this literature, the front DFB laser section is directly modulated by a 40 M Symbol s^{-1} 32-QAM signal with 6 GHz carrier. After 50 km of transmission, a better result is achieved in that the EVM of the system is reduced from 5.25% to 2.94% when the MI-MISL is switched from free-running state to injection-locked state ($I_s = 50$ mA, $I_m = 150$ mA). This decrease clearly indicates that the received constellation diagram under the injection-locked state is more superior to that of the free-running state. Furthermore, the comparison diagram of the measured EVMs with different received optical powers between injection locked state and free-running state is also investigated. It can be concluded that the received 32-QAM signal is improved significantly by utilizing the mutual injection state in the case of the same received optical power. All the outstanding performances above prove the feasibility and reliability of the monolithic integrated multi-stage lasers in the application of ROF link. The internal schemes of optical feedback-induced PPR effect as well as mutual injection can account for this variation.

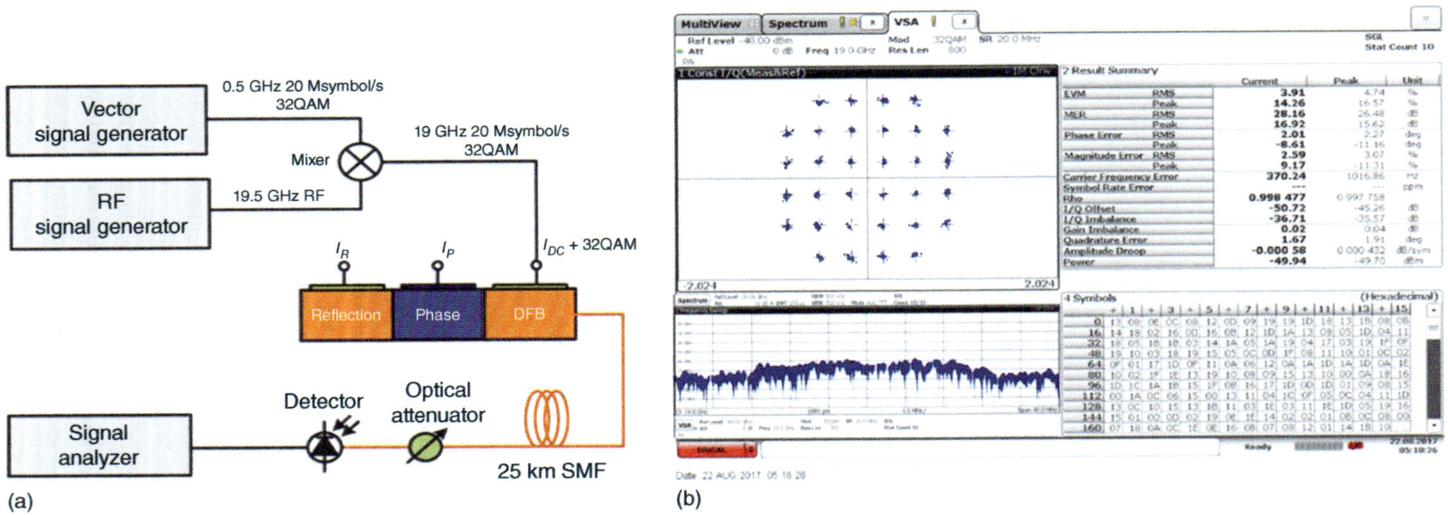

Figure 7.8 (a) Schematic of ROF link using the laser module, and (b) analysis of the received signal.

7.3.2 Nonlinearity Reduction

Nonlinear distortions are crucial factors in the electro-optic conversion characteristics, and they also determine the transmission performance of the directly modulated semiconductor lasers [27, 68]. With respect to this specific subject, this has also been extensively discussed in the MI-MSSLs [63–65]. Some issues mainly including the 1-dB compression point, the second harmonic distortion (2HD), the third-order intermodulation distortion (IMD3) as well as spurious free dynamic range (SFDR) are well covered in the literature. All of them are essentially generated from the interaction between electrons and photons. Especially, as the modulating frequency approaches the relaxation resonance frequency, the interaction becomes more and more severe. Now, attributing to the mutual optical injection and optical feedback-induced enhanced modulation bandwidth, the nonlinear distortions can be suppressed in the MI-MSSLs.

Commonly, the output of RF power is in direct proportion to the input RF power. But, once the input power is too high, the output RF power will deviate from the original linear curve, attributing to the RF link gain compression. Generally, the 1-dB compression point, where the output RF power decreases by 1 dB from the linear curve, is proposed to describe the linearity work region of the semiconductor lasers. It has been reported that the 1-dB compression point of the MI-MISLs had a remarkable improvement of more 3 dBm than that in free-running state [64]. Besides, high-order harmonics, covering the second and third-order as well as the fourth-order harmonics, were also investigated in [63–65], when the microwave signal with the frequency close to the relaxation resonance frequency is modulated on the MI-MISLs [65]. The power of the second and the third-order harmonics are reduced by 17 and 23 dB, respectively, indicating a great improvement in nonlinear distortions. Moreover, it is worthy of being noted that the noise near the fundamental frequency is also suppressed by about 4 dB due to the reducing RIN under mutual injection locking, which defines the fluctuations in the optical power of lasers and mainly determines the ground noise level. The entire evolution process of RIN under different current states can be also found in [56] with a reference to an AFL, which is the same as that in MI-MISLs. As the current of the reflector section increases, the RIN peak position moves to higher frequencies and the corresponding RIN value is reduced. This phenomenon can be explained by the increasing relaxation resonance frequency induced by PPR effect.

As one of the most vital factors for directly modulated lasers in analog optical links, the IMD3 of two closely spaced RF signals determines the linearity of the device. A two-tone signal with the frequency gradually close to the relaxation resonance frequency was modulated to investigate whether the IMD3 can be improved in MI-MISLs [64]. The spectra of the IMD3 signal in free-running state and injection-locked state are given, respectively, in Figure 7.9, showing a significant improvement of 24.54 dB in the suppression of IMD3 after utilizing the injection-locked state. Another report demonstrated the reduction of IMD3 at various frequencies ranging from 0 to 4 GHz due to mutual injection in detail in [67], clearly revealing how nonlinearity is reduced to a greater extent as the

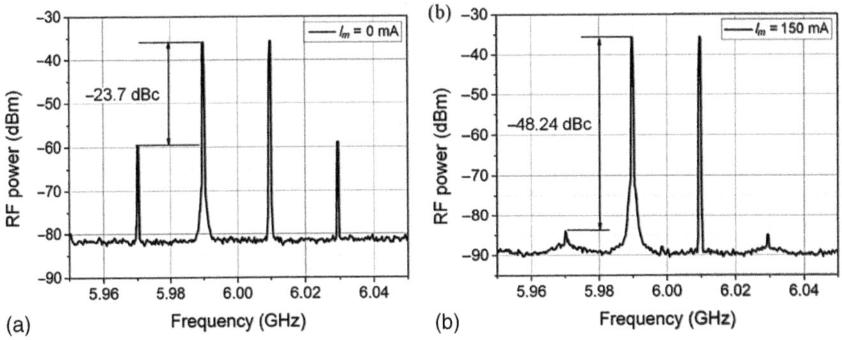

Figure 7.9 IMD3. Source: Zhang et al. [64].

performed frequency approaches the nonlinear operation range of the SL. For further investigation, the SFDR of the multi-section laser is also measured by the functions of fundamental and IMD3 signals respective to the input RF signal power [56, 64, 65]. Compared with the free-running condition, SFDR is improved by about 9.1 dB from 80.7 to 89.8 dB $Hz^{2/3}$ [64], attributing to the combined effect of nonlinear distortion reduction and RIN suppression. Conceptually identical work, utilizing a similar method, was proposed by [56] based on the AFL, in which perfect SFDR is calculated to be 102.8 dB $Hz^{2/3}$ around 19 GHz.

7.3.3 Chirp Suppression

In semiconductor lasers, frequency modulation (FM) is inherently accompanied by intensity modulation as a result of the carrier and photon density variation as well as the subsequent change of refractive index. The properties of frequency chirp induced by the current modulation play an important role in an optical link since it can heavily limit the transmission distance due to the effect of power fading, especially in dispersive systems [23]. According to whether the photon density is changed with time, we can separate the chirp process into transient and adiabatic chirp. It was reported in [47] that both the transient and adiabatic chirps can be managed and suppressed by adjusting the feedback strength and the feedback phase. Just like the high-speed digital transmission, the chirp also exists in analog transmission links [65]. The same phenomenon of chirp reduction can be also achieved in the case of the MI-MISLs as a result of the injection locking. One 40-km-long SSMF was inserted in an analog optical link to test the performance of the integrated laser under mutual injection locking. The measured results depicted in Figure 7.10 manifest that an improvement of more than doubled frequency-length product was achieved for the fiber analog optic link, indicating a perfect suppression in the frequency chirp.

7.4 Photonic Microwave Generation

Microwave waveform generation has received intense attention in recent years. It is found that high-frequency microwave signals can support a large range of

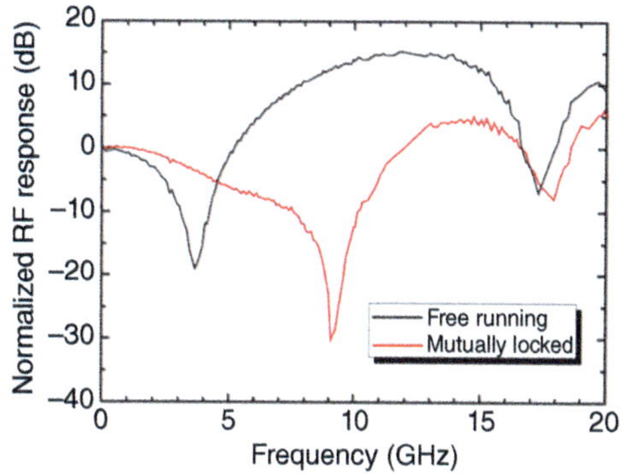

Figure 7.10 RF power fading after 40 km SMF.

applications in radar systems, 60-GHz wireless communication access, and so on. Lots of types of microwave generators have already been spurred and developed. Traditionally, high-frequency microwave signals we desired are achieved by means of frequency multiplication based on complicated and costly electronic circuits. However, due to the limitations of the electronic bottlenecks, once the signal frequency is beyond about 10 GHz, the phase noise performance of microwave waveforms will deteriorate sharply at a rough rate of 6-dB reduction with the frequency doubling, leading to a large degree of restriction in the maximum of the microwave frequency. Moreover, the high loss of copper cables is also unfriendly to long-haul transmission. Aiming at these existing questions in electrical domain, a large number of corresponding photonic microwave generation techniques based on single-section DFB lasers, which mainly include optical heterodyne technique, external modulation, optoelectronic oscillators (OEOs), and P1 oscillation state, have been proposed to generate high-frequency microwave signals with the help of the photon's unique dominances such as high speed and wide bandwidth. Among these methods, optical waves at different frequencies, which are provided by two laser sources, are detected to generate microwave signals through one photodetector in the optical heterodyne technique. The approach can generate no matter how high frequency you want in the absence of considering the maximum response frequency of the photodetector. The beating frequency up to several hundred gigahertz bands could be easily realized without a hitch. Yet, the phase noise of the microwave signal is terrible because the two lights are not phase correlatively. External modulation is another important photonic technique to achieve microwave signals by heterodyning the different optically carried microwave sidebands derived from modulation. The generated microwave signals are not only high-frequency but also of a relatively low phase noise as a result of the totally correlated optical spectral sidebands. However, due to the existence of external modulator, the operation schemes are often subject to high insertion loss and large volume. The approach of OEO systems is also attractive for the fact that they can generate high-quality microwave signals with extremely low phase noises, but it is a pity that the similar question of the complex structures is also exposed in OEO schemes. Besides, the technique of utilizing the nonlinear behavior of semiconductor lasers themselves, subject to optical injection and optical feedback systems is of great interest and has been explored thoroughly, like self-pulsations, P1 oscillation state, P2 oscillation state, during the last 10 years. Especially, when adjusting the optically injected semiconductor laser system to operate in P1 state, a broadly tunable microwave waveform, whose frequency is supposed to be up to 300 GHz far exceeding the intrinsic relaxation resonance frequency of the semiconductor laser, can be realized with no difficulty. However, in previous studies, the realization of optical injection and optical feedback systems always required at least two discrete single-section laser sources besides one optical isolator or circulator or one DFB laser with an additional mirror, and thus these methods are facing the common challenges of costly and complex system structures. Moreover, the precise control of polarization state is not well acquired.

Recently, microwave signal generation realized via MI-MSSLs has attracted much research interest for the superior performances of more compactness, lower power consumption, lower cost, and higher reliability than the conventional microwave generation techniques based on single-section lasers. In this section, the previous and current attempts at photonic microwave generation techniques based on MI-MSSLs are described. In special, some of the most pertinent papers in the subject field are reviewed and listed in Table 7.3. Besides the single-tone microwave signals, the generation of frequency-modulated microwave waveforms, such as F-H and LCMWs, which are widely applied in radar and solar systems, are described as well. In the end, one optimizing method of boosting the signal quality and narrowing the linewidth is demonstrated.

7.4.1 Tunable Single-Tone Microwave Signal Generation

7.4.1.1 Free-Running State

The frequency heterodyning of two optical modes coming from two independent DFB lasers and separated at a desired microwave frequency is a well-suited microwave generation method in the MI-MISLs due to their natural two DFB lasers structure. In this approach, both DFB lasers operate in the free-running state and oscillate on their own. By independently altering the corresponding drive currents to change the wavelength deviation between the two oscillating optical modes, the beating microwave signal frequency can be changed correspondingly according to the following equation.

$$f = c\Delta\lambda/\lambda^2 \tag{7.9}$$

Here, c is the light velocity in vacuum, λ is the Bragg wavelength, and $\Delta\lambda$ is the difference between the two modes. Based on this method, the continuous adjustment of optical microwave signals from 13 to 42 GHz has been reported with the help of two parallel DFB lasers integrated with Y-branch waveguide coupler, and the corresponding wavelength differences when fixing one injected current and scanning another one are depicted in Figure 7.11 [82]. Another electrical signal with frequency tuning range from microwave (~10 GHz) to millimeter wave (~50 GHz) was also reported in [83] based on the same Y-branch laser structure. The frequency of the generated microwave signal can be extremely high if not considering the bandwidth of the measurement setup. The signal quality is also ensured to some extent for the

Table 7.3 Overview of diverse experimental results optical generation technique based on different types of MI-MSSLs.

Integrated laser structure	Optical generation technique	Frequency tuning range	Signal quality (3 dB linewidth, SSB phase noise)	Year and reference
MI-MISLs	Free-running state	10–50 GHz	~10 MHz	1999 [83]
MI-MISLs	Free-running state	13–42 GHz	~10 MHz	2007 [82]
AFL	MB-SPs	17–35 GHz	—	2004 [94]
AFL	MB-SPs	28–41 GHz	~8 MHz (28–41 GHz)	2002 [88]
AFL	MB-SPs	12–45 GHz	—	2003 [90]
AFL	MB-SPs	11–40 GHz	—	2014 [97]
AFL	MB-SPs	46.8–72 GHz	—	2015 [96]
AFL	MB-SPs	75–109 GHz	<4.5 MHz	2016 [95]
AFL	MB-SPs + sideband injection locking	19.87–26.3 GHz	30 kHz @ 22.70 GHz	2009 [99]
AFL	MB-SPs + direct modulation	15–33 GHz	−77.8 dBc Hz^{-1} @ 10 kHz @ 26 GHz (original 13-GHz driven RF signal (−84 dBc Hz^{-1} @ 10 kHz))	2015 [102]
AFL	MB-SPs + dual all-optical feedback loop	30–38 GHz	<2 kHz (30–38 GHz)	2014 [118]
AFL	MB-SPs + single all-optical loop	29–37.6 GHz	1.8–4.0 kHz	2014 [117]
AFL	MB-SPs + OEO + self-injection locking	32–41 GHz	<−97 dBc Hz^{-1} @ 10 kHz (32–41 GHz) −101.6 dBc Hz^{-1} @ 10 kHz @ 36 GHz	2015 [115]
AFL	MB-SPs + OEO + self-injection locking	28–41 GHz	<−106 dBc Hz^{-1} @ 10 kHz (28–41 GHz) <−115.93 dBc Hz^{-1} @ 10 kHz @ 33.5 GHz	2015 [133]
MI-MISLs	Sideband injection locking	42 GHz (eight times frequency)	−94.6 dBc Hz^{-1} @ 10 kHz, −98.5 dBc Hz^{-1} @ 100 kHz	2009 [101]
MI-MISL	Self-pulsation + OEO	8 GHz	−108 dBc Hz^{-1} @ 5 kHz	2012 [114]
MI-MISLs	P1	14.15–35 GHz	~3 MHz	2014 [61]
MI-MISL	OEO	17.3–21.7 GHz	−115.3 dBc Hz^{-1} @ 10 kHz @ 20.3 GHz −92.8 dBc Hz^{-1} @ 1 kHz @ 20.3 GHz	2019 [116]

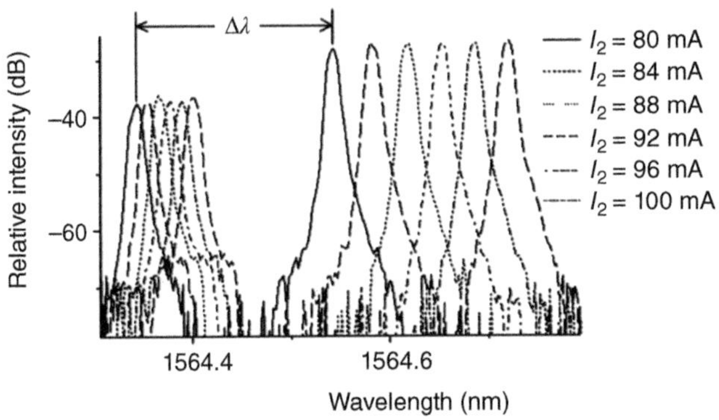

Figure 7.11 Wavelength difference when fixing one injected current and scanning another one.

two lasing wavelengths sharing the same laser cavity in MI-MISLs when compared with that generated based on discrete DFB lasers. However, an important constraint on all the work discussed in this method is that a Lorentzian-shaped microwave signal with a linewidth on the order of 10 MHz can be still observed in the electrical spectrum due to the lack of phase correlation between the two DFB lasers. Good consistency has been observed with a linewidth of 10.7 MHz full width half maximum (FWHM) at 15.187 GHz in [86], which exceeds the minimum requirement in many photonic microwave applications.

7.4.1.2 Mode-Beating Self-Pulsations (MB-SPs)

The state of the MB-SPs in discrete optical feedback semiconductor laser systems has been studied intensely both in theory and experiment in the last two decades for their application in high-frequency microwave generation. Similarly, MB-SPs can be also obtained in MI-OFLs when one section is operated as a saturable absorber. High-speed MB-SPs in one two-section DFB laser were observed firstly in the gigahertz range in 1992 [87]. By adjusting an appropriate current injection, the MB-SPs of MI-MSSLs can be established reproducibly from a common gain medium, in which two compound cavity modes coexist and share the same threshold gain [95]. The frequency separation of the two main lasing optical modes defines the frequency of the beating signal.

For a certain IFB laser system, the beating frequency is mainly determined by two main parameters, i.e. the feedback strength K and the delay time τ. The corresponding function between them is listed as followed according to [90, 92].

$$K = K_0 \left| \frac{\pi f \tau}{\sin(\pi f \tau)} \right| = \frac{1}{2C_f} \frac{L_{DFB}}{L_{EC}} \left| \frac{\pi f \tau}{\sin(\pi f \tau)} \right| \tag{7.10}$$

The factor K_0 represents the minimum feedback strength, which is the basic requirement for the emergence of MB-SPs. Its size is determined by three parameters, including the effective lengths of DFB laser (L_{DFB}) and external cavity (L_{EC}) as well as the feedback sensitivity C_f of the DFB section [92]. The delay time τ is decided by L_{EC} and the group index (n_g). One thorough image analysis of the relationships between beating frequencies and variables such as K/K_0 and L_{EC} is shown

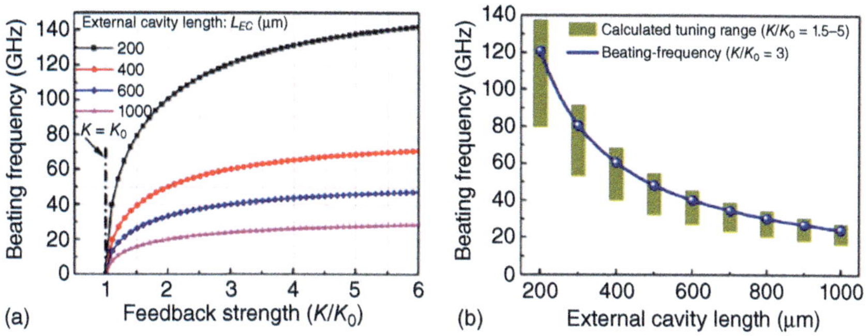

Figure 7.12 Calculated beating frequency as functions of feedback strength (a) as well as external cavity length (b).

in Figure 7.12. It can be seen that the beating frequency increases with the growing feedback strength under the condition of a fixed external cavity length. Moreover, when feedback strength is fixed, the tuning range of the beating frequency is gradually smaller with an increasing L_{EC}. In especial, not only the maximum available beating frequency ($f_{max} \propto 1/\tau = c/(2n_g L_{EC})$), but the minimum beating frequency is determined by L_{EC}. Hence, it can be concluded that a short external cavity length is more attractive to result in high-frequency microwave signals as well as a large tuning bandwidth [92, 93]. An optimized laser structure with a very short feedback cavity of 230 μm was successfully designed to obtain high frequencies around 100 GHz in [95], almost covering the entire W-band (75–110 GHz).

As the simplest MI-OFLs, PFL is firstly discussed to realize the MB-SPs. However, the fact of relatively low feedback strength is an inevitable shortcoming, heavily limiting the development of the PFLs in terms of high-frequency generation. Generally, the microwave signals generated based on MB-SPs are always at low frequencies below 20 GHz in PFLs. One possible solution is to substitute the universal anti-reflection (AR) coating at the side of the feedback section with an appropriate high-reflection coating to promote the frequency of the generated microwave signals through the improved feedback strength. Yet, the shortcoming of not controlling the feedback strength separately is still remaining in PFLs, which is the fundamental factor in limiting the size of the feedback strength.

AFLs, in which an additional amplifier section is integrated into the feedback cavity, are proposed to compensate for the optical losses and obtain a higher feedback strength, and hence higher beating frequencies. Especially, the independent control of the feedback strength and the feedback phase can be allowed. In an AFL, it has been found that the oscillation frequency decreases with the increase of the phase current, attributing to the decrease feedback strength induced by free carrier absorption in the phase tuning section. Also, when the current of the amplifier section is fixed, the MB-SP oscillations occur periodically with the phase current, which results from the change in the feedback phase due to the refractive index change in the feedback cavity [94]. In special, with the increasement of the current of amplifier section, the injection current in phase section has a more vital effect on the MB-SP frequency. More importantly, the areas, where MB-SP states occur, increase with the increasing amplifier section current, and the extra changes of carrier density acting in the added amplifier section induced a self-adjusting effect of the optical feedback amplitude and phase, which can account for this phenomenon. Hence, AFLs working under MB-SP state can provide not only higher frequencies but also a wider tuning range of the generated frequency, offering a more stable method to generate high-frequency microwave pulse signals compared with the PFLs. It has been shown experimentally that a continuous tuning of the self-pulsation frequency from 28 to 41 GHz can be adjusted via changing only the feedback level, i.e. by changing the current in the active section [88]. Next year, the later work of [90] further optimized the frequency tunability ranging from 12 to 45 GHz based on the same method. Similar work has also been pursued by others [96] in which beating-frequency tuned from 46.8 to 72 GHz was recorded. The frequency tuning range can cover most frequencies of U-band (40–60 GHz) and V-band (50–75 GHz), which is very suitable for the future millimeter wave over fiber networks.

Obviously, MB-SPs oscillation state can provide a possibility to generate microwave signals at different pulse frequencies only utilizing a single DFB laser device. However, the method is flawed for that in the operation of a MI-OFL, a precise and careful adjustment of the optical path length in the laser cavity is supposed to be guaranteed for that the MB-SPs phenomena are sensitively dependent on the feedback phase, or it will be difficult for us to observe these pulsations during the experiment. Multiple studies have confirmed the problem. In addition, most of the available experiments in literature carry notable limitations related to the 3-dB linewidth of the generated microwave signal, which is wide at MHz level [95]. Thus, the quality of microwave signals is worthy of being further optimized, and it is recognized to reduce the linewidth to the kilohertz or even millihertz level by adding extra optoelectronic feedback or all-optical feedback loops.

7.4.1.3 Period-One (P1) Oscillation

Besides the MB-SPs in the optical feedback systems, the P1 oscillation is also an attractive microwave generation technique based on the nonlinearity of the semiconductor laser itself. The so-called P1 state means that the temporal waveform shows constant intensity and the pulsing period is consistent with the beating signal frequency. In the OFL systems, the physical mechanism of the P1 state is dispersion-Q-switching (DQS) self-pulsation. The microwave frequency generated by this method is heavily limited, approximately equivalent to the relaxation oscillation frequency of the laser device [97]. Accordingly, P1 state is more commonly studied for microwave generation in optical injection systems, because the generated microwave frequency employing this method far exceeds the intrinsic relaxation resonance frequency of the semiconductor laser. However, one matter should be noted that most investigations of P1 state have been made based on unidirectional optical injection in a discrete optically injected semiconductor laser system, which originates from the beating between the injected optical mode provided by the ML and the shifted cavity mode of SL. While in MI-MISLs, due to the shortage of optical isolation/isolator, the two laser sections are mutually injected. As a result, it is inferred that the P1 oscillation in this condition attributes

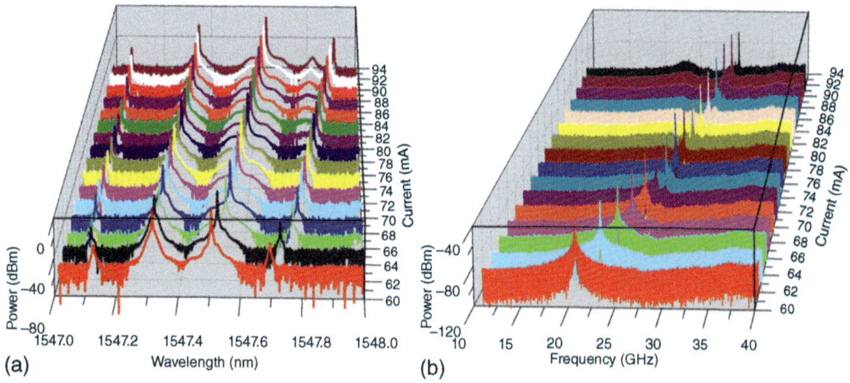

Figure 7.13 (a) Measured optical spectrum with I_{DC2} being varied from 60 to 92 mA while I_{DC1} being fixed at 77 mA, (b) measured electrical spectrum with I_{DC2} being varied from 60 to 92 mA while I_{DC1} being fixed at 77 mA.

to the beating between cavity modes of both two coupled DFB laser sections [61]. One microwave signal generation at the frequency ranging from 14.15 to 35 GHz was observed experimentally, and the measured optical spectrum with I_{DC2} varying from 60 to 92 mA and I_{DC1} fixed at 77 mA is demonstrated in Figure 7.13a [56]. The largest frequency can be up to hundreds of gigahertz, if not limited by the bandwidth of the spectrum analyzer as well as the photodetector. However, the linewidth of the generated microwave signal is terrible at a level of MHz level just like the discrete optical injection systems, as shown in Figure 7.13b. Furthermore, it is a pity that to the best of our knowledge, the study about applying the P1 oscillation to single side band (SSB) transmission has not been reported so far, which is recognized as a potential application prospect in ROF systems, especially with the advent of 5G time.

7.4.1.4 Sideband Injection Locking

Among various photonic methods for mm-wave generation, sideband injection locking technique is a promising approach for generating high-frequency microwave signals. The technique of sideband injection locking has two advantages. The first, attributing to the optically carrier sidebands are derived from the same optical source, different optical modes are totally correlated in phase and thus, the phase noise of the generated microwave signals by beating multi-stage carrier sidebands is of great improvement. Secondly, microwave frequency multiplication can be achieved with the corresponding beating frequency at $n \cdot f$ when the ML is modulated with a low frequency of f and the oscillating wavelength of the SL is adjusted to be locked to the high-order ($+n$) sideband of the ML. As a result, high-frequency microwave signals with low phase noises can be easily accomplished [83].

External Injection Locking Photonic microwave generation utilizing external injection locking is an important method of realizing high-frequency photonics. Attributing to the optically carrier sidebands from external modulation are derived from the same optical source, different optical modes are totally correlated in phase after modulation. However, in the conventional discrete DFB laser structures, the spectrum purity of the microwave signals generated by the sideband injection locking is always suffering from phase decorrelation because, after the modulator, the two beating optical signals undergo different optical paths between the ML and the two SLs [84, 98]. Now, this problem can be settled when the two discrete SLs are substituted by one MI-MISL because the optical path difference can be neglected for the short distance between the two DFB lasers integrated on one single chip. Microwave carrier generation based on external injection locking has been experimentally reported in [85]. In this system, the ML is a tunable laser source and the Y-branch integrated dual-wavelength laser plays a role of the SL. The multi-optical carrier sidebands are generated by utilizing a $LiNbO_3$ modulator. When injecting the modulated signals to the SL, i.e. an integrated dual-wavelength laser diode, the oscillated lights of two DFB laser diodes (LDs) in dual-wavelength laser diode can be respectively locked to the +1st and −1st order modulation sidebands of the ML by tuning control currents. Thereby, one frequency-doubled microwave signal of 24 GHz was generated after applying a 12-GHz RF signal into external modulator. Compared to the generated microwave signal in the free-running state in MI-MISLs, the high spectral purity of the microwave carrier was also ensured due to the sideband injection locking that induced high correlation between the two heterodyning lights. In addition, if the two oscillating optical modes in the Y-branch integrated laser are tuned to be locked to the higher stage modulation sidebands of external injection signal, a greater frequency signal with a low phase noise can be achieved incidentally. Similar work has also been pursued by others [99] based on an AFL in which by utilizing the external optical injection locking, the linewidth of the locked RF was 30 kHz while the linewidth of the unlocked mode was 3 MHz. Simultaneously, quadruple frequency at 22.88 GHz microwave signal was obtained by injecting the 5.72 GHz modulated signal. However, prior work does not address the issue of complex structures and high cost due to the existence of the external modulator as well as an additional ML.

Self-Injection Locking Besides the external injection locking, direct modulation of the MI-MSSLs to realize the self-injection locking is a simpler and lower-cost method for eliminating the need for a ML and an external modulator. Attributing to the combined effect of amplitude modulation (AM) and FM, a series of optical sidebands can be generated on both sides of the optical carrier when one of the DFB lasers is modulated. One pure 36-GHz microwave signal with high power and high fidelity was generated in [100] by RF modulation of the SL section with a 12-GHz RF signal. The third sideband of the modulated SL was locked to the ML, and thus the noise induced by random phase of two DFB oscillating modes was eliminated, leading to a low phase noise. To our knowledge, it is the first time to utilize monolithic sideband injection locking of a two-section DFB laser without any external optical element to generate microwave signals. Another microwave signal with a frequency up to 42 GHz has been experimentally recorded as well, corresponding to eight-times the modulation frequency by utilizing the Y-branch

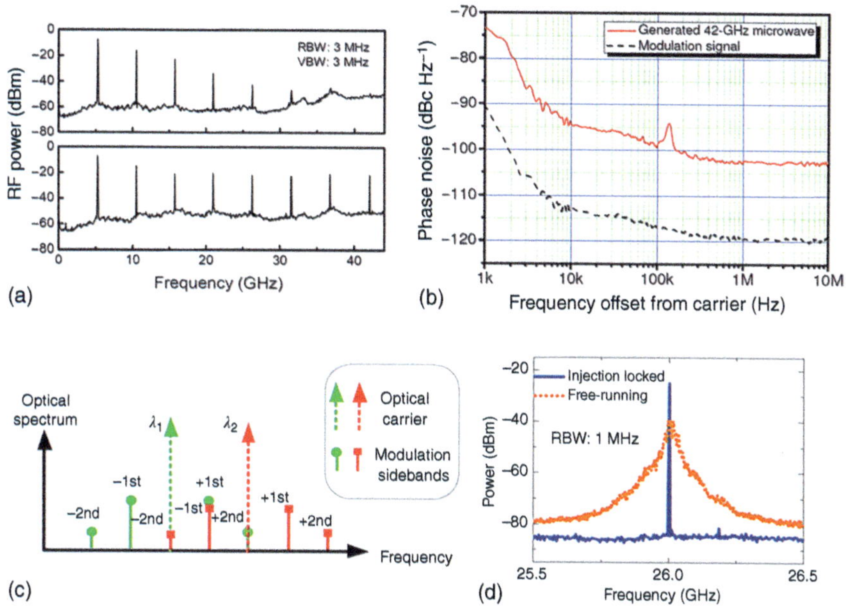

Figure 7.14 (a) Electrical spectra measured at the Y-branch facet (with/without injection locking) [101]. (b) Phase noise spectra of the optically generated 42-GHz carrier and the modulation signal. Source: Huang et al. [101]. (c) Illustration of the operation of the electrical injection locking scheme [102]. (d) The RF spectra of the free-running AFL and the injection-locked AFL. (a, b) Source: J. Huang, et al., [101]. (c, d) Source: Yu et al. [102]/ with permission of Elsevier.

integrated dual-wavelength laser diode, as Figure 7.14a shows [101]. The result of obtaining a high-frequency mm-wave signal from one low-frequency modulation signal was accomplished by taking advantage of the enhanced modulation nonlinearity around the relaxation resonance frequency. Compared with the modulation signal, the phase noise of optically generated mm-wave signal at 42 GHz exhibited only about 18 dB degradation, which was in accord with eight-time frequency multiplication as Figure 7.14b depicts. The phase noise at 10 kHz offset is $-94.6\,\text{dBc Hz}^{-1}$ and no Lorentzian-shaped noise floor was observed, indicating excellent phase correlation between the two lasers. However, this result is limited by the bandwidth of the electrical spectrum analyzer (ESA), or higher frequencies at multiples of the modulation frequency can be observed, though the signal intensity decreases with the increasing frequencies. The drawback of this method here is that due to the limited modulation response intensity, large modulation current is often required to realize high-order frequency multiplication, which may bring about abundant nonlinear dynamics, such as period doubling and chaos, and finally severely degrade the quality of the generated microwave signal.

Frequency tunability is also an important aspect in practical applications, which can be also allocated in the technique of self-injection locking. By varying the initial modulation signal frequency and adjusting the detuning frequency between the two DFB lasers, the frequency of the microwave signals can be easily tuned accordingly [20]. There has been reported that a tuning range of over 4 GHz was realized by adjusting the modulation frequency of the ML and the bias currents of both DFB lasers in [101]. In addition, no significant variations in the RF power as well as the spectral purity of the generated microwave signals were observed, indicating high stability and tunability in the high-frequency microwave generation. Furthermore, it is believed that the tunable range can be made wider by improving the DFB laser structure to increase the flatness around the relaxation frequency.

Utilizing the sideband injection locking technique to suppress the chaos induced by the large-signal direct modulation in MI-MISLs is another novel and simple method for high-frequency signal generation [60]. In conventional single-section laser, such large-signal modulation-induced chaos not only limits the intensity of the high-order harmonics but also severely degrades their spectral purity. But in MI-MISLs, the broadband chaos can be greatly reduced, and due to optical coupling, the rear DFB laser section is also under modulation, though the modulation signal is only applied into the front DFB laser section. Thus, the modulation sidebands of the two DFB lasers are mutually locked, and the front DFB laser is pulled from chaos to a stable state. After the photodetector, high-order microwave signals with great spectral purity can be significant by carefully controlling the key parameters. It has been reported that after the optical injection, about 20 sidebands with comparable intensities are mutually locked due to the sideband injection locking. The chaos was significantly suppressed, and high-frequency signals with many times the modulation frequency began to stand out gradually. Harmonics at 10 times the modulation frequency were generated, and one 33-GHz mm-wave signal, corresponding to sixfold multiplication of the 5.5 GHz modulation signal with the phase noise of $-95.2\,\mathrm{dBc\,Hz^{-1}}$ at 10 kHz offset and $-100\,\mathrm{dBc\,Hz^{-1}}$ at 100 kHz offset was recorded experimentally.

Identically, the technique of sideband injection locking different optical modes to improve the quality of beating signals can be applied equally in the dual-wavelength AFL [102]. The operation principle of the injection locking scheme is depicted in Figure 7.14c. It can be seen that with the driven electrical signal injected into the AFL, whose frequency corresponds to the 1/2 subharmonic frequency of the mode spacing of the AFL under free-running state, the λ_1-mode optical carrier will be locked by the −2nd order sideband λ_2-mode optical carrier, while λ_2-mode optical carrier will be locked by the +2nd order sideband λ_1-mode optical carrier. Thus, after the sideband injection locking, the dual modes of the AFL are synchronized in phase, and a beating signal with low phase noise will be obtained. Meanwhile, frequency-doubled photonic microwave waveforms are generated. Results demonstrate that the beating frequency ranging from 15 to 33 GHz was achieved. In addition, microwave signals with the frequency of 26 GHz were recorded, as Figure 7.14d shows, in which after injection locking, a dramatic beating linewidth narrowing from a few megahertz to hertz level was observed.

7.4.2 Frequency-Modulated Microwave Signal Generation

LCMWs are extensively applied in radar and sonar systems due to their unique characteristics in long-distance detection, high degree of resolution, and potential imaging recognition of the target. Especially in the military field, LCMWs with a wide bandwidth and large time-bandwidth product (TBWP) are in more urgent demand for advancing the performance of radar and sonar systems. Conventionally, lots of electric and photonic methods based on discrete devices have been proposed to research how to advance the performances of LCMWs in the past decades, such as voltage-controlled oscillators (VCO) [103], surface acoustic wave (SAW) [104], direct digital synthesizer technique (DDS) [107], and spectral shaping and frequency-to-time mapping technique (SS-FTT) [108]. Due to the limitations of electric bottleneck, all these electric methods usually suffer from a large degree of restrictions in the maximum of the central frequency, bandwidth, and TBWP. As for the photonic-based approaches, although these methods have the advantage of a good tunability in any central frequency, bandwidth, and high compression ratio, the most obvious conundrum in these approaches is that the systems used are bulky and costly because of the application of numerous discrete photonic components [107, 108]. Moreover, the stability of the systems in many aspects, such as polarization and temperature, is supposed to be seriously considered.

Recently, some solutions based on MI-MSSLs have been put forward to solve these problems. An LCMW with a pulse duration of 1 μs and bandwidth of 3.3 GHz centered around 31.5–34.8 GHz was realized by using an AFL working on the MB-SPs state [109]. By controlling the chip to work in the region where beating frequency changes linearly with the feedback strength, a linear chirped microwave signal will be obtained if a sawtooth sweeping signal is applied to the amplifier section because the feedback strength is linear to the input signal voltage. The method is simple and cost-effective due to the photonic integration component. However, the generated LCMWs only have a small bandwidth of 3.3 GHz due to the limited linear region. To improve the detection ability of LCMWs, which is inversely proportional to the bandwidth, another method based on the intrinsic P1 oscillation state of the MI-MISL was proposed [56]. This method also takes the advantage of the changes in the relationship between the injection current imposed on the front DFB section and the generated microwave signal frequency in essence. After optical-to-electrical conversion in photodetector, the generated LCMW reaching up to a bandwidth of 6.7 GHz together with a temporal duration of 77 μs can be realized by injecting optimized sweeping signal, whose bending tendency is compensated with the help of inverse fitting. As Figure 7.15a shows, the corresponding TBWP achieved as high as 5.159×10^5. The excellent results give the credit to the advantage of P1 state in large tunability in frequency. However, the phase noise of the generated LCMW by above two approaches is supposed to be improved further. Fourier-domain mode locking (FDML) technique is considered as the best candidate to narrow the linewidth and boost the signal quality to obtain significant arbitrary waveforms we desired, which has been successfully applied in the discrete devices [110, 111]. In addition, the bandwidth of the LCMW requires a further increasement for the early realization in future radar systems.

Besides linearly chirped microwave signals, triangular microwave signal is also extremely attractive in radar and electronic warfare, for it can not only measure the relative distance of the target but also can judge the relative velocity of the object.

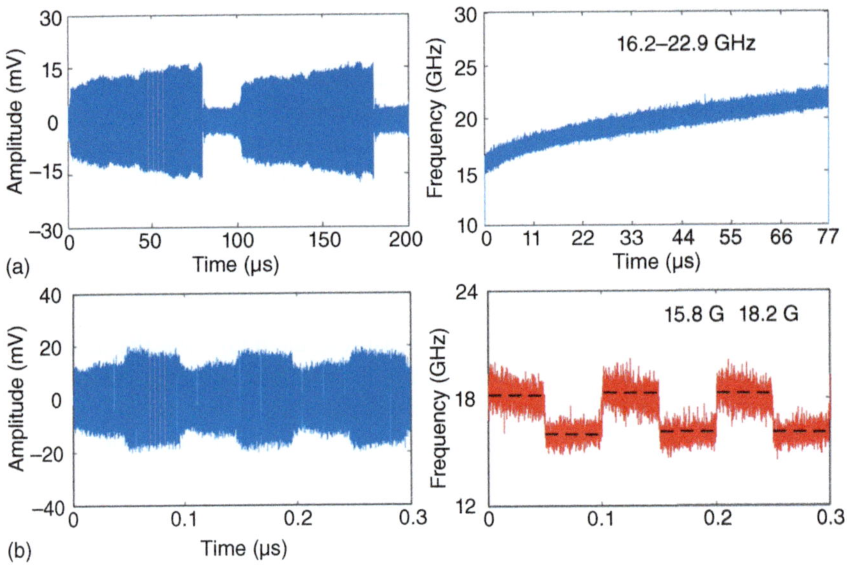

Figure 7.15 (a) Measured LCMW in one period as well as the calculated instantaneous frequency. Source: Li et al. [56]. (b) Measured two-mode FH microwave signal. Source: Adapted from Li et al. [113].

One triangular wave signal was experimentally achieved in [109] when the injection sweeping current was triangular wave signal. The adopted principle is just the same as that of the first method of generating an LCMW mentioned before. In the meantime, based on the P1 principle, the F-H microwave waveforms generation can be achieved successfully as well, which is widely used in military radar systems for its strong anti-jamming capability [112, 113]. Two or more frequency modes with a hopping rate of 1 MHz have been reported in [113] based on the integration mode and one two-mode F-H signal is shown in Figure 7.15b. Furthermore, it is convinced that utilizing the two methods referred before can also achieve perfect F-H microwave signal as well as triangular microwave signal with only need to alter the pattern of injection current. More importantly, there are also reasonable grounds to believe that more complex arbitrary waveform generation can be available without any difficulty.

7.4.3 High-Performance Microwave Signal Generation Optimizing Technique

OEOs, which can generate high-quality microwave signals with low phase noises and narrow linewidths, have attracted lots of attention among photonic techniques of microwave generation. Conventionally, OEOs are incorporated with one continuous-wave laser source, an external modulator, a long-delay-fiber loop, RF amplifiers, as well as an electronic band-pass filter (EBPF). The quality of the generated microwave signal is dependent on the fiber length, and the oscillator frequency of the generated microwave signal is determined by the EBPF. However, due to the existence of external modulators, most of OEOs are suffering from high costs and complex configurations. Moreover, an RF amplifier with a gain of at least 60 dB is typically necessary to compensate for the link losses including the optical-to-electrical and electrical-to-optical conversion losses and the fiber propagation loss in the feedback loop. Only while the loop gain is larger than all of the losses, the loop begin to oscillate, and eventually, a low-phase-noise microwave signal will be generated. In addition, the frequency tuning property is limited heavily by the EBPF.

Recently, simple OEOs are proposed to generate widely tunable microwave signals with superior performance by utilizing directly modulated MI-MISLs to replace the synergistic functions of both laser sources and external modulators. In 2012, an 8-GHz OEO signal with a phase noise of −108 dBc Hz^{-1} at a 5-kHz offset based on the self-pulsation state was reported successfully [114]. The OEO configuration is greatly simplified for the reason that the need for the external modulator is eliminated. Additionally, due to the advantage of enhanced laser modulation response around the relaxation resonance frequency subject to monolithic optical injection, the RF threshold gain for loop oscillations and the requirement of RF amplifier are both greatly reduced simultaneously. One should be noted that the beating of the two main oscillating optical modes will greatly promote the initial oscillation of the OEO, while in the traditional OEOs, the oscillation is built up from the lower transient noise in the loop constantly [115]. However, this scheme is still equipped with an EBPF and thus the output microwave signal has only a single-tone frequency, which limits the tuning range of the generated microwave signals largely. More importantly, when considering high frequencies, it is not easy to find appropriate EBPFs with sufficiently narrow 3-dB bandwidth and wide frequency tuning range to define the desired frequencies. In addition, there was a ramp-up of the phase noise at multiples of around 20-kHz offset in the spectrum image as a result of the side modes from the free spectral range of the 10-km-long fiber loop. To further improve the performance of the OEO, another simple frequency-tunable OEO based on an integrated multi-section DFB semiconductor laser with the help of dual-loop structure was proposed [116]. By adding an extra optical-to-electronic loop of different fiber length, one high-performance microwave signal at the frequency of 20.3 GHz was generated through vernier caliper effect, whose SSB phase noise of the generated signal at the frequency offset of 10 and 1 kHz were −115.3 and −92.9 dBc Hz^{-1}, respectively. Here, the two parallel fiber loops with different lengths were used to suppress the unwanted side modes, as Figure 7.16a depicts. Longer fiber can result in a high Q-factor, leading to a smaller linewidth of the laser, and the shorter fiber loop was adopted to filter the desired mode to guarantee the single-mode oscillation. The SMSR was 60.94 dB, which had a 40-dB improvement compared with a single loop. In this method, a built-in MPF based on the sideband injection locking effect of the rear and front section lasers under mutual injection was adopted to replace the EBPF without no additional loss except for the coupling loss, making the structure simpler and more flexible besides the direct modulation. The central frequency of MPF can be tuned electrically through current tuning of the MI-MISL. Thus, microwave signals with diverse frequencies can be achieved easily only by adjusting the bias currents of two DFB laser sections. Limited from the electrical amplifier, the

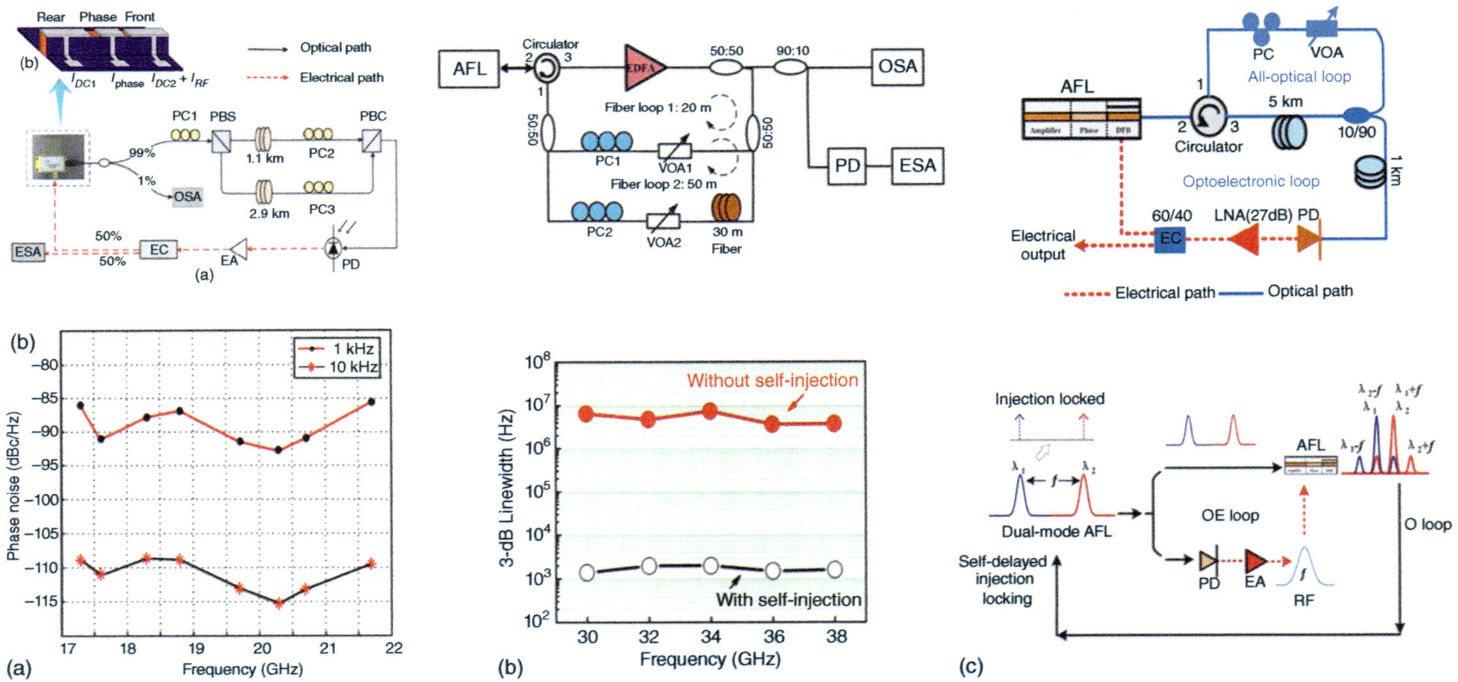

Figure 7.16 Different schematic diagrams with optical-to-electrical/all-optical feedback loops as well as their outstanding results, or the principle (a) Source: Zhang et al. [116]/Optica Publishing Group, (b) Source: Pan et al. [118], (c) Source: Adapted from Pan et al. [133].

OEO was tuned ranging from 17.3 to 21.7 GHz. In addition, the phase noise is heavily restricted for using two cascaded power amplifiers with high noise figure and it can be foreseeable that the phase noise will get further improved if the low-noise-figure electrical amplifier is used. However, there are two deficiencies in this OEO scheme. First, two polarization beam splitters and three polarization controllers are required to make sure the polarization alignment due to the utilization of dual fiber feedback loop. Second, the link loss is increased with the increasement of the total fiber length, which will make an unreasonable demand upon the RF amplifier.

Besides the optical-to-electronic loop, the method of improving the quality of the generated microwave signal through increasing an additional all-optical feedback loop is also attractive for reducing the linewidth of each oscillating wavelength and, in the meantime, avoiding the conversion loss between optical domain and electric domain. A photonic microwave generation scheme based on monolithic dual-wavelength AFL under single all-optical feedback loop has been reported. Optical microwave generation tunable from 29 to 37.6 GHz with a linewidth reduced from 3.7–7.5 MHz to 1.8–4.0 kHz was experimentally demonstrated [117]. However, a narrow-bandpass microwave filter is required to achieve single-mode output. To further compress the linewidth of each wavelength of the AFL and suppress the unwanted side modes accompanied with a single external fiber loop, Pan et al. proposed the dual optical feedback loops scheme, as shown in Figure 7.16b, in which the two fiber loops functioned as the external cavity. In the tuning range from 30 to 38 GHz, all of the linewidth was below 2 kHz, indicating more than 3 orders of magnitude reduction compared with free-running state [118]. Though the significant linewidth narrowing characteristics, the all-optical feedback approach is still troublesome for the real application due to the following imperfections. Firstly, polarization controllers are necessary in each all-optical feedback loop to adjust the polarization state of the signal injected to the AFL, largely increasing the complexity of systems. Secondly, due to the lack of synchronization between the two oscillating wavelengths, the beating microwave signal will fluctuate with time during the whole operation. For cases where stability is a priority consideration, the all-optical feedback loop scheme is not suitable. To make the two wavelengths stay in phase with each other and increase the phase correlation between the two beating optical modes, a widely tunable OEO based on a self-injection-locked monolithic dual-mode AFL has been proposed and the schematic diagram is shown in Figure 7.16c [133]. Roughly same as above scheme [115], the directly modulated AFL functions not only as a seeding laser source but also can be equivalently recognized as an active tunable MPF, an optical amplifier, and an intensity modulator for the OEO, and therefore, it is obvious that neither a high-speed external modulator nor a tunable EBPF is required. By sending the modulated lights back to the directly modulated AFL, two kinds of injection locking will happen inside the AFL. One is the self-injection locking of an individual mode by its own fiber-delayed replica, and another is the mutual injection locking between the laser emitting modes and subsequently injected modulation sidebands. With the optical loop closed, the combination of two kinds of effects will result in two coherence laser modes oscillating in AFL with narrowed linewidth down to kHz and finally lead to a much better beating signal than conventional OEO structures with one optical–electrical loop. Moreover, by taking advantage of the PPR effect, the beating frequency will be far beyond the intrinsic modulation bandwidth of the laser itself. Therefore, stable microwave signals that were continuously tunable from 28 to 41 GHz with SSB phase noise below $-106\,\text{dBc}\,\text{Hz}^{-1}$ @10 kHz and threshold electronic gain of no more than 27 dB were obtained by tuning the injection currents of the laser sections [133]. The minimum phase noise of $-115.93\,\text{dBc}\,\text{Hz}^{-1}$ was also achieved at 33.5-GHz frequency.

Though adopting extra optical-to-electronic or optical-to-optical feedback loop can greatly improve the quality of generated microwave waveforms, it should be noted that this original multimode operation stems from MB-SPs state or P1 oscillation dynamic in the MI-MSSLs at least in the current programs. Thus, optical-to-electronic or optical-to-optical feedback loop can be only considered as one of optimizing measures of overcoming original schemes' limitation in linewidth at MHz level. High-quality microwave signals with kHz-scale linewidth will be achieved through continuous cycle oscillation in the feedback loop.

It is a pity to find that until now, the investigations concerning the theoretical analysis of the phase noise characteristics of the generated microwave waveforms have not been reported. The phase noise of the microwave waveforms signal is extremely important in deciding whether it can be successfully applied. It is supposed to be the key question of the next research. Though the optimizing method has been utilized to promote the phase noise performances of the generated microwave signals in optical approaches, there still exists a big gap compared with that generated in electronic methods. As we know, high-quality microwave signals generated by OEwaves can already reach at least $-140\,\text{dBc}\,\text{Hz}^{-1}$ at 10 kHz offset by traditional electronic method. Merely, the carrier frequency is only tuned from 10 to 12 GHz and the electronic bottlenecks remain a problem hard to jump. Thus, the advantages of MWP are quite obvious when at high frequencies. According to the experimental results reported, many factors affect phase noise. First, no matter whether in MI-MISLs or MI-OFLs, the coupling/feedback strength and the coupling/feedback phase should be adjusted carefully, in which different combinations would lead to different phase noise performances just like in discrete systems. Especially in MI-MISLs, phase synchronization is supposed to be stressed. Although there is some correlation due to the coexistence in one integrated chip, it is not completely synchronous. Second, the RIN and the phase noise of the laser itself are also non-negligible factors, which are influenced by the laser design parameters.

7.5 Microwave Photonic Filter (MPF)

Besides the photonic generation of microwave waveforms, frequency-tunable and high-selectivity microwave filter is another key function device in the MWP systems. Conventionally, to achieve any microwave frequency we desired and dislodge undesired clutter, an EBPF is of the essence. However, with the increasement of the favorite frequency, continuing to utilize EBPFs to pick out microwave signals means higher central frequency, leading to more expensive cost. In addition, the tunability

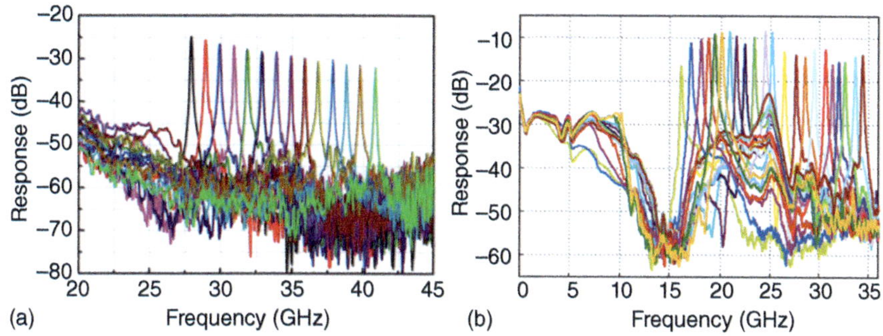

Figure 7.17 Experimentally measured microwave photonic filters (a) Source: Pan et al. [133], (b) Source: Zhang et al. [116].

of the whole MWP system is degraded once the EBPF is fixed. The method of utilizing photonic technique to realize filter function is an attractive orientation for a large tunable range. Within the scope of research on MI-MSSLs, there are two main built-in MPF generation techniques, according to their diverse principles at present.

One built-in MPF is realized by taking advantage of the PPR effect, in which the central frequency is defined by the beating frequency [133]. Due to the PPR oscillation, the modulation response will exhibit strong resonance frequency and resonance peak, and hence when the MI-MSSL is modulated around the frequency at the resonance peak point, higher modulation efficiency will be achieved, corresponding to a higher filter frequency response at the frequency near relaxation resonance frequency compared with other frequency signals. Since the phenomenon only happens for those frequencies around the resonance frequency, a corresponding single-passband MPF with a 3-dB bandwidth of MHz level can be obtained, whose amplitude response shape strictly follows the modulation response curve and the homologous central frequency is dependent on the resonance peak frequency. Moreover, as inferred before, the modulation response resulting from the PPR effect can be changeable when the currents imposed on the multi-sections are altered. There has been reported that one tunable MPF with the central frequency ranging from 28 to 41 GHz just by injecting different combinations of currents into the laser sections, as Figure 7.17a shows.

Another photonic method of generating the MPF is based on the sideband injection locking process of two DFB laser sections in MI-MISLs [116]. As mentioned before, the cavity modes of the two DFB laser sections will move to the longer wavelength, respectively, and oscillate at their new red-shifted cavity resonance frequencies due to the anti-guidance effect of the semiconductor laser gain medium with the increasing currents. Meanwhile, in the vicinity of new red-shifted cavity frequencies, there will appear an ultra-narrow gain spectrum within the region of the injection locking and the pulling bandwidth. In this area, the sideband would get a resonance amplification from the injecting mode due to the ultra-narrow gain spectrum. Thus, when the external RF signals are modulated to the front DFB section and the weak +1st sidemode of the optically modulated signal exactly right fall into the range of the gain spectrum by tuning the corresponding current, the +1st sidemode will get amplified and at the same time, undesired sidemodes are set outside the bandwidth of the gain spectrum. Subsequently, the red-shifted rear section cavity mode will be locked by the amplified sidemode, attributing to the small frequency difference. Because the phenomenon only happens for those frequencies falling in the gain spectrum, a corresponding single-passband MPF can be generated afterward. The amplitude response of the MPF exhibits the same shape as the gain spectrum, which can be tuned by adjusting the injection currents of active DFB lasers. The experimental results, which are shown in Figure 7.17b, reveal that one MPF with a large frequency tuning range of 16.9–34.7 GHz is generated based on the mutually injected three-section laser, indicating about 20-GHz tuning bandwidth. Moreover, the 3-dB bandwidth of the MPF at different central frequencies is excellently limited to around 10 MHz ranging from 6 to 18 MHz.

These methods to generate MPFs are based on the characteristics of MI-MSSLs themselves, suggesting novel photonic measures to realize the function of MPF with extensive simplicity and flexibility.

7.6 Laser Arrays

WDM-ROF networks, which combine advantages of ROF link and the WDM technique, can efficiently improve the carrying capacity of wireless access networks and thus have become an ideal wired platform for the development of high-speed and large-capacity information systems. While in WDM-ROF networks, large-scale laser arrays such as DFB multi-wavelength semiconductor laser arrays (DFB-MW-SLAs) are considered as the predominant light sources for their notable figure-of-merits of small size, low energy consumption, high reliability, and so on compared with several discrete laser sources. They are accomplished by integrating multiple lasers with different wavelengths onto a single chip. Hence, multi-photonic microwave transmission links can be achieved by only requiring one laser array. However, in spite of their tremendous potential, the implementation of DFB-MW-SLAs has been hampered by a number of issues, particularly a poor wavelength precision, and a low SLM yield because the production of the laser array will rapidly decrease as the channel number in one laser array gradually increases [51, 52, 57]. Thus, both of them are essential parameters, which are supposed to be highly guaranteed, or the entire laser array will break down. Usually, phase-shift gratings with antireflection/antireflection (AR/AR) coatings as well as the REC fabrication technique are employed to guarantee DFB-MW-SLAs precise wavelengths and great SLM yields. Thus, high reproducibility can be obtained at a remarkably low fabrication cost. One packaged eight-channel directly modulated DFB-MLA transmitter module in [134] has been successfully reported. For the corresponding small-signal modulation response of each laser in the laser array. It can be observed that for all eight lasers, the 3-dB modulation bandwidth is approximately 10 GHz when setting the whole biasing currents of directly modulated lasers to be 55 mA, indicating excellent operation for multi-wavelength analog ROF link transmission. Nevertheless, it is difficult to obtain high-efficiency working state due to the symmetrical laser structures inducing 50% optical output. HR coatings are often applied to one of the facets to break the balance and obtain high efficiency operation. However, the wavelength precision as well as the SLM yield will be deteriorated unavoidably with the cleavage process, which attributes to the random phase change of the HR-coated facet [135]. Changing the position of phase shift to the side of the output port is another option to advance the optical output efficiency, but the movement distance is limited to a narrow range due to the degraded SMSR as the asymmetry increases. Hence, simple-effective methods to fabricate DFB-MW-SLAs with improved wavelength accuracy, good SLM performance as well as high output efficiency are highly desired.

On account of the great significant performance improvement from the multi-section DFB laser, for further optimization of the characteristics of integrated laser arrays, a novel approach of utilizing MSSLs instead of conventional single-segment DFB semiconductor lasers to be integrated onto a single chip is a bright alternative. A four-channel DFB-MLA integrated with passive grating reflectors has been reported in [52]. Due to the increasement of additional GR section, the symmetrical structure of each multi-section laser is naturally broken and an improved laser output efficiency can be ensured without changing the pattern of DFB laser section. Meanwhile, as a result of the fact that the reflectivity of the GR for the main oscillating mode is much higher than that of the other side modes, the SMSR will be promoted dramatically and the lasing wavelength can be controlled with high accuracy. As Figure 7.18 shows, after incorporating a grating reflector, 40% increase of slope efficiency, about 10 mA decrease of threshold current, and 7 dB increase of SMSR are achieved. Compared to the design value of 3.2 nm between the adjacent lasers, the average error and the maximum deviation of wavelength spacing are separately 0.027 and 0.07 nm. Moreover, the SMSRs of all the lasers are higher than 60 dB. Furthermore, in order to study the dynamic characteristics of the two-section DFB laser array in one WDM-ROF link, the laser array is packaged using a commercial 10-GHz butterfly housing. Compared with conventional one-section DFB lasers, the 3-dB modulation bandwidth of the two-section DFB lasers is 1.9 GHz larger than that of the one-segment DFB lasers, indicating an improved modulation bandwidth. Moreover, the modulation response curves are much flatter, and the relaxation resonance peaks are suppressed significantly under different control currents, which is of great benefit to WDM-ROF transmission systems. In addition, SFDR is also considered. Two-tone frequency signals of 10 and 10.02 GHz are generated to modulate the DFB lasers. The experimental result clearly shows the SFDR of the two-section DFB laser and the conventional single-section DFB laser are 92 and 87 dB $Hz^{2/3}$, respectively, indicating a better performance in terms of suppressing the nonlinear distortion. Finally, one 50-MSymbol s^{-1} 64-QAM signal accompanied by a 10 GHz carrier is modulated on one of the lasers in the laser array and the corresponding schematic is just like that shown in Figure 7.8a. After a 40-km SMF transmission, the average EVM of the received signal is 2.97%, indicating an excellent constellation diagram. These results manifest that good characteristics can be achieved in the four-channel DFB-MLA integrated with passive grating reflectors, which may provide new ideas

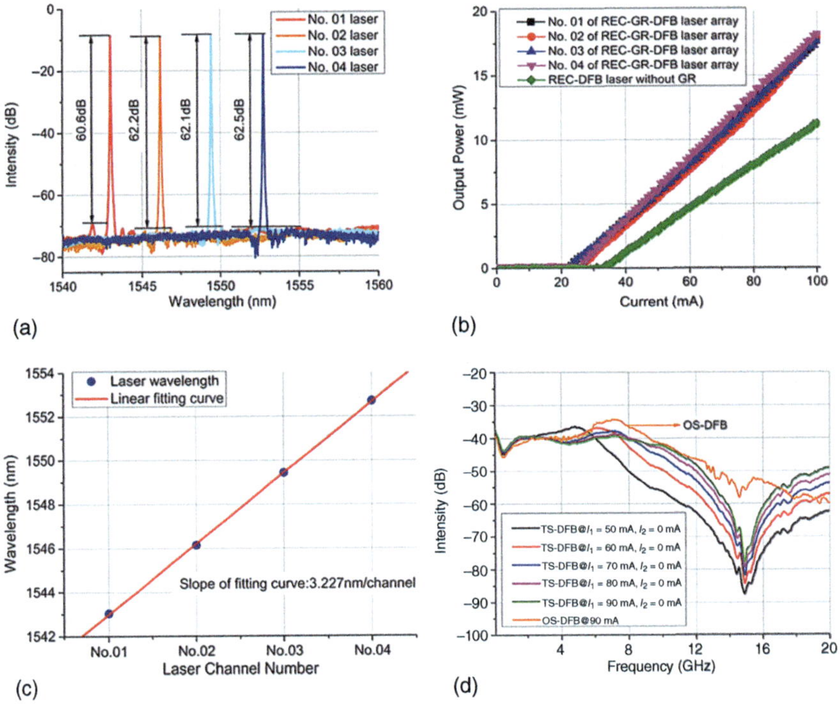

Figure 7.18 Improved characteristics of the four-channel TS-DFB laser, (a) SMSR, (b) *P*–*I* curve, (c) wavelength spacing, (d) frequency response, (e) schematic of ROF link using the TS-DFB laser, (f) analysis of the received signal when the injection current of the TS-DFB laser.

and methods for solving the drawbacks in existing schemes. More importantly, it is convinced that the analog performances of the two-section DFB laser array can be improved greatly by optimizing the key parameters in each laser, such as the cavity length, grating and waveguide structure, package, and epitaxial material.

It is also found that wavelength spacing accuracy can be further optimized by tuning the injection current of the GR section [51, 136]. Due to the thermal crosstalk between the adjacent lasers in a laser array, when changing the control current of the No. 1 laser of the four-channel two-section DFB laser array, not only the wavelength of the No. 1 laser, but also the oscillating wavelength of the adjacent section is red-shift as Figure 7.19 depicts. The internal mechanism is that the extra heat produced by the increasing current of the GR section of the No. 1 laser changes the temperature of the adjacent laser. Consequently, the refractive index of the adjacent laser is varied, resulting in the variation of the lasing wavelength. Meanwhile, precisely because of the difference in slope efficiencies, the wavelength spacing accuracy can be altered slightly by changing the control current of the adjacent laser, which is beneficial to further improve the wavelength spacing accuracy. In the proposed laser arrays, the maximum wavelength residual reduction to 0.096 nm is realized.

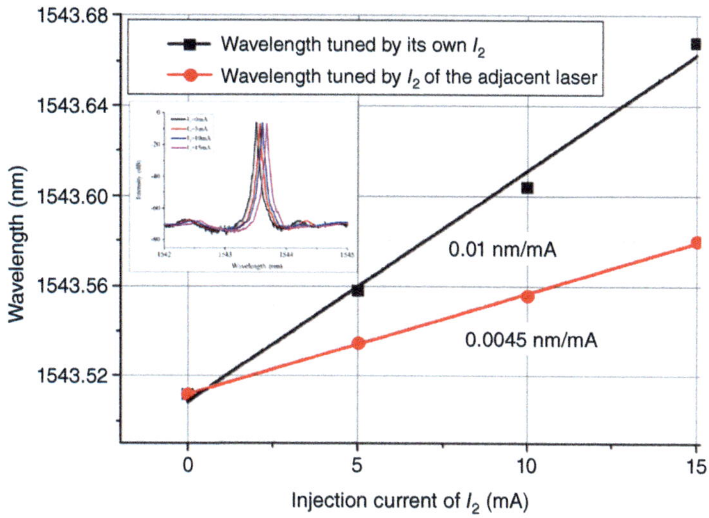

Figure 7.19 The wavelengths of the No. 01 laser in the four-channel TS-DFB-LAs as a function of its own I_2 and I_2 of the adjacent laser.

7.7 Conclusion

In this chapter, the three main MI-MSSLs types, including AFLs, PFLs, and MI-MISLs, are described. The advantages and restrictions of each laser structure are identified and discussed. Their key properties relevant to practice applications, such as high-efficiency electro-optic conversion characteristics, photonic microwave signal generation, frequency-variable MPFs, and the DFB-MW-SLAs for WDM-ROF systems, are demonstrated as well.

The focus of this chapter is on modulation properties enhancement utilizing the PPR effect. Lots of experiments cited in Section 7.3 have reported that the modulation bandwidth based on the PPR frequency can be enhanced up to a range of 30–40 GHz, far beyond three times that determined by CPR frequency in the conventional single-section semiconductor lasers. Moreover, the nonlinearity reduction and chirp suppression are discussed. Another topic of this chapter is the photonic microwave generation based on the nonlinear dynamics of MI-MSSLs, mainly consisting of four dynamic states, namely free-running state, MB-SPs state, P1 oscillation state, as well as sideband injection locking. The photonic approaches to generating tunable microwave signals utilizing the MI-MSSLs are beneficial from compact devices and flexible control when compared with conventional single-segment semiconductor lasers. Furthermore, common radar waveforms such as F-H signals, LCMWs, as well as triangular microwave waveforms are discussed by injecting the corresponding control currents with specific patterns into the MI-MSSLs. Last but not least, extra optical-to-electronic or all-optical feedback loops are described to further decrease the phase noise and significantly improve the quality of the generated microwave signal. Considering the characteristics of the nonlinear dynamics of MI-MSSLs and the urgent demands required for potential applications, several future prospects are forecasted as follows:

(1) To further improve the laser packaging technology and advance the electrical isolation between the adjacent sections in MI-MSSLs are sophisticated problems, especially in DFB-MW-SLAs, which have a great effect on the performance of the direct modulated laser module. Usually, to ensure the MI-MSSLs convenient use and inhibit oxidation of aluminum surface layer, MI-MSSLs are usually packaged in the form of butterfly. However, laser characteristics, especially direct modulation response, are limited and heavily subject to RC parasitic. Hence, for future wide-bandwidth application scenarios, especially in WDM-ROF networks, a more excellent laser packaging technology is suggested. On the other hand, excellent electrical isolation is of great significance in deciding the independent work characteristics of each section in MI-MSSLs, but in the actual manufacturing process of the device, the isolation resistance between the adjacent areas is not infinite, leading to an unavoidable offset current leakage. More importantly, the leak current at mA level is high enough to change the working state of the device. Therefore, striving to improve the isolation resistance between two areas is also an urging problem to be addressed.
(2) To promote the MI-MSSLs from laboratory research stage to being commercially available is need to be further explored. Many aspects attribute to this consequence. The first, MI-MSSLs, sharing the same laser structure, but with different section lengths could present specific characteristics. Hence, it is an unavoidable problem about how to balance multi properties (cavity length, materials of laser, etc.) of MI-MSSLs and make the produced lasers to be excellent as far as possible for the different scene demands. The second, limited to the high precision requirements throughout the entire fabrication process, it is difficult to achieve the MI-MSSLs with expected parameters such as the SLM operation, suffering from a low yield. Luckily, one new manufacturing technology of waveguide grating, i.e. REC technique, which can achieve the fine structure of nanometer level resolution by the large micron grade operation is proposed and is recognized to effectively alleviate this problem [51, 52, 56, 57, 64, 116]. Attributing to its superiority in reducing the high requirement in process dimension, it is an important method to fabricate complex grating structure nowadays. Therefore, the frequency stability and repeatability of microwave signals can be ensured. Thirdly, though we can achieve plenty of nonlinear dynamics in using well-designed MI-MSSLs, there still exists an undeniable fact that people do not have a complete and constant verdict concerning how to accurately control the multi-section lasers to work in exactly right dynamic states we desired by tuning relevant variables including drive currents as well as working temperature to concrete values.
(3) To further understand, study, and make full use of the nonlinear dynamics of MI-MSSLs and expand their applications in the future, IMWP is a very important aspect. Despite a large range of accomplishments that have been realized based on the discrete devices, extremely limited applications have

been achieved based on the MI-MSSLs due to the limited depth of integration, fabrication process, cognitive deficit, and so on. A large range of MWP functions, such as up converter, channelized reception, and frequency measurement, is still in the blanks and can be worthy of being studied. In addition, with the development of 6G research as well as the consumer demands for bandwidth-intensive applications, the photonic microwave frequency is expected to be extended into the THz band and the ROF link is expected to be replaced with the THz over fiber (TOF) link. Novel photonic microwave techniques, which can efficiently and effectively generate and transmit high frequencies into the THz range based on the MI-MSSLs are also an important area of research.

(4) To effectively integrate the laser chips with other optical chips, such as silicon chips, would be a challenging and important work. For specific applications, mature PICs are more attractive toward the future IMWP because of their higher integration and more stability. For instance, HI-Q™ OEOs patented by OEWAVE company can generate microwave-to-millimeter signals with the same low-spectral purity at all RF frequencies, whose size is only analogous to a coin. However, in some applications, such as optical link transmission networks, especially in WDM-ROF systems, MI-MSSLs are the better choices as light sources. In the future, how to tie these two technologies together to realize more perfect photonic microwave functions is a promising area of future research.

Funding Information

National Natural Science Foundation of China (NSFC) (61974165).

Disclosures

The authors declare no conflicts of interest.

References

1 Capmany, J. and Novak, D. (2007). Microwave photonics combines two worlds. *Nat. Photonics* 1: 319–330.

2 Seeds, A.J. and Williams, K.J. (2007). Microwave photonics. *J. Lightwave Technol.* 24: 4628–4641.

3 Yao, J. (2009). Microwave photonics. *J. Lightwave Technol.* 27: 314–335.

4 Li, M. and Zhu, N. (2016). Recent advances in microwave photonics. *Front. Optoelectron.* 9: 160–185.

5 Marpaung, D., Roeloffzen, C., Heideman, R. et al. (2013). Integrated microwave photonics. *Laser Photonics Rev.* 7: 506–538.

6 Marpaung, D., Burla, M., and Capmany, J. (2018). Editorial special issue on advances in integrated microwave photonics. *IEEE Photonics Technol. Lett.* 30: 1813.

7 Zou, X., Zou, F., Cao, Z. et al. (2019). A multifunctional photonic integrated circuit for diverse microwave signal generation, transmission, and processing. *Laser Photonics Rev.* 13: 1800240.

8 Chow, W., Koch, S.W., and Sargent, M. (1994). *Semiconductor-Laser Physics*. Berlin: Springer.

9 Ohtsubo, J. (2017). *Semiconductor Lasers and Theory*. Berlin: Springer.

10 Simpson, T.B. and Liu, J.M. (1997). Enhanced modulation bandwidth in injection-locked semiconductor lasers. *IEEE Photonics Technol. Lett.* 9: 1322–1324.

11 Lau, E.K., Wong, L., and Wu, M.C. (2009). Enhanced modulation characteristics of optical injection-locked lasers: a tutorial. *IEEE J. Sel. Top. Quantum Electron.* 15: 618–633.

12 Hwang, S.K., Liu, J.M., and White, J.K. (2004). Characteristics of period-one oscillations in semiconductor lasers subject to optical injection. *IEEE J. Sel. Top. Quantum Electron.* 10: 974–981.

13 Hwang, S.K. and Liang, D.H. (2006). Effects of linewidth enhancement factor on period-one oscillations of optically injected semiconductor lasers. *Appl. Phys. Lett.* 89: 061120.

14 Lee, C.H. and Shin, S.Y. (1993). Self-pulsing, spectral bistability, and chaos in a semiconductor laser diode with optoelectronic feedback. *Appl. Phys. Lett.* 62: 922–924.

15 Tang, S. and Liu, J.M. (2001). Chaotic pulsing and quasi-periodic route to chaos in a semiconductor laser with delayed opto-electronic feedback. *IEEE J. Quantum Electron.* 37: 329–336.

16 Zhao, Y.G. (1992). Simulation studies of frequency locking, quasi-periodicity, subharmonic bifurcation, and chaos in modulated semiconductor lasers. *IEEE J. Quantum Electron.* 28: 2009–2015.

17 Lin, F.Y. and Liu, J.M. (2002). Harmonic frequency locking in a semiconductor laser with delayed negative optoelectronic feedback. *Appl. Phys. Lett.* 81: 3128–3130.

18 Kovanis, V., Gavrielides, A., Simpson, T.B., and Liu, J.M. (1995). Instabilities and chaos in optically injected semiconductor-lasers. *Appl. Phys. Lett.* 67: 2780–2782.

19 Eriksson, S. and Lindberg, A.M. (2001). Periodic oscillation within the chaotic region in a semiconductor laser subjected to external optical injection. *Opt. Lett.* 26: 142–144.

20 Zhu, N.H., Shi, Z., Zhang, Z. et al. (2018). Directly modulated semiconductor lasers. *IEEE J. Sel. Top. Quantum Electron.* 24: 1–19.

21 Mogensen, F., Olesen, H., and Jacobsen, G. (1985). Locking conditions and stability properties for a semiconductor laser with external light injection. *IEEE J. Quantum Electron.* 21: 784–793.

22 Simpson, T.B., Liu, J.M., and Gavrielides, A. (1995). Bandwidth enhancement and broadband noise reduction in injection-locked semiconductor lasers. *IEEE Photonics Technol. Lett.* 7: 709–711.

23 Lin, C. and Mengel, F. (1984). Reduction of frequency chirping and dynamic linewidth in high-speed directly modulated semiconductor lasers by injection locking. *IEEE Electron Device Lett.* 20: 1073–1075.

24 Iwashita, K. and Nakagawa, K. (1982). Suppression of mode partition noise by laser diode light injection. *IEEE Trans. Microwave Theory Tech.* 82: 1657–1662.

25 Mogensen, F., Olesen, H., and Jacobsen, G. (1985). FM noise suppression and linewidth reduction in an injection-locked semiconductor laser. *IEEE Electron Device Lett.* 21: 696–697.

26 Chrostowski, L., Chang, C.H., and Chang-Hasnain, C. (2003). Reduction of relative intensity noise and improvement of spur-free dynamic range of an injection locked VCSEL. In: *The 16th Annual Meeting of the IEEE Lasers and Electro-Optics Society*, Tucson, AZ, USA (27–28 October 2003), 706–707. Piscataway, NJ, USA: IEEE.

27 Meng, X.J., Chau, T., and Wu, M.C. (1999). Improved intrinsic dynamic distortions in directly modulated semiconductor lasers by optical injection locking. *IEEE Trans. Microwave Theory Tech.* 47: 1172–1176.

28 Wang, J., Haldar, M.K., Li, L., and Mendis, F.V.C. (1996). Enhancement of modulation bandwidth of laser diodes by injection locking. *IEEE Photonics Technol. Lett.* 8: 34–36.

29 Simpson, T.B., Liu, J.M., Huang, K.F. et al. (1995). Cavity enhancement of resonant frequencies in semiconductor lasers subject to optical injection. *Phys. Rev. A: At. Mol. Opt. Phys.* 52: 4348–4351.

30 Sung, H.K., Lau, E.K., and Wu, M.C. (2007). Optical properties and modulation characteristics of ultra-strong injection-locked distributed feedback lasers. *IEEE J. Sel. Top. Quantum Electron.* 13: 1215–1220.

31 Sung, H.K., Lau, E.K., and Wu, M.C. (2007). Optical single sideband modulation using strong optical injection-locked semiconductor lasers. *IEEE Photonics Technol. Lett.* 19: 1005–1007.

32 Hong, C., Zhang, C., Li, M. et al. (2010). Single-sideband modulation based on an injection-locked DFB laser in radio-over-fiber systems. *IEEE Photonics Technol. Lett.* 22: 462–464.

33 Xiong, J., Wang, R., Pu, T. et al. (2013). A novel approach to realizing SSB modulation with optimum optical carrier to sideband ratio. *IEEE Photonics Technol. Lett.* 25: 1114–1117.

34 Chan, S., Hwang, S., and Liu, J. (2007). Period-one oscillation for photonic microwave transmission using an optically injected semiconductor laser. *Opt. Express* 15: 14921–14935.

35 Sung, H. (2006). *Strong Optical Injection Locking of Edge-Emitting Lasers and Its Applications*, vol. 158. University of California.

36 Jung, T., Liou, K., Sung, H., et al. (2005). Semiconductor injection locked lasers and method. US 2005/0013337 A1. filed 27 May 2004 and issued 20 January 2005.

37 Lau, E.K., Zhao, X., Sung, H. et al. (2008). Strong optical injection-locked semiconductor lasers demonstrating >100-GHz resonance frequencies and 80-GHz intrinsic bandwidths. *Opt. Express* 16: 6609–6618.

38 Qi, X. and Liu, J. (2011). Photonic microwave applications of the dynamics of semiconductor lasers. *IEEE J. Sel. Top. Quantum Electron.* 17: 1198–1211.

39 Chow, W.W., Yang, Z.S., Vawter, G.A., and Skogen, E.J. (2009). Modulation response improvement with isolator-free injection-locking. *IEEE Photonics Technol. Lett.* 21: 839–841.

40 Kreissl, J., Vercesi, V., Troppenz, U. et al. (2012). Up to 40 Gb/s directly modulated laser operating at low driving current: buried-heterostructure passive feedback laser (BH-PFL). *IEEE Photonics Technol. Lett.* 24: 362–364.

41 Matsuda, M., Uetake, A., Simoyama, T. et al. (2015). 1.3-μm-wavelength AlGaInAs multiple-quantum-well semi-insulating buried-heterostructure distributed-reflector laser arrays on semi-insulating InP substrate. *IEEE J. Sel. Top. Quantum Electron.* 21: 241–247.

42 Hou, L., Tan, M., Haji, M. et al. (2013). EML based on side-wall grating and identical epitaxial layer scheme. *IEEE Photonics Technol. Lett.* 25: 1169–1172.

43 Akram, M.N., Silfvenius, C., Kjeborn, O., and Schatz, R. (2004). Design optimization of InGaAsP–InGaAlAs 1.55 μm strain-compensated MQW lasers for direct modulation applications. *Semicond. Sci. Technol.* 19: 615–625.

44 Reithmaier, J., Kaiser, W., Bach, L. et al. (2005). Modulation speed enhancement by coupling to higher order resonances: a road towards 40 GHz bandwidth lasers on InP. In: *International Conference on Indium Phosphide and Related Materials*, Glasgow, UK (08–12 May 2005), 118–123. Piscataway, NJ, USA: IEEE.

45 Radziunas, M., Glitzky, A., Bandelow, U. et al. (2007). Improving the modulation bandwidth in semiconductor lasers by passive feedback. *IEEE J. Sel. Top. Quantum Electron.* 13: 136–142.

46 Kreissl, J., Troppenz, U., Rehbein, W., et al. (2008). 40 Gbit/s directly modulated passive feedback laser with complex-coupled DFB section. *International Conference on Indium Phosphide & Related Materials.* 2008 20th International Conference on Indium Phosphide and Related Materials, Versailles, France (25-29 May 2008). 1-4, IEEE, Piscataway, NJ, USA.

47 Möhrle, M. and Schell, M. (2011). 40 Gbit/s directly modulated lasers: physics and application. *Proc. SPIE Int. Soc. Opt. Eng.* 7953: 79530F-1–79530F-10.

48 Troppenz, U., Bornholdt, C., Kreissl, J. et al. (2009). 1.3 μm passive feedback laser for 28 Gb/s and 40 Gb/s transmission over uncompensated SSMF links. In: *2009 35th European Conference on Optical Communication*, Vienna, Austria (20–24 September 2009), 1–2. Piscataway, NJ, USA: IEEE.

49 Bornholdt, C., Troppenz, U., Kreissl, J. et al. (2008). 40 Gbit/s directly modulated passive feedback DFB laser for transmission over 320 km single mode fiber. In: *2008 34th European Conference on Optical Communication*, Brussels, Belgium (21–25 September 2008), 1–2. Piscataway, NJ, USA: IEEE.

50 Uusitalo, T., Virtanen, H., Bardella, P., and Dumitrescu, M. (2017). Analysis of the photon–photon resonance influence on the direct modulation bandwidth of dual-longitudinal-mode distributed feedback lasers. *Opt. Quantum Electron.* 49: 1–14.

51 Zhang, Y., Zheng, J., Shi, Y. et al. (2015). Study on two-section DFB lasers and laser arrays based on the reconstruction equivalent chirp technique and their application in radio-over-fiber systems. *IEEE J. Sel. Top. Quantum Electron.* 21: 232–240.

52 Zhang, Y., Zheng, J., Zhang, F. et al. (2015). Study on DFB semiconductor laser array integrated with grating reflector based on reconstruction-equivalent-chirp technique. *Opt. Express* 23: 2889–2894.

53 Zhao, G., Sun, J., Xi, Y. et al. (2016). Design and simulation of two-section DFB lasers with short active-section lengths. *Opt. Express* 24: 10590–10598.

54 Liu, G., Zhao, G., Sun, J. et al. (2018). Experimental demonstration of DFB lasers with active distributed reflector. *Opt. Express* 26: 29784–29795.

55 Yu, L., Guo, L., Lu, D. et al. (2015). Modulated bandwidth enhancement in an amplified feedback laser. *Chin. Opt. Lett.* 13: 051401.

56 Li, J., Pu, T., Zheng, J. et al. (2018). Photonic generation of linearly chirped microwave waveforms using a monolithic integrated three-section laser. *Opt. Express* 26: 9676–9685.

57 Zhang, Y., Zheng, J., Shi, Y., et al. (2015). High efficiency and SMSR distributed feedback laser array integrated with passive grating reflector based on reconstruction-equivalent-chirp technique. *2015 Optical Fiber Communications Conference and Exhibition*, Los Angeles, CA, USA (22–26 March 2015). IEEE, Piscataway, NJ, USA.

58 Jung, T., Sung, H., Wu, M.C., et al. (2003). Demonstration of monolithic optical injection locking using a two section DFB laser. *Conference on Lasers and Electro-Optics/Quantum Electronics and Laser Science Conference*, Technical Digest (Optica Publishing Group, 2003), paper CThC4.

59 Sung, H., Lau, E.K., Wu, M.C., et al. (2005). Large-signal analog modulation response of monolithic optical injection-locked DFB lasers. *Conference on Lasers and Electro-Optics/Quantum Electronics and Laser Science and Photonic Applications Systems Technologies*, Technical Digest (CD) (Optica Publishing Group, 2005), paper CTuV7.

60 Liu, D., Sun, C., Xiong, B., and Luo, Y. (2013). Suppression of chaos in integrated twin DFB lasers for millimeter-wave generation. *Opt. Express* 21: 2444–2451.

61 Liu, D., Sun, C., Xiong, B., and Luo, Y. (2014). Nonlinear dynamics in integrated coupled DFB lasers with ultra-short delay. *Opt. Express* 22: 5614–5622.

62 Lau, E.K., Xiaoxue, Z., Chang-Hasnain, C., and Wu, M.C. (2008). 80-GHz intrinsic 3-dB bandwidth of directly modulated semiconductor lasers under optical injection locking. In: *2008 IEEE 21st International Semiconductor Laser Conference*, Sorrento, Italy (14–18 September 2008), 171–172. Piscataway, NJ, USA: IEEE.

63 Sung, H.K., Jung, T., Wu, M.C. et al. (2003). Modulation bandwidth enhancement and nonlinear distortion suppression in directly modulated monolithic injection-locked DFB lasers. In: *International Topical Meeting on Microwave Photonics*, Budapest, Hungary (10–12 September 2003). Piscataway, NJ, USA: IEEE.

64 Zhang, Y., Li, L., Zhou, Y. et al. (2017). Modulation properties enhancement in a monolithic integrated two-section DFB laser utilizing side-mode injection locking method. *Opt. Express* 25: 27595–27608.

65 Sun, C., Liu, D., Xiong, B. et al. (2015). Modulation characteristics enhancement of monolithically integrated laser diodes under mutual injection locking. *IEEE J. Sel. Top. Quantum Electron.* 21: 628–635.

66 Karar, A.S., Rezania, M.A., Deb, N., and Cartledge, J.C. (2013). 5.3 GHz modulation passband for fiber lengths up to 100 km using a directly modulated passive feedback laser. In: *39th European Conference and Exhibition on Optical Communication*, London (22–26 September 2013), 1–3. Savoy Place, London, England: IET.

67 Browning, C., Shi, K., Latkowski, S. et al. (2011). Performance improvement of 10 Gb/s direct modulation OFDM by optical injection using monolithically integrated discrete mode lasers. *Opt. Express* 19: 289–294.

68 Chrostowski, L., Zhao, X., and Chang-Hasnain, C.J. (2006). Microwave performance of optically injection-locked VCSELs. *IEEE Trans. Microwave Theory Tech.* 54: 788–796.

69 Hong, C., Li, M., Zhang, C., et al. (2008). Millimeter-wave frequency tripling based on four-wave mixing in sideband injection locking DFB. *2008 International Conference on Microwave & Millimeter Wave Technology*, Nanjing (21–24 April 2008). IEEE, Piscataway, NJ, USA.

70 Juan, Y.S. and Lin, F.Y. (2011). Photonic generation of broadly tunable microwave signals utilizing a dual-beam optically injected semiconductor laser. *IEEE Photonics J.* 3: 644–650.

71 Zhu, H., Wang, R., Xiang, P. et al. (2017). A novel approach to photonic generate microwave signals based on optical injection locking and four-wave mixing. *Opt. Commun.* 400: 101–105.

72 Fan, L., Wu, Z., Deng, T. et al. (2014). Subharmonic microwave modulation stabilization of tunable photonic microwave generated by period-one nonlinear dynamics of an optically injected semiconductor laser. *J. Lightwave Technol.* 32: 4058–4064.

73 Suelzer, J.S., Simpson, T.B., Devgan, P., and Usechak, N.G. (2017). Tunable, low-phase-noise microwave signals from an optically injected semiconductor laser with opto-electronic feedback. *Opt. Lett.* 42: 3181–3184.

74 Hung, Y. and Hwang, S. (2015). Photonic microwave stabilization for period-one nonlinear dynamics of semiconductor lasers using optical modulation sideband injection locking. *Opt. Express* 23: 6520–6532.

75 Zhuang, J. and Chan, S. (2013). Tunable photonic microwave generation using optically injected semiconductor laser dynamics with optical feedback stabilization. *Opt. Lett.* 38: 344–346.

76 Sung, H.K., Zhao, X., Lau, E.K., and Parekh, D. (2009). Optoelectronic oscillators using direct-modulated semiconductor lasers under strong optical injection. *IEEE J. Quantum Electron.* 15: 572–577.

77 Chen, G., Lu, D., Guo, L. et al. (2018). Frequency-tunable OEO using a DFB laser at period-one oscillations with optoelectronic feedback. *IEEE Photonics Technol. Lett.* 30: 1593–1596.

78 Zhou, P., Zhang, F., Zhang, D., and Pan, S. (2018). Performance enhancement of an optically injected semiconductor laser based optoelectronic oscillator by subharmonic microwave modulation. *Opt. Lett.* 43: 5439–5442.

79 Wang, P., Xiong, J., Zhang, T. et al. (2015). Frequency tunable optoelectronic oscillator based on a directly modulated DFB semiconductor laser under optical injection. *Opt. Express* 23: 20450–20458.

80 Zhang, T., Xiong, J., Wang, P. et al. (2015). Tunable optoelectronic oscillator using FWM dynamics of an optical-injected DFB laser. *IEEE Photonics Technol. Lett.* 27: 1313–1316.

81 Zhuang, J., Li, X., Li, S., and Chan, S. (2016). Frequency-modulated microwave generation with feedback stabilization using an optically injected semiconductor laser. *Opt. Lett.* 41: 5764–5767.

82 Xie, H., Wang, L., Zhao, L. et al. (2007). Optical microwave generation using two parallel DFB lasers integrated with Y-branch waveguide coupler. *Chin. Phys.* 16: 1459.

83 Laperle, C., Svilans, M., Poirier, M., and Tetu, M. (1999). Frequency multiplication of microwave signals by sideband optical injection locking using a monolithic dual-wavelength DFB laser device. *IEEE Trans. Microwave Theory Tech.* 47: 1219–1224.

84 Huang, J., Sun, C., Xiong, B. et al. (2008). Phase noise characteristics of optically generated microwave by sideband injection locking. In: *2008 International Nano-Optoelectronics Workshop*, Shonan Village (02–15 August 2008), 301–302. Piscataway, NJ, USA: IEEE.

85 Luo, Y., Huang, J., Sun, C., and Xiong, B. (2010). Optical generation of low phase noise microwave by Y-branch integrated dual wavelength laser. In: *The 19th Annual Wireless and Optical Communications Conference*, Shanghai, China (14–15 May 2010), 1–4. Piscataway, NJ, USA: IEEE.

86 Laperle, C., Svilans, M., Poirier, M., and Tetu, M. (1997). Microwave generation with monolithic dual-wavelength DFB lasers. In: *1997 Digest of the IEEE/LEOS Summer Topical Meeting*, Montreal, QC, Canada (11–13 August 1997), 17–18. Piscataway, NJ, USA: IEEE.

87 Mohrle, M., Feiste, U., Horer, J. et al. (1992). Gigahertz self-pulsations in 1.5 μm wavelength multisection DFB lasers. *IEEE Photonics Technol. Lett.* 4: 976–979.

88 Bauer, S., Brox, O., Kreissl, J., and Sahin, G. (2002). Optical microwave source. *Electron. Lett.* 38: 334–335.

89 Lang, R. and Kobayashi, K. (1980). External optical feedback effects on semiconductor injection laser properties. *IEEE J. Quantum Electron.* 16: 347–355.

90 Brox, O., Bauer, S., Radziunas, M. et al. (2003). High-frequency pulsations in DFB lasers with amplified feedback. *IEEE J. Quantum Electron.* 39: 1381–1387.

91 Bauer, S., Brox, O., Kreissl, J. et al. (2004). Nonlinear dynamics of semiconductor lasers with active optical feedback. *Phys. Rev. E* 69: 016206.

92 Tager, A. and Petermann, K. (1994). High-frequency oscillations and self-mode locking in short external-cavity laser diodes. *IEEE J. Quantum Electron.* 30: 1553–1561.

93 Favre, F. (1987). Theoretical analysis of external optical feedback on DFB semiconductor lasers. *IEEE J. Quantum Electron.* 23: 81–88.

94 Yee, D., Leem, Y., Kim, S. et al. (2004). Loss-coupled distributed-feedback lasers with amplified optical feedback for optical microwave generation. *Opt. Lett.* 29: 2243–2245.

95 Pan, B., Guo, L., Zhang, L. et al. (2016). Widely tunable monolithic dual-mode laser for W-band photonic millimeter-wave generation and all-optical clock recovery. *Appl. Opt.* 55: 2930.

96 Pan, B., Lu, D., Yu, L. et al. (2015). Widely tunable dual-mode amplified feedback laser with mode separation covering 60-GHz band. In: *2015 14th International Conference on Optical Communications and Networks (ICOCN)*, Nanjing, China (03–05 July 2015), 1–3. Piscataway, NJ, USA: IEEE.

97 Yu, L., Lu, D., Pan, B. et al. (2014). Monolithically integrated amplified feedback lasers for high-quality microwave and broadband chaos generation. *J. Lightwave Technol.* 32: 3595–3601.

98 Huang, J., Sun, C., Song, Y. et al. (2009). Influence of master laser's lineshape on the optically generated microwave carrier by injection locking. *Appl. Phys. Express* 2: 072502.

99 Sun, Y., Pan, J., Zhao, L. et al. (2009). Widely frequency tunable amplified feedback laser as 20 GHz optical microwave source. *J. China Univ. Posts Telecommun.* 16: 50–53.

100 Hyuk, K., Jung, T., Wu, M.C. et al. (2003). Optical generation of millimeter-waves using monolithic sideband injection locking of a two-section DFB laser. In: *The 16th Annual Meeting of the IEEE Lasers and Electro-Optics Society*, Tucson, AZ, USA (27–28 October 2003), vol. 2, 1005–1006. Piscataway, NJ, USA: IEEE.

101 Huang, J., Sun, C., Xiong, B., and Luo, Y. (2009). Y-branch integrated dual wavelength laser diode for microwave generation by sideband injection locking. *Opt. Express* 17: 20727–20734.

102 Yu, L., Lu, D., Sun, Y., and Zhao, L. (2015). Tunable photonic microwave generation by directly modulating a dual-wavelength amplified feedback laser. *Opt. Commun.* 345: 57–61.

103 Kwon, H. and Kang, B. (2005). Linear frequency modulation of voltage-controlled oscillator using delay-line feedback. *IEEE Microwave Wireless Compon. Lett.* 15: 431–433.

104 Johnson, D.G. and Brooker, G.M. (2008). Wide band linearization of a millimetre-wave, linear frequency modulated radar employing a surface acoustic wave, delay line discriminator. In: *Smart Sensors and Sensing Technology* (ed. S.C. Mukhopadhyay and G.S. Gupta), 153–164. Springer.

105 Adler, E., Viveiros, E., Ton, T., and Kurtz, J. (1995). Direct digital synthesis applications for radar development. In: *IEEE International Radar Conference*, Alexandria, VA, USA (08–11 May 1995), 224–226. IEEE, Piscataway, NJ, USA.

106 Wang, C. and Yao, J. (2009). Chirped microwave pulse generation based on optical spectral shaping and wavelength-to-time mapping using a Sagnac loop mirror incorporating a chirped fiber Bragg grating. *J. Lightwave Technol.* 27: 3336–3341.

107 Zhou, P., Zhang, F., Ye, X. et al. (2016). Linearly chirped microwave waveform generation with large time-bandwidth product by optically injected semiconductor laser. *Opt. Express* 24: 18460–18467.

108 Zhou, P., Zhang, F., Guo, Q. et al. (2017). Reconfigurable radar waveform generation based on an optically injected semiconductor laser. *IEEE J. Sel. Top. Quantum Electron.* 23: 1.

109 Guo, L., Lu, D., Pan, B. et al. (2017). Linearly chirped microwave generation using a monolithic integrated amplified feedback laser. *IEEE Photonics Technol. Lett.* 29: 1915–1918.

110 Cen, Q., Dai, Y., Yin, F. et al. (2017). Rapidly and continuously frequency-scanning opto-electronic oscillator. *Opt. Express* 25: 635.

111 Zhou, P., Zhang, F.Z., Guo, Q.S., and Pan, S.L. (2017). Linear frequency-modulated waveform generation based on a tunable optoelectronic oscillator. In: *International Topical Meeting on Microwave Photonics*, Beijing, China (23–26 October 2017), 1–4. Piscataway, NJ, USA: IEEE.

112 Zhou, P., Zhang, F.Z., Ye, X.W. et al. (2016). Flexible frequency-hopping microwave generation by dynamic control of optically injected semiconductor laser. *IEEE Photonics J.* 8: 5501909.

113 Li, J., Pu, T., Zheng, J., et al. (2018). Frequency hopping microwave signal generator based on three-stage integrated semiconductor laser. *18th National Conference on Optical Fiber Communication and 19th National Conference on Integrated Optics*, Changchun, Jilin, China (13–16 July, 2018).

114 Cho, J. and Sung, H. (2012). Simple optoelectronic oscillators using direct modulation of dual-section distributed-feedback lasers. *IEEE Photonics Technol. Lett.* 24: 2172–2174.

115 Lu, D., Pan, B., Chen, H., and Zhao, L. (2015). Frequency-tunable optoelectronic oscillator using a dual-mode amplified feedback laser as an electrically controlled active microwave photonic filter. *Opt. Lett.* 40: 4340–4343.

116 Zhang, X., Pu, T., Zheng, J. et al. (2019). A simple frequency-tunable optoelectronic oscillator using an integrated multi-section distributed feedback semiconductor laser. *Opt. Express* 27: 7036–7046.

117 Pan, B., Lu, D., Zhang, L., et al. (2014). Tunable photonic microwave generation using self-injection-locked monolithic dual-wavelength semiconductor laser. *2014 Asia Communications & Photonics Conference*, Shanghai, China (11–14 November 2014). IEEE, Piscataway, NJ, USA.

118 Pan, B., Lu, D., Sun, Y. et al. (2014). Tunable optical microwave generation using self-injection locked monolithic dual-wavelength amplified feedback laser. *Opt. Lett.* 39: 6395.

119 Chan, S., Liu, Q., Wang, Z., and Chiang, K. (2011). Tunable negative-tap photonic microwave filter based on a cladding-mode coupler and an optically injected laser of large detuning. *Opt. Express* 19: 12045–12052.

120 Xiong, J., Wang, R., Pu, T. et al. (2014). A novel approach to realizing a widely tunable single passband microwave photonic filter based on optical injection. *IEEE J. Sel. Top. Quantum Electron.* 21: 1500706.

121 Hsieh, K., Hwang, S., and Yang, C. (2017). Photonic microwave time delay using slow- and fast-light effects in optically injected semiconductor lasers. *Opt. Lett.* 42: 3307–3310.

122 Fu, X., Cui, C., and Chan, S. (2010). Optically injected semiconductor laser for photonic microwave frequency mixing in radio-over-fiber. *IEEE J. Electromagn. Waves Appl.* 24: 849–860.

123 Chan, S., Hwang, S., and Liu, J. (2016). Radio-over-fiber AM-to-FM upconversion using an optically injected semiconductor laser. *Opt. Lett.* 31: 2254–2256.

124 Li, M., Deng, Y., Tang, J. et al. (2016). Reconfigurable optical signal processing based on a distributed feedback semiconductor optical amplifier. *Sci. Rep.* 6: 19985.

125 Mørk, J., Tromborg, B., and Mark, J. (1992). Chaos in semiconductor lasers with optical feedback: theory and experiment. *IEEE J. Quantum Electron.* 28: 93–108.

126 Heil, T., Fischer, I., Elsäër, W., and Gavrielides, A. (2001). Dynamics of semiconductor lasers subject to delayed optical feedback: the short cavity regime. *Phys. Rev. Lett.* 87: 1–4.

127 Lang, R. (1982). Injection locking properties of a semiconductor laser. *IEEE J. Quantum Electron.* 18: 976–983.

128 Tartwijk, G. and Lenstra, D. (1995). Semiconductor lasers with optical injection and feedback. *Quantum Semiclassical Opt.* 7: 87.

129 Korneyev, N., Radziunas, M., and Wuensche, H. (2003). Bifurcations of a DFB laser with short optical feedback: numerical experiment. *Proc. SPIE Int. Soc. Opt. Eng.* 4986: 480–489.

130 Radziunas, M., Wünsche, H., Brox, O., and Henneberger, F. (2002). Excitability of a DFB laser with short external cavity. *Proc. SPIE Int. Soc. Opt. Eng.* 4646: 420–428.

131 Simpson, T., Liu, J., Huang, K., and Tai, K. (1997). Nonlinear dynamics induced by external optical injection in semiconductor lasers. *Quantum Semiclassical Opt.* 9: 765.

132 Kobayashi, W., Ito, T., Yamanaka, T. et al. (2013). 50-Gb/s direct modulation of a 1.3-μm InGaAlAs-based DFB laser with a ridge waveguide structure. *IEEE J. Sel. Top. Quantum Electron.* 19: 1500908.

133 Pan, B., Lu, D., Zhang, L., and Zhao, L. (2015). A widely tunable optoelectronic oscillator based on directly modulated dual-mode laser. *IEEE Photonics J.* 7: 1–7.

134 Wang, J., Liu, Y., Chen, X. et al. (2014). Compact packaging for multi-wavelength DML TOSA. *Chin. Sci. Bull.* 20: 29–32.

135 Orfanos, I., Sphicopoulos, T., Tsigopoulos, A., and Caroubalos, C. (1991). A tractable above-threshold model for the design of DFB and phase-shifted DFB lasers. *IEEE J. Quantum Electron.* 27: 946–956.

136 Klepser, B. and Hillmer, H. (1998). Investigations of thermal crosstalk in laser arrays for WDM applications. *J. Lightwave Technol.* 16: 1888–1894.

Index

Optical Imaging and Sensing: Materials, Devices, and Applications, First Edition.
Edited by Jiang Wu and Hao Xu.

t

v